John Flint South

Memorials of the Craft of Surgery in England

John Flint South

Memorials of the Craft of Surgery in England

ISBN/EAN: 9783337365844

Printed in Europe, USA, Canada, Australia, Japan

Cover: Foto ©berggeist007 / pixelio.de

More available books at **www.hansebooks.com**

MEMORIALS

OF THE

CRAFT OF SURGERY

IN ENGLAND.

FROM MATERIALS COMPILED BY

JOHN FLINT SOUTH,

TWICE PRESIDENT OF THE ROYAL COLLEGE OF SURGEONS OF ENGLAND, AND SURGEON TO ST. THOMAS'S HOSPITAL.

EDITED BY

D'ARCY POWER, M.A. OXON., F.R.C.S. ENG.

WITH INTRODUCTION BY

SIR JAMES PAGET, BART., F.R.S.,

SERJEANT SURGEON TO HER MAJESTY THE QUEEN.

WITH COLOURED PLATES AND ENGRAVINGS.

CASSELL & COMPANY, LIMITED:

LONDON, PARIS, NEW YORK & MELBOURNE.

1886.

THIS BOOK IS

Dedicated

IN AFFECTIONATE GRATITUDE BY

D'ARCY POWER

TO

HIS MASTER, THE PRESIDENT,

AND TO

HIS FATHER, A VICE-PRESIDENT

OF

THE ROYAL COLLEGE OF SURGEONS OF ENGLAND.

INTRODUCTION,

BY

SIR JAMES PAGET, Bart., F.R.S.

The history of Surgery had long been a favourite study of Mr. South's, but he was past seventy when he began in earnest to collect the materials from which he intended to write a history of the Royal College of Surgeons of England. Unable to resist that fascination of searching and collecting which, I suppose, none feel more keenly than they who love to gather the curious facts that are in old records, he accumulated far more than he could arrange. "I read," he said, "no end of useless things in hope of finding something to my purpose;"[1] and very many of these "useless things" he not only read, but, with his usual steadfastness of purpose and laborious accuracy, copied and preserved. Still, together with the useless, he gathered much that was excellent and well selected; and when Mrs. South, affectionately anxious that his industry should not be fruitless, sent his volumes of manuscripts to the College of Surgeons, and, at the request of the Council, I examined them, I could not doubt that there were in them materials which, if well

[1] "Memorials of John Flint South," by the Rev. C. L. Feltoe, p. 193. 8vo. 1884.

arranged, as they have been by Mr. D'Arcy Power, would be a valuable contribution to a part of the history of Surgery in England. Many of them, indeed, may have a wider value, in that they minutely relate facts which may help the general historians of our country to estimate the methods by, and the measure in, which social progress has been influenced, not by the prominent few who are the heroes of our history, but by the constant impetus of the mental powers and the ambitions of the commonalty.

Surgeons and students of surgery may certainly find in these Memorials many facts of great interest, and, I think, some useful lessons. Only, let it be borne in mind that they are not full records of the progress of surgery in the present usual meaning of the words. They do not tell much of its progress as either a science or an art, though they may help to the understanding of this part of its history if read together with the writings of the successively contemporary surgeons. They are, properly, as the title says, "Memorials of the *Craft* of Surgery," that is, of its business and of the corporate life and government of those who, in successive centuries, practised it. We can trace in them, especially in the last century, an earnest desire for the promotion of surgical knowledge; but, generally, they record events which were of personal or corporate interest; they tell of the acquirement and defence of civil rights, the maintenance of dignity and discipline, the repression of rivalry, the settlement of disputes.

Many of the things thus told must seem to us very strange and, in this sense, amusing; especially if we only think of them as if they were occurring now, and under the same conditions as we are living under. A more careful reading will show that the "strange" things were, usually, fitted to the times and the circumstances in which they happened; and that, like the organs and the changes in an embryo, and in spite of many errors and defects of human management, they were in the progress towards better things.

Whoever will study these Memorials as the history of a development may find in them abundant interest.[1] Especially, he may trace the progress of medical education onwards from the teaching of apprentices, who were to be comely and to be able to read and write, and to wear no beards, and to be well punished for their faults; or the progress of the teaching of anatomy, from the custom of public demonstrations once or twice a year onwards to the methods of our medical schools. Or he may try to imagine, for he can hardly trace, the contrast between the conjoint examination of the fellowships of physicians and surgeons in the first half of the fifteenth century and that which their descendants in the Royal Colleges have happily arranged in the last few years. He may wonder why so good a plan should have lapsed for more than four hundred years, and may find the bad reason for this and many other errors in the maintenance of vested rights, as if they were better

[1] The several subjects and facts referred to may be found with the help of the index.

than the promotion of knowledge. Or he may wonder that women were licensed to practise surgery in the fourteenth century, and hindered in the nineteenth; or that in the sixteenth century licences were granted for the separate practice of specialities; and then, as in the study of a development, he may consider whether the abolition of those usages was like the timely cessation of processes that had only an embryonic use.

Thus and in many other ways these Memorials may be usefully, and, I think, happily studied. But I will suggest attention to only one more subject, the relations that used to exist between barbers and surgeons. They will be found, I believe, more fully illustrated here than hitherto they have been, and the facts recorded may correct some errors commonly prevalent.

The union of the barber and the surgeon in one person, or even in the same corporation, may seem very strange to us now, and in England. But it was quite natural that when bleeding was deemed necessary for the cure of most ailments, and even for the prevention of many, and when the medical ecclesiastics were forbidden to shed blood, they should turn for assistance to their barbers, whom they knew to be dexterous with sharp instruments, and with basins and towels; and, of course, when the barbers were thus admitted to the practice of one piece of surgery, they constantly ventured further, and, after a time, practised many parts of minor surgery independently of the ecclesiastics. Something of this kind was really necessary, because of the small number of surgeons not being ecclesiastics who were in

London; for the whole fellowship in 1491 consisted of only eight, and in 1513 of only twelve, members. And the utility of the barber-surgeon may still be observed in several parts of Europe; especially in Russia, where the fully educated surgeons are far too few for the vast and wide-spread population, and many of the Feldshers, who are generally educated to be military hospital attendants, become barber-surgeons in villages and in the poorer parts of towns, and do good work in both divisions of their calling.

But, however useful the union of surgery and "barbery" may have been in long past times in this country, or may even now be in others, it is an error to suppose that English surgeons are, in any fair sense, the descendants of barbers. These Memorials will show that, from first to last, and even during their temporary conjunction with the Barbers' Company, the real surgeons held themselves apart as a distinct body. The surgeons from whom we and our College can trace an uninterrupted descent were not barbers. When the barbers were first incorporated, early in the fourteenth century, there were, and doubtless long had been, surgeons who practised separately from them; many of whom had served in the army, and had, probably, learned much of their art abroad. In and after 1368, and, probably, for some time previously, these surgeons formed a separate guild or fellowship, with license from the City authorities; and about the year 1421 they combined with the physicians, an evidence of their good repute, for the physicians were then, as always, among the most educated persons of their time. And this evidence of their station is confirmed by the fact that, like the physicians, the surgeons then, and for a long time afterwards, were licensed to practise by the Bishop of London or the Dean of St. Paul's; for it was, I think, only over the classes deemed learned that the Church had such jurisdiction as these licences imply.

It does not appear how long the combination of the physicians and surgeons lasted; but, after it was dissolved, the surgeons still continued separate from the barbers and the barber-surgeons or barbers practising surgery. Each formed a distinct "fellowship," and the incorporation of the Barbers and Surgeons' Company in 1540 scarcely affected the separation. The surgeons did not, in any sense, become barbers or barber-surgeons. That which is often called a union was, really, only an official junction. It was, probably, a convenient means of putting an end to the disputes as to what the few barbers in or near London who still practised any kind of surgery might do; and it gave sanction and authority to the agreement made between the barbers and surgeons in 1493, of which the chief effect was to give the surgeons control over the practice of surgery by the barber-surgeons.

There was no fusion of the two callings. The Company had two distinct sections, and two names, carefully maintained, Barbers *and* Surgeons. In the one section were the barbers, a few of whom practised some simple parts of surgery, and many of whom, it is probable, were then, as are all the present members of the Company, barbers only in name. In the other section were the surgeons. The surgeons were not allowed to practise shaving; the barber-surgeons were not allowed to do more than draw teeth; and if any of them became surgeons it was after such education as, in some measure at least, fitted them for surgical practice.

This division of the Company was maintained by every rule and custom. The surgeons were constantly asserting their superiority in all things relating to anatomy or surgery; they resisted all encroachments from the barbers' side; they resisted the control of the physicians, and gradually superseded them as teachers of anatomy in their own hall; and, when any one was to be admitted into the joint Company who had not been apprenticed to any of its members, it was only as a barber that he could be admitted.

At last, as the surgeons became more skilled and influential, and surgery became a science as well as an art, even the appearance of the union became intolerable; and although the barbers were an influential body of citizens in various lines of business, and had always, as they said, "with the greatest deference, submitted to the surgeons in all matters peculiar to them," yet the surgeons insisted on separation. They gave up all claim to any share in the property or other treasures of the Company, and obtained for themselves the separate Charter which preceded that by which their successors are incorporated in the Royal College.

Thus these "Memorials of the Craft of Surgery"

will show that there is nothing discreditable or ridiculous in its pedigree. Surgery, we may believe, has always been an occupation of men who might be deemed well educated and who held good social rank. All the more, therefore, should it be maintained and taught as a science and an art, of which both the study and the practice may employ the strongest and most honest minds.

EDITOR'S PREFACE.

After the death of Mr. South in 1882, Mrs. South sent to the Royal College of Surgeons the various manuscripts which her husband had laboriously collected during the last ten years of his life. These papers were examined by Sir James Paget, who considered them worthy to be classified with a view to their subsequent publication. He asked me to undertake this work, and the result appears in the following pages.

So far as the present Memorials are concerned, I found that in the first quarter of the present century Mr. Carwardine, to whom we owe the discovery and preservation of the original midwifery forceps manufactured by the Chamberlens, had projected a history of surgery in England. From inability to obtain access to the Barbers' records, however, he was compelled to desist. Many years subsequently his manuscript appears to have fallen into the hands of Mr. South, and it probably suggested to him the idea of carrying out Mr. Carwardine's design. Mr. South entitled his work "Historical researches into the rise and progress of surgery in England, illustrated by a review of our early medical literature." In pursuance of this design the outline was slightly sketched in as far as the year 1450, an epoch here and there being more fully worked out.

These elaborated portions appear to have formed part of the Hunterian oration, which he delivered before the Royal College of Surgeons in the year 1844.

More fortunate than his predecessor, Mr. South gained unrestricted access to the valuable records of the Corporation of Barbers and Surgeons, which are now in the custody of the Barbers' Company at their hall in Monkwell Street. These records he made the centre of his work; and the most important, viz. the Court Minute Books, he, with immense labour and indefatigable industry, copied word by word, letter by letter, and abbreviation by abbreviation, a toil which can only be duly appreciated by those who have had experience in deciphering the crabbed writings of clerks living under the Tudors and Stuarts. The materials thus obtained were supplemented by extracting from the Letter Books and Repertories belonging to the City of London,[1] as well as from the various state papers in

[1] The Letter Books (Stow's "Books of Remembrances") are described by the late Mr. H. T. Riley in his "Memorials of London and London Life in the Thirteenth, Fourteenth, and Fifteenth Centuries" as "a series of folio volumes (the early ones of somewhat smaller size than those of later date) in manuscript on parchment; containing entries of the current matters of the day in which the City has been in any way interested or concerned, downwards from the early part of the reign of King Edward the First. . . . Though for convenience' sake styled 'Letter Books' since the latter part of the reign of Edward the Third, the earliest volumes were originally known as the 'Lesser Black Book,' the 'Greater Black Book,' the 'Red Book,' and the 'White Book,' from the respective colours of their original bindings, no doubt. From November, 1416, the proceedings in which the mayor and aldermen alone have taken part, have ceased to be entered in the Letter Books, the Journals having then superseded them for that purpose. At a later date the 'Repertories' were devoted to the record of proceedings in the Court of Aldermen."

the Record office, all that could be gleaned relating to surgeons and their practice of surgery. It is of this period of his life that Mr. South writes in a letter dated Sept. 11th, 1868: "I . . . . pay visits daily to the Record office, and read no end of useless things in hope of finding something to my purpose."[1] In regard to the City records, Mr. South appears to have worked independently of the late Mr. Riley, as I find no mention of the latter author's name; and the translations of the earlier notices of the Barbers and Barber-Surgeons differ in many important respects from the versions given by Mr. Riley in the "Memorials of London and London Life." I am the more surprised at this, as from the date of the above letter Mr. South and Mr. Riley must have been working at the Guildhall almost contemporaneously.

In editing the following pages from the vast mass of material left by Mr. South, I found that the first two chapters had been finished, and were apparently ready for the press. They are therefore printed almost as they stood, and Mr. South must be held answerable for the opinions expressed as to the scientific knowledge possessed by the Druids and their immediate successors. For the rest, I have followed as closely as possible upon the lines which Mr. South had laid down for his own guidance, though I have not hesitated to deviate from them when by so doing it has been possible to give greater effect to the work, or to render it more interesting

1 "Memorials of John Flint South," collected by the Rev. C. L. Feltoe, M.A., p. 193. Lond. 1884.

to the general reader. For this purpose, I have made numerous extracts from the Court Minute Books of the Barbers and Surgeons' Company in regard to matters which, though not strictly relevant to the issue, appeared to be of too great interest, as illustrating the manners of the age, to be passed over in silence. These extracts I have copied out *verbatim et literatim*, only writing in full the contractions, and marking the omitted letters by italics. I have throughout followed Mr. Riley [1] in using the word "craft" or "trade" as a translation of "mestera" or "mestier." For, as Mr. Riley points out, "the word 'mystery,' or its old-fashioned and now obsolete form 'mistery,' as signifying a trade, ... is in no way connected with 'mysterium,' a 'secret' which the use of the misguiding English word 'mystery' as its representative might easily lead the purely English reader to suppose." In the translations of the older ordinances, of which the originals are written in abbreviated Latin or Norman-French, I have been careful to compare Mr. South's rendering with that of Mr. Riley.

The history of the Barbers and Surgeons is throughout the history of the development of a guild into a company. In the years succeeding the Norman Conquest, the government of the City of London was oligarchical. The aldermen were the hereditary governors, and their wards belonged to them by right of purchase. In the names of many of the wards the remembrance of this state of affairs is still left to us:

[1] *Op. cit.* p. 1.

thus Bassishaw, or Basinghall Ward, is the ward belonging to the family of Basing; and Farringdon Ward takes its name from its last hereditary alderman and owner, William de Farringdon, one of whose descendants is mentioned as Sir Nicholas de Farÿngdon in the Barber's oath.[1] Cornhill and its ward, according to Mr. Loftie,[2] owes its name to the family of Cornhill, who once possessed it. From these powerful landowners, who were in all cases wealthy merchants, many of the proudest families of the English peerage have taken their origin.[3]

At an exceedingly remote period, not only in England, but throughout Germany, the followers of trades had banded themselves together into fellowships or guilds, partly for trade purposes, and partly for the observance of religious duties. The guilds, at first in complete subjection, gradually acquired power until, after a long and very severe struggle with the hereditary oligarchy, they conquered, and in London at least converted the civic government into an elective representation. The details of this contest are but little known, but it was less bitter in England than in many of the German towns. It was virtually concluded in 1377, the year of the death of King Edward the Third, the guilds of the more important and wealthier trades becoming incorporated as companies, from whose members alone the higher City

[1] Page 14.

[2] "Hist. of London," vol. i. p. 160. Ed. ij.

[3] The Coventrys, Earls of Coventry, and, at a much later date, the Osbornes, Dukes of Leeds, are cases in point.

officers could be selected. About the same time the guilds of the meaner trades obtained, upon payment of annual dues, the right to nominate their own officers, and a formal confirmation of their craft-ordinances, which transferred the management of all trade-concerns and the settling of all trade-disputes to the guild with which each was associated. A slight advance had been made in this direction of guild-independence as early as 1272, when charters of incorporation were granted by Walter Harvy, the Mayor, but the oligarchical party was too strong, and the guilds relapsed for a time into their former subordinate position.

The history of the Barbers as a guild can only be traced back to the year 1308, when, in the reign of Edward the Second, Richard was chosen supervisor of the Barbers. He probably held the position of Guild-alderman or graceman, a post which is not to be confounded with that of *ward* alderman. Similar guilds of Barbers existed in York, in Exeter, and in other large towns, where they were associated partly for trade purposes and partly for purposes of religion. They assembled on a special saint's day, and, after attending mass, dined together, and subsequently elected their officers for the ensuing year. The death or marriage of any of their members also caused them to meet. In the Barbers' Guild, as in nearly all other trade-guilds, women as well as men might become members, although they were not eligible to any of the higher offices. In 1375 some of the Barbers had come to be Barber-Surgeons and the Company consisted of two portions, the Barbers who practised shaving, and the Barbers who practised surgery. A similar separation took place in the great guild of Weavers, who branched off into woollen drapers and linen armourers, the latter afterwards becoming the powerful company of Merchant Taylors.

In the year 1415 the City authorities, "to prevent dispute," nominated the wardens of the Barber-Surgeons' Guild, and it is therefore probable that the guilds had not as yet fully emancipated themselves. This must have been one of the last years of the supremacy of the City oligarchy, however, for shortly afterwards the Barber-Surgeons succeeded in getting their ordinances enrolled, and obtained the right of using a particular *livery*. The livery had long been a distinguishing feature of the older guilds; but in the reign of Richard the Second a determined but ineffectual attempt had been made to limit its use. It was therefore only confirmed to those who could claim it by prescriptive right. The Barber-Surgeons, as being one of the less important or less wealthy guilds, did not obtain their incorporation as a company so early as many of the other guilds; and it was only in 1460, the year of the accession of Edward the Fourth, that they took their place as one of the livery companies of the City of London.

Side by side with the Barbers the Guild of Surgeons existed. They were at first merely incorporated as a confraternity, but at a later period they obtained a charter. This guild appears to have been of comparatively recent origin, and to have sprung out of an association of the military surgeons who had been trained in the hundred years' (1337-1444) war with France. It was always a small body, and it was never a guild in the older and truer sense of the word, like that of the Barbers. Their smallness of numbers, however, was more than counterbalanced by the influence of the members, who were the élite of the surgeons of the time. After an attempted alliance with the physicians, the Surgeons amalgamated with the Barbers' Company, the two being united by a charter granted in the year 1540 by Henry the Eighth.

The United Company of Barbers and Surgeons thus formed appears to have been peculiar amongst the other City companies in the fact that non-freemen and strangers were admitted either permanently or for a time to the privileges of the Company upon payment of somewhat higher fees than were required of those who obtained admission by servitude in the ordinary way, such strangers, unless actually licensed as surgeons, being accounted as belonging to the Barbers' side of the United Company. The journeymen of the craft, too, under the title of yeomanry formed a subordinate body within the Company, systematically organised, and possessing wardens with powers similar to those exercised by the wardens of the Company itself.

In 1745, after a union of more than two hundred years, the Surgeons severed themselves from the Barbers, and established a new body called the Surgeons' Company, founded on the exact lines of the pre-existing united corporation. From small beginnings the Surgeons' Company rapidly acquired considerable influence. By a foolish blunder in 1796, the Charter was forfeited, and failing to obtain an Act of Parliament for the reconstitution of their Company, the Surgeons were incorporated by a charter of George the Third in the opening year of the present century as the Royal College of Surgeons in London.

Exception may, perhaps, be taken to the title of the present work, inasmuch as the history relates rather to the craft of surgery in London than in England. The following facts, however, will show that the general regulations by which surgery was practised were the same for England and Ireland, and probably, too, for Scotland. Barbers' guilds, as has been already said, existed in several towns in England, notably in London, York, and Exeter. In some of these towns the guild remained as a religious body; whilst in others, as at York, they allied themselves with the Surgeons, and were active in managing the matters relating to their craft. A manuscript volume of ordinances belonging to the York guild is preserved in the Egerton collection at the British Museum. The various details of this guild appear to have been based upon the regulations of the London Company, and to be so far identical that what is here written of the one may be held as true of the other: even the arms are similar.

In Dublin the Barber-Surgeons were incorporated as a guild by a charter granted to them by Henry the Sixth as early as 1446 (25 Henry VI.). Surgeons not members of the guild associated themselves in the same way as they did in London. In 1576 these Surgeons were amalgamated by Queen Elizabeth with the Barber-Surgeons and periwig-makers under the title of the "Master, Wardens, and Fraternity of Barbers and Chirurgeons of the Guild of St. Mary Magdalene within our City of Dublin."[1] The united fraternity for some time used the arms granted to the Barbers and Surgeons' Company of London "with some small difference, being a note of diminution or subordination," but they subsequently obtained from the Ulster King of Arms a separate grant, thereby showing themselves to be independent of their neighbours. The united confraternity was dissolved in opposition to the wishes of the Barbers' side in the year 1784, and the College of Surgeons was founded immediately afterwards.

In Edinburgh the Barbers and Surgeons were united in the year 1505, but I have met with no further details of their history.

In conclusion, I must offer my best thanks to the gentlemen who have materially assisted me in preparing the present volume, and especially to Mr. Sidney Young, of the Barbers' Company. Although he is engaged upon a somewhat similar work, he has invariably replied to my numerous questions with the greatest readiness and courtesy, and has on many occasions devoted to me

[1] Since the above was printed, a fuller account of this Guild is announced in "The History of the Royal College of Surgeons in Ireland," by Sir C. A. Cameron, but I have had no opportunity of seeing the work.

a far larger share of his valuable time than I as a stranger could have claimed, or he need have granted. His help has been invaluable, especially in matters of detail relating to the Barbers' Company; and his answers have in many cases thrown light upon passages which, in my ignorance of City customs, were obscure. I have further to thank Mr. Young for giving me several extracts relating to the Surgeons which he met with in his reperusal of the books belonging to the Barbers' Company, and which appear to have escaped the notice of Mr. South.

To Dr. Norman Moore, the Warden of the College in St. Bartholomew's Hospital, I must tender my gratitude for the kindness with which he read the proof-sheets, correcting many minor errors.

Mr. Horace Noble has laid me under an obligation by the care with which he has revised the paragraphs and appendices upon the Company's heraldry, a science of which, I regret to say, I am profoundly ignorant.

Mr. Joseph Mills kindly lent me the plate from which the interior of the Barbers and Surgeons' Hall has been re-engraved.

Lastly, Sir James Paget, by the readiness with which he undertook, at my request, to write the introduction, has added a valuable feature to the work, and has rendered it of greater and more permanent value than I could otherwise hope to have made it.

D'ARCY POWER.

May, 1883.

CHRONOLOGICAL TABLE.

A.D. 1308. First mention of the Barbers in the City records; the Barbers' oath.
,, 1354. First mention of the Surgeons in the City records.
,, 1368. Masters of the Guild of Surgeons (?) sworn.
,, 1375. Untrained Barbers intermeddle in Barbery and Surgery.
,, 1387. The Barbers a livery guild.
,, 1389. Masters of the Guild of Surgeons (?) sworn.
,, 1392. Thomas Stodeley, Master of the Surgeons, sworn.
,, 1409. Barbers authorised to practise surgery.
,, 1415. Overseers of the Barber-Surgeons appointed.
,, 1416. Overseers of the Barber-Surgeons again appointed.
,, 1423. Union of the Guild of Surgeons with the Physicians.
,, 1424. Rector of Medicines sworn.
,, 1435. Ordinances of the Guild of Surgeons engrossed.
,, 1450. The Guild of Barbers confirmed in the practice of surgery.
,, 1452. Grant of arms to the Guild of Barbers.
,, 1462. Charter of Edward the Fourth, making the Guild of Barbers a Company, with power to govern all Surgeons acting as Barbers in the City of London.
,, 1492. Grant of arms to the Guild of Surgeons.
,, 1493. Alliance of the Barbers' Company with the Guild of Surgeons.
,, 1540. Act of Parliament incorporating the Barbers' Company and the Guild of Surgeons as the Company of Barbers and Surgeons.
,, 1745. Separation of the Surgeons from the Barbers.
,, 1745. Formation of the Surgeons' Company.
,, 1796. The Surgeons' Company dissolved.
,, 1800. The Royal College of Surgeons in London established by Royal Charter.

CONTENTS.

CHAPTER I.

CHAPTER II.

CHAPTER III.

CHAPTER IV.

CHAPTER V.

CHAPTER VI.

CHAPTER VII.

CHAPTER VIII.

CHAPTER IX.

CHAPTER X.

CHAPTER XI.

CHAPTER XII.

APPENDIX A.

APPENDIX B.

APPENDIX C.

APPENDIX D.

APPENDIX E.

APPENDIX F.

APPENDIX G.

APPENDIX H.

APPENDIX I.

APPENDIX J.

APPENDIX K.

APPENDIX L.

APPENDIX M.

APPENDIX N.

APPENDIX O.

APPENDIX P.

APPENDIX Q.

APPENDIX R.

LIST OF PLATES.

MEMORIALS

OF THE

CRAFT OF SURGERY.

CHAPTER I.

MEDICINE IN ENGLAND FROM THE TIME OF THE DRUIDS TO THE SEPARATION OF MEDICINE AND SURGERY IN THE THIRTEENTH CENTURY—THE SCHOOL OF SALERNUM.

The rise of the medical profession in England.

THE eleventh century may be taken as the starting point from which to trace the rise and progress of the medical profession in England. Its separation into distinct branches has been the result either of the wants and expectations of the people, or of the jealous vigilance and narrow policy of public companies invested with exclusive privileges under the power and authority of charter and incorporation; whilst its position has been modified by the state and usages of society at different periods. A slight sketch of the condition of surgery before the incorporation of the fellowships of Barbers and Surgeons will not be devoid of interest, and will serve as a fitting introduction to these memorials.

The Druids, according to Pliny, were at once priests, poets, and physicians. In the latter capacity they acted partly, as became their priestly office, by prayer and the laying on of hands, by divination, and by charm, and partly in a more scientific manner. The numerous human sacrifices which they must of necessity have witnessed would lead to their acquiring a fair amount of anatomical knowledge of the truest and best kind, since it was derived from the actual inspection of the bodies of men. They were well versed also in medical botany, and their veneration for the mistletoe is too well known to require more than a passing notice.[1] The marshwort and vervain were held in high esteem, whilst the Britannica, whether the great water dock or scurvy grass,[2] was known to the whole civilised world and derived its name from the island whence the supply was obtained. The Druids, moreover were no mean pharmacists, since they could extract the

Medicine amongst the Druids.

[1] In the words of Pliny (book xvi. chap. 44), "They call it in their language All Heale (for they have an opinion of it that it cureth all maladies whatsoever). And when they are about to gather it, after they have well and duly prepared their sacrifices and festivall cheare under the said tree, they bring thither two young bullocks, milke white, such as never yet drew in yoke at plough or waine, and whose heads were then, and not before, bound by the horne; which done, the priest, arraied in a surplesse or white vesture, climbeth up into the tree, and with a golden hook or bill cutteth it off, and they beneathe receive it in a white souldiour's cassocke or coat of armes. Then fall they to kill the beasts aforesaid for sacrifice, mumbling many oraisons and praying devoutly, That it would please God to blesse this gift of His to the good and benefit of all those to whome he had vouchsafed to give it. Now this persuasion they have of Misselto thus gathered, That what living creature soever (otherwise barraine) doe drinke of it, will presently become fruitfull thereupon; also, that it is a soveraign countrepoison or singular remedie against all vermine" (= venom). Pliny, "Natural History," translated by Philemon Holland, p. 497 *D*. 1601.

[2] See, upon this point, Paris' "Pharmacologia," p. 54, note 2. Ed. 8.

juices of herbs and plants by bruising and steeping them in cold water, whilst they prepared tinctures by infusing in wine the juices thus obtained, and made potions and decoctions by boiling the herbs in water. Pliny further states that they administered drugs by fumigation, and that they were well acquainted with the art of making salves and ointments.

Medicine amongst the Saxons.

The Danish and Saxon leeches were perhaps more ignorant than the Druids. As early as the seventh century, however, there were men who made the science of medicine a study and who practised it as a profession. This advance was owing to the influence of the clergy, who not only introduced books from Rome, but often had in a monastery a member of the community who was consulted by the neighbourhood as a physician. Physicians are mentioned by Bede, and amongst the letters of Boniface there is one from a Saxon, desiring some books *de medicinalibus*. He says that they had plenty of such works in England, but that the foreign drawings in them were unknown to his country and difficult to acquire. The Saxon "leechdoms," published in the Rolls series,[1] enable us to form some idea of the degraded condition of medicine during the earlier part of this period.

Gradually, however, these darker days passed away, and we have a splendid instance of the attention which was at a later time bestowed upon medical knowledge in

[1] "Leechdoms, Wort-cunning, and Star-craft of Early England," collected and edited by the Rev. Oswald Cockayne. Lond. 1864.

the Saxon treatise described by Wanley.[1] This treatise may have been written in the time of Alfred, but Mr. Cockayne attributes it to the former half of the tenth century. The first part of it contains eighty-eight remedies against various diseases; the second part adds sixty-seven more, and in the third part are seventy-six prescriptions.[2] Little is known of the surgical attainments of the leeches of this period, but they do not seem to have exceeded those common operations which every people somewhat removed from barbarism cannot fail to know and use. Venesection was employed, but in a rude and unskilful manner which led to many accidents. The lancet was known as the "œder seax" or vein knife, and its use was governed less by necessity than by superstitions of the idlest kind.

The School of Salernum.

In the years immediately preceding the accession of William the Conqueror a stimulus was given to the study of medicine by the medical schools of Salernum, Naples, and Montpellier, which were frequented by students from all parts of Europe, who carried the methods and practice which they had learnt to their various homes, and by imparting their knowledge to others assisted greatly in the spread of scientific culture. The School of Salernum is now best known by the Schola Salernitana, which for many years served as a text-book of hygiene and dietetics in western Europe. This work appeared very

[1] "Catalogue of the Harleian MSS.," vol. j. 585. 4 ed. 1808.

[2] Sharon Turner's "History of the Anglo-Saxons," vol. iii. p. 445 (London, 1830), and transcribed in full in the "Leechdoms," vol. ii.

early in an English form; the original is believed to have been dedicated to the eldest son of William the Conqueror, though some authorities maintain that Edward the Confessor is the king named in the first lines of the work, which run as follows:

> "Anglorum Regi scribit schola tota Salerni:
> Si vis incolumem, si vis te reddere sanum
> Curas tolle graves, irasci crede profanum."

The reason for the dedication may have been that Duke Robert was under treatment at Salernum for a sinus in his right arm, due to a wound received at the siege of Jerusalem. The work appears to have been edited by John of Milan in the name of the whole community. It is written in verse, and exercised a most extensive influence upon what has been termed the "folk medicine" of this country. There is hardly a scrap of proverbial wisdom handed down to us from our ancestors upon the inexhaustible subject of what is wholesome or unwholesome in diet, etc., which may not be traced in one form or another to the "Regimen Sanitatis." The reputation which it acquired was so great that more than twenty editions appeared in Latin within a century after the invention of printing, of which the earliest was published in 1480.

The School of Salernum, from which the work emanated, was perhaps the greatest medical school of the period. It conferred after examination a licence to practise, a privilege which was not possessed by any other body, for even at Naples the college could only recommend their students to the king or his chancellor as fit persons for the necessary licence. The statutes of the school are worthy of attention, as well on account of their antiquity as of their propriety. Licences to practise as a physician, and, it appears from the following statute, as a surgeon also, were granted by this college as early as the eleventh century, whilst special measures were taken for the supervision of the apothecaries. The statute in reference to the licence in surgery is, that "the person examined must be twenty-one years of age, and must bring testimonials of having studied physic for five years; if to be admitted in surgery he must learn anatomy for one year; he must swear to be true and obedient to the society, to refuse fees from the poor, and to have no share of gains with the apothecaries."

The results of this long training were not, however, wholly satisfactory to the public, if we may trust what John of Salisbury, living in the twelfth century, says upon the subject. He writes[1] of the physicians of his own time, that "the professors of the *theory* of medicine are very communicative; they will tell you all they know, and perhaps out of their great kindness a little more. From them you may learn the nature of all things: the cause of sickness and of health; how to banish the one and preserve the other, for they can do both at pleasure. They will describe to you minutely the origin, the beginning, the progress, and the cure of all diseases. In a word, when I hear them harangue I am charmed; I think them not

Medicine under the Angevins.

[1] Johan. Salisburiensis; Policraticus lib. ii. c. 29.

inferior to Mercury or Æsculapius, and almost persuade myself that they can raise the dead. There is but one thing which makes me hesitate: their theories are as directly opposite to one another as light to darkness. When I reflect on this I am a little staggered. Two contradictory propositions cannot be true. But what shall I say of the *practical* physician? I must say nothing amiss of them. It pleaseth God, for the punishment of my sins, to suffer and fall too frequently into their hands. They must be soothed and not exasperated, that I may not be treated roughly in my next illness. I dare hardly allow myself to think in secret what others proclaim aloud."

In another work, however, the writer plucks up more courage, and speaks his mind of the practical physician as freely as he had before done of his theoretical brother. "They soon return from college, full of flimsy theories, to practise what they have learned. Galen and Hippocrates are continually in their mouths. They speak aphorisms on every subject, and make their hearers stare at their long, unknown, and high-sounding words. The good people believe that they can do anything because they pretend to all things. They have only two maxims which they never violate: 'Never mind the poor; never refuse money from the rich.'"

The clergy as early physicians.

The clergy were for many years almost the only persons who taught and practised physic as well as the other sciences, and there are but few names celebrated in the annals of medicine at this period which are not those of ecclesiastics.

This profession became so lucrative, and so many monks applied themselves to the study and practice of it, deserting their monasteries and neglecting their own religious duties, that the eighth canon promulgated by the Council of Tours, in A.D. 1163, prohibited monks from staying out of their monasteries above two months at one time, and forbade them to teach or practise physic.[1] No restraint was at first laid upon the secular clergy, and many of the bishops and other dignitaries of the Church acted as physicians in ordinary to kings and princes, a service by which they acquired both riches and honour. These very reverend physicians had received their education at Salernum, and derived much of their medical knowledge from the writings of Rhazes, Avicenna, Avenzoar, and other Arabian writers whose works had been translated into Latin at an earlier date by Constantine, a monk of Mount Casino, near Salernum.

It is not improbable that the scientific method of teaching and studying physic, which was introduced into the medical schools of the eleventh and twelfth centuries, gave rise to the separation in this country of physicians and surgeons into two distinct classes of practitioners. Be this as it may, however, the separation was complete by the end of the twelfth century, when a contemporary poet, in describing the attempts made to cure the wound received by Richard the First

[1] This ordinance was only one of a number to like purpose. The second Council of Lateran in 1139 had previously enjoined a similar duty; and by a decree of Henry III. in 1216, the clergy and monks were prohibited from exercising the profession of advocates and physicians.

before the castle of Chalons, in 1199 A.D., clearly distinguishes the two professions, and assigns to each its peculiar duties.[1] About this period too, in England, some applied themselves more particularly to the study of the *materia medica*, and to the composition of medicines, and were on that account termed apothecaries. In the annals of the Church of Winchester, Richard Fitz Nigel (de Ely), who died Bishop of London in 1198, is said to have been apothecary to Henry II.[2] Nor was the post of apothecary at this time a light one, if we may judge from the complicated and artificial mixtures which were prescribed. Theriac itself, which was often combined with other substances, contained more than fifty ingredients.

Spoliation of the monasteries.

The clergy exercised the learned professions, and the monasteries became the depositories of our literature, the spark of learning being kept alive amongst us in those religious foundations. But it was like the story of the lamp burning in a sepulchre which expired when the sanctuary was burst open by the rude violence of Henry's agents. How ample, in bulk at least, were these records of science and history, has been shown by Bale, who says, "I have been also at Norwich, our second city of name, and there all the library monuments are turned to the use of their grocers, candlemakers, soapsellers, and other worldly occupiers."[3]

[1] Interea regem circumstant undique mixtim,
Apponunt medici fomenta, secantque chirugi
Vulnus, ut inde trahant ferrum leviore periclo.

[2] Wharton's "Anglia Sacra," pars i. p. 304. Lond. 1691.

[3] "The Laborious Journey in Search of J. Leland." Lond. 1549.

And again, very much to the same purpose:[1] "Covetousnesse was at that time so busie about private commodity that public Wealth . . . . was not anywhere regarded. A number of them, which purchased those superstitious mansions, reserved of those Library-books, some to serve their jakes, some to scour their candlesticks, and some to rub their boots; some they sold to the Grocers and Sope sellers, and some they sent over sea to the Book binders, not in small number, but at times whole ships full. . . . I know a Merchant-man (which shall at this time be namelesse) that bought the contents of two noble Libraries for fourty shillings price, a shame it is to be spoken. This stuffe hath he occupied instead of gray paper by the space of more than these ten years, and yet he hath store enough for as many years to come." Fortunately, however, printing had been in use for more than half a century prior to the dissolution of the monasteries, and the majority of those works were preserved to us which were considered as the most worthy of being communicated to the public.

The Jews as physicians.

It will be necessary to retrace our steps after this digression, in order to show more clearly the nature and origin of such an apparently incongruous association of ecclesiastics and physicians, barbers and surgeons, grocers and apothecaries, as we find at a somewhat later period. The Jews, next to the clergy, perhaps possessed the largest share of learning. The vagrant life to which this extraordinary

[1] Fuller's "Church History," book vi. p. 335. Lond. 1655. Bale *op. cit.* fol. B. i. 6.

people was reduced afforded them an intercourse with the different nations of the world, and rendered them in some measure a medium of communication both in literature and science throughout the western hemisphere. Benjamin of Tudela, in his Itinerary, written about A.D. 1165,[1] enumerates the cities in which the Jews had any settlement, and mentions their numbers in each place. He names many who were physicians, and who practised not only amongst their own tribes, but also amongst the Moors and Christians. This learned Israelite, upon his return from his travels over the greater part of the then known world, commends the school at Salernum as the best seminary of physic amongst the "Sons of Edom," as he calls the western Christians.[2]

The priests, as might be expected, looked with a jealous eye upon the encroachments of the Jewish physicians and of lay surgeons. To exclude the former from any participation in the honours and emoluments of medical practice, they obtained, through their interest at Rome, a formal excommunication against all who committed themselves to the care of a Jewish physician, and by the canon law no Jew might give physic to any Christian. Yet so celebrated had these practitioners

[1] Translated from the Hebrew into Latin by Benedict Arias Montanus. The first printed edition was in Hebrew: it was published at Constantinople 1543. Of these travels and of this translation, however, Isaac D'Israeli wrote: "He describes a journey which, if ever he took it, must have been with his nightcap on, being a perfect dream. . . . The Travels of which we have a curious translation must have been apocryphal." —*Curiosities of Literature; Art., Literary Impostures.*

[2] Freind, "History of Physic," ed. j. part ii. p. 227.

become, and so ardent was the desire for health and long life, that even the power of Rome was ineffectual in excluding them from practice. The simple fact that the Jews were masters of the Arabic language, at a time when no translation of Hippocrates or Galen could be procured in Europe, was sufficient to ensure the employment of physicians belonging to this religion at nearly every court in Christendom.

Surgery divorced from physic.

As the efforts of the clergy to restrain the practice of the Jewish physicians were to a great extent ineffectual, so they had a still more difficult task to perform with respect to the lay surgeons, who were at this time very numerous. A mastery over the principles of medicine as well as of surgery is necessary to form a perfect practitioner in either, and in England the priests appear to have been disposed to preserve the two branches united. The popes, however, jealous of such an interruption in the duties of the clergy, and looking upon the manual part of surgery as derogatory, made several attempts to prohibit priests from the performance of surgical operations. In 1215 the ecclesiastics were debarred by an ordinance of Pope Innocent III. from undertaking any operation involving the shedding of blood, on the plea that the Church "abhorret a sanguine."

By two subsequent decrees, the one issued by Pope Boniface the Eighth at the close of the thirteenth century, and the second by Pope Clement the Fifth about the beginning of the fourteenth century, surgery was formally separated from physic, and the priests were absolutely forbidden to practise the art. These measures must have abandoned surgery entirely to the laity, who were as yet a wholly illiterate race. The priests, however, still kept their hold upon the art by making use of their servants the barbers, who, having been employed to shave the heads of the priests, and to perform the minor operations in surgery, were now instructed to work entirely under the direction of their masters. These men, qualifying themselves by the instruction of the clergy, assumed the title of barber-surgeons, and became a confraternity or fellowship. The more enlightened of the barber-surgeons again, in the march of knowledge, by attending lectures and practising dissection, began to spurn such a degrading conjunction, and at last, freeing themselves from the barbers, became a college of surgeons.

CHAPTER II.

THE DIVISION OF BARBERS INTO THOSE PRACTISING SURGERY AND THOSE REMAINING AS BARBERS—THE FELLOWSHIP OF SURGEONS A DISTINCT BODY.

First notice of barbers.

THE first notice of the barbers which I have found in the City Records is rather indicative of the doubtful character of their morals; an imputation which is the less unlikely, if it be borne in mind that in addition to their occupation of shavers and hair and beard trimmers, they were also professional bathers. The extract runs as follows, and may be headed the Barber's Oath.[1] "Richard the barber, living opposite the church of All Hallows the Less, was elected and presented by the barbers of London on the Tuesday after the Feast of St. Lucy the Virgin [Dec. 13th N.S.] in the second year of the reign of King Edward son of King Edward [*i.e.* the Second, 1308], to Nicholas de Faryngdon, then Mayor of London, John de Wengrave, and the other Aldermen, for the purpose of keeping order amongst the barbers. And he was admitted and sworn to make diligent search through the whole of his craft every month, and if he shall find any brothel keeper or other disreputable

[1] Letter Book C. fol. 96*b*. See also Riley's "Memorials of London and London Life," p. 67. In regard to the Letter Books, see Editor's Preface.

folk to the scandal of the craft, he shall detain them and cause them to be brought before the chamber." It appears from this notice, and also by a journal which was formerly kept in the Town Clerk's office of the City of London, that a Barbers' Guild or confraternity existed as early as 1308, though it did not attain to the rank of a company for many subsequent years. No records remain of its foundation, and we only find scattered and incidental notices of its existence.

As early as 1354 the following inquisition by surgeons as to the treatment of a wound occurs in the City records:[1] "Be it remembered, that on the Monday next after the Feast of St. Matthias the Apostle [24th February N.S.] in the 28th year, etc., the Prior of Hogges, Master Paschal, Master Adam de la Poletrie, and Master David de Westmerland, surgeons, were sworn before the Mayor, Aldermen, and Sheriffs, to certify them as to a certain enormous and horrible hurt on the right side of the jaw of Thomas de Shene appearing; whether or not such injury was curable at the time when John le Spicer of Cornhulle took the same Thomas under his care to heal the wound aforesaid. Who say upon their oath that if the aforesaid John le Spicer at the time when he took the said Thomas under his care had been expert in his craft or art, or had called in counsel and assistance to his aid, he might have cured the injury aforesaid; and they further say, that through want of skill on the part of the said John le Spicer the said injury under his care became apparently incurable." In 1368, however, the

[1] Letter Book G, fol. xviii., quoted from Riley; *op. cit.* p. 273.

surgeons are first noted as a distinct body who, like all other persons following trades and professions, were required to appear before the authorities of the City of London to be by them licensed to practise within their jurisdiction. This licence was conceded under promise "well and truly to serve the people in their cures, to take of them reasonable fees,[1] to exercise their mystery faithfully, to report to the Mayor and Aldermen any surgeon neglecting his patients, and to give information to the City officers of the hurt, wounded, or otherwise in peril of death."

The admission of Master Surgeons.

The following document is endorsed, "Admission of Master Surgeons." In it the persons who came before the Mayor and Aldermen to be sworn as *Master Surgeons* of the City of London are simply termed *Surgeons*. From this it is doubtful whether these persons only became Masters in Surgery by the authority of the court, after having been previously licensed to practise surgery by the Bishop of London or the Dean of St. Paul's, as was the common custom at this time, for the legal practice of the art: or whether they were sworn as Masters of the Guild of Surgery within the City of London having authority over their brethren of the craft. The latter hypothesis seems the more probable, inasmuch as in the

[1] The licence to practise surgery granted by the University of Oxford is still very similar; it runs: "Primo, scilicet, quod quatuor saltem pauperes gratis et intuitu caritatis (quumprimum sese occasio tulerit) cures. . . . Secundo, quod fines artis tuæ non excedas, aut medicinam practices. Tertio, quod nimium pro salario non exigas; aut curationem aliquam retardes uberioris lucri intuitu."—*Stat. Univ. Oxon.* 1874; vi. (ix.) vii. 8, p. 151.

foreign universities at this period the names of Doctor and Master were only beginning to be known as specific titles of honour, and were still used in their original significations for teachers and persons skilled in their art.[1] In either case it is evident that all surgeons practising in the City were compelled to appear before the Court of Aldermen to be sworn as surgeons.

The form of admission is also of interest, as showing the care taken by the civic authorities for those maimed or wounded, who must have been so numerous when guild fights were of almost daily occurrence. The formula runs as follows: "On the Monday after the Feast of the Purification of the blessed Mary [Feb. 2nd], in the forty-third year [1368] of the reign of King Edward the Third after the conquest, Master John Dunheued, Master John Hyndstoke, and Nicholas Kyldesby, surgeons, were admitted at full hustings before Simon de Morden [Mayor] and the Aldermen, and were sworn as Master Surgeons of the City of London, to deserve well and truly of the people in doing their cures,[2] to take from them reasonable payment, and truly to practise their craft, and to report as often as need be to the Mayor and Aldermen the faults of those who undertook cures. To take charge of the hurt or wounded, and to give true information to the officers of the City about such persons whether they be in danger of death or not, etc.,[3] and to act

[1] "Regimen Sanitatis Salernitanum;" Sir Alex. Croke. Oxford, 1830, p. 15.

[2] Attending their cases.

[3] Mr. Riley remarks that the abbreviation *etc.* frequently occurs at the end of a passage: it seems to have been used as a matter of course,

uprightly in all other things belonging to their calling."[1]

The next admission of Master Surgeons is dated 1389, and although nearly identical in form with the preceding, it presents two points of interest. In the first place there is a distinct recognition of the practice of women; and secondly, amongst the Master Surgeons admitted is the name of Master John Hynstok. The practice of medicine by the opposite sex was not of recent date, since there were several women whose writings were held in high esteem at Salernum, as early as the eleventh century. Dr. William Moore,[2] of Dublin, has shown that in the Irish Guild of Barber-Surgeons, founded in 1446, women are recognised as distinctly sharing in the privileges of the foundation. The name of Master Hyndstoke is recorded in the previous admission twenty years earlier. The John Hynstok here mentioned may be either the same person or a relative. If the two are identical it would give weight to the supposition that the persons admitted as *Magistri Surgici* were not merely Master Surgeons, but were actually Masters or Aldermen of the Surgeons' Guild, and were thus publicly placed in authority over their brethren. Office bearers of this kind will by-and-by be found to be actually

in many instances without being needed by the context, but rather as a sort of saving clause to cover any omission that might possibly have been made.

[1] Letter Book G, fol. 219. See also Riley; *op. cit.* p. 337.

[2] "Dublin Quarterly Journal of Medical Science," vol. vi. pp. 76 and 101; and vol. viii. p. 232.

existing and to be elected time after time, though not at regular intervals. After an interval of twenty years, however, the second John Hynstok might have been the son or other relative of the previous one, and the question as to the exact position of Magistri chirurgici remains unsettled.

The document runs as follows: "On Monday the tenth day of April, in the thirteenth year of King Richard the Second, Master John Hynstok, Master Geoffrey Grace, Master John Brademore, and Master Henry Suttone surgeons were admitted in the court of Guildhall in London, before William Venour, Mayor, and the Aldermen. They were sworn as Masters Surgical of the aforesaid city, well and truly to serve the people in working their cures, taking of them reasonable recompense, etc. To practise truly their trade, and to make faithful oversight of all others, both men and women, occupied in cures or using the art of surgery, presenting their lack both in practice and medicines so often as needs be to the aforesaid Mayor and Aldermen. They shall be ready when warned thereto to take charge of the hurt or wounded, etc., and to give faithful information to the servants of the City of such hurt or wounded as are in danger of death or not."[1]

Unskilful barber-surgeons.

At a somewhat earlier period than this is a petition to the Mayor and Aldermen from the Barbers of London, who, in 1375, complain that "men barbers from Uppeland little skilled

[1] Letter Book H, fol. 248*b*. See also Riley; *op. cit.* p. 519.

in their craft, come into the City from day to day, take houses, and intermeddle with barbery, surgery, and to cure other maladies. Whereas they have not known nor ever were taught how to do such things, to the great danger and cheating of the people, and grievous disgrace of all honest barbers of this City."[1]

It appears from this that the barbers were practising surgery as part of their craft as early as the reign of Edward the Third, though it is not clear how far this practice extended. It is certain, however, that they were desirous of preventing any person using barbery until, as their petition set forth, "they had been found able and skilled in the said art by trial and examination before certain barbers of the City." To this petition the Mayor and Aldermen gave assent, and it was enrolled as an ordinance in the Chamber of London.

In the year 1392 occurs the first record of a "Magister chirurgorum," or Master of the Surgeons, in the person of Thomas Stodeley, who was sworn on the 7th of May, before Will: Standon, the Mayor.[2]

Barbers petition against unwarranted supervision.

The barbers, although considering surgery as part of their craft, since it was recognised as such by the ordinance of 1375, either failed to prosper in their surgical practice or found their authority of small avail, in consequence of the existence of surgeons who were not shavers, and over whose actions they had in consequence no control.

[1] This document, on account of its quaintness and importance, I have translated as literally as possible from the Norman French in which it is written. See Appendix A. Letter Book H, fol. 27*b*. Riley, *op. cit.* p. 393.

[2] Letter Book H, fol. 276.

In 1409, therefore, the barbers again petitioned the Court of Aldermen, appearing before them by counsel. The Court granted their petition, which set out more fully the surgical aspect of the barbers' profession. It appears that the Master Surgeons, thinking themselves empowered to do so by their oath before the Mayor and Aldermen, had interfered with those who practised surgery, among whom were probably some of the barbers.

The minute of this petition and its confirmation runs thus. "On Friday, the 7th day of the month of March, in the eleventh year of the reign of King Henry the Fourth from the Conquest (1409), there came before the honourable man, Richard Merlawe, who was the Mayor of the City of London, good and honest folk, barbers of the said City, by their counsel, John Weston, in the private chamber. There were then present John Shadwithe, Robert Chichylly, John Waryner, William Norton, Thomas Fauconer, Walter Cotton, Henry Pountfreyt, Stephen Spillman, Henry Barton, William Chichylly, Thomas Pyk, Aldermen, and John Lane, one of the Sheriffs of the aforesaid City. And did present a copy or transcript of a certain petition formerly offered by their predecessors, the barbers of the City of London, to the lords, the Mayor, and Aldermen, in the 49th year of the reign of King Edward the Third after the Conquest, John Ward being then Mayor (1375).[1] . . . . And after their petition, or rather ordinance, had been read before Richard, the Mayor, and the Aldermen and Sheriff,

[1] See Appendix A, page 297.

and had been well considered by them, and had been approved, ratified, and confirmed in each of its articles. It was further unanimously agreed by the whole court that the barbers, who are for themselves and their successors barbers of the City of London, should for ever peaceably enjoy the privileges contained in the ordinance without scrutiny of any person of other craft or trade than barbers. And this neither in shavings, cutting, bleeding, nor other thing in any way pertaining to barbery or to such practice of surgery as is now used or in future to be used within the craft of the said barbers."[1]

Appointment of Masters of the Barber-Surgeons.

The powers thus confirmed and increased to the barbers at the expense of the practitioners in surgery do not seem to have been efficient, even so far as their own members were concerned, for in 1415 "it was tumultuously reported to Thomas Fauconer Mayor and the court of Aldermen that certain barbers of the city of London, inexperienced in the art of surgery, very frequently take charge of sick and wounded persons with the intent of fraudulently acquiring their goods; whereby the sick were often worse off at their departure than at their incoming, and on account of the unskilfulness of these barbers were oftentimes maimed, to the scandal of the skilled, and the manifest harm of the people of our Lord the King." Wherefore the Mayor and Aldermen, desirous of putting down such scandals and damage, and also to apply fitting remedy, determined that "by the

[1] Letter Book I, fol. 94.

majority of all the barbers practising surgery, and dwelling within the liberties of the City of London, there should be chosen two, the ablest, wisest, and most discreet of all the barbers practising the surgical faculty, to minister what in their opinion was wanting in cases of death and maim,"[1] where ignorant persons had the control.

These masters, as often as they were elected, were to present themselves to the Mayor and Aldermen to be by them sworn. To prevent any dispute in regard to this new procedure the Court of Aldermen took the first election into their own hands. On the third day of May, 1415, the Mayor and Aldermen, after careful deliberation and counsel, "caused to be brought before them the name of every barber practising the art of surgery, and dwelling within the liberty of the City." After due inquiry had been made "the names of Simon Rolf and Richard Wellys, citizens and barbers of London, practising the faculty of surgery, were commended above all others, as well for their science and probity as for the divers difficult cures[2] wisely treated by them throughout. The witness to their acts being based on sound and undamaged information and good faith." Precepts were therefore issued by the Mayor and Aldermen to one of their sergeants that the said Simon and Richard should present themselves before the court in the chamber of the Guildhall on Monday, the sixth day of May.

Upon this Monday the two elect appeared together

[1] Letter Book I, fol. 149*b*. Riley, *op. cit.* p. 606. [2] cases.

before the Mayor, and were accepted and sworn upon the Holy Gospels of God over all the barbers practising the faculty of surgery, and dwelling within the City. They took further oath well and faithfully to oversee and superintend their brethren, to keep the rules and ordinances of the craft, to spare no one for love, favour, lucre, or hate; diligently and without concealment to present to the Chamberlain of the City all faults which they might detect. At all times when called upon so to do to superintend, as in duty bound, all wounds, bruises, maims, and other ills, without asking aught for their trouble. Finally, to do all other things which "are fit and proper for masters or overseers to perform;"[1] in good sooth a comprehensive oath.

Fauconer's ordinance, however, does not seem to have been more effectual than that of Merlawe. In a very short time after this arrangement had been made it is noted that the barber-surgeons "pretending that they were wiser than the overseeing masters," refused to call them to consultation. This may not improbably have been a fact, as it is not unreasonable to believe that the impressment of surgeons for the war which Henry V. was waging in France[2] not only included surgeons, but also those barber-surgeons who had attained to any repute for their skill. Those who had escaped the pressgang would, from their scarcity, be held in high esteem by the neighbours requiring their services. They would thus be liable to exaggerate their own importance, and set the overseeing masters at defiance.

[1] Letter Book I, fol. 149*b*.

[2] See page 47.

Be this as it may, however, the offences and pretences of these recalcitrant barber-surgeons and the penalties enacted with a view to securing their better behaviour for the future are amusingly set forth in the following ordinance dated July 4th, 1416. "It was stated on the authority of some trustworthy and discreet citizens of the Craft of Barbers practising surgery, as well as other able, skilful, and worthy persons, that, in defiance of the ordinance, very many unskilful persons of the said Craft of Barbers indiscreetly presume and presumptuously pretend that they are wiser than the overseeing masters and with still less reason utterly disdain to call them to any ailments, or to be by them discreetly advised or diligently overlooked. Whence they desist not from daily taking under their care persons sick and in instant peril of death or maim, without shewing such sick person and their ailment or danger to the overseeing masters. Such sick, therefore, often fall into the greatest danger of death or maim on account of the presumption and unskilfulness of the barbers aforesaid." Whereupon the Lord Mayor and Aldermen are prayed that they would "so far as possible deign to provide a sure remedy as well for the common good of the whole realm as for the special honour of the City of London."[1]

Presentation of the sick in danger of death or maim.

The result of this representation was that "the Mayor and Aldermen being well disposed to the petition, both for its justice and reasonableness, after due

[1] Letter Book I, fol. 149*b*.

consideration, and seeing that many persons in these times dread the loss or payment of money more than the rule of honesty and of a safe conscience, ordain and establish that no barber practising the surgical faculty within the liberty of the City should presume to take under his charge any sick person in actual danger of death or maim without shewing him to the overseeing masters. If such presentations were not made within three days after receiving such sick person the barber so offending was to pay a penalty of six shillings and eightpence for each occasion on which he was found acting contrary to the ordinance. Five shillings of the fine to go to the use of the Chamber in the Guildhall and twentypence to the use of the barbers." The system of the presentation of the more serious cases here enforced was a jealously guarded privilege until a late period in the history of the united and incorporated barbers and surgeons. It was productive of much benefit to the science of surgery; and a somewhat similar plan, as we shall see later on, is still pursued in one of the large metropolitan hospitals in this kingdom.[1] On the 2nd of October, 1416, John Parker and Simon Rolf were sworn in as the overseeing masters of the faculty.[2]

Barber-surgeons who had ceased entirely to act as barbers must from henceforth be considered as a distinct class in the guild of barber-surgeons. The control over them, as well as the oversight of all who professed to practise surgery, devolved upon the two masters appointed by the ordinance of the Court of

[1] See chapter viii. page 144. [2] Letter Book I, fol. 166*b*.

Aldermen. In like manner masters of the barbers practising barbery were annually selected. But the two officers in no case interfered with each other, and in the records of the annual swearing in of the masters of the barbers' guild (a fellowship which, as we have seen, preceded by many years the foundation of the Barber-Surgeons' Company, and is in no way to be confounded with it) the two masters who are to govern the barbers practising surgery are always designated as *Magistri barbitonsorum cirurg. facult. exercent.*, and as such are distinct from the *Magistri barbitonsorum.* The *Magistri chirurgorum* or Masters of the Guild of Surgeons were sworn separately, so they clearly had no connection direct or indirect with the Guild or confraternity of Barbers.

CHAPTER III.

ENGLISH PRACTITIONERS OF THE FOURTEENTH CENTURY—JOHN OF GADDESDEN AND JOHN OF ARDERNE.

Two practitioners of the fourteenth century.

IN the reign of Edward the Third there lived in England two very remarkable men belonging to our profession. The one was John of Gaddesden, a physician, the other John of Arderne, a surgeon. They appear to have been men of a very different type of character. Gaddesden was a follower of Galen, and his book is written upon Galenian principles, with comparatively few cases, whilst Arderne refers but seldom to Galen or any other writer, and his work consists chiefly of the cases which he had himself treated.

John of Gaddesden.

John of Gaddesden, or Johannes Anglicus, concerning whose life but few authentic facts have reached us, flourished about 1320; he was a member of Merton College, Oxford, a doctor of physic, as well as an ecclesiastic; he was the first Englishman employed at Court as a physician, and he appears to have obtained the prebendal stall of Ealdland in St. Paul's Cathedral. As physician to Edward the Second and his son, Gaddesden may have met Chaucer, who was one of Prince Lionel's personal attendants in the year 1357, and was acting as valet to the king in

1367, whilst John was a *young* doctor of physic in 1320. It would be interesting to imagine that the poet and the physician had met; in any case, however, the reputation of Gaddesden had reached the author of the Canterbury Tales, who thus mentions him in his Prologue[1] amongst the classical authors of medicine in describing the learning of his doctor of physic:

> "Wel knew he the olde Esculapius,
> And Deiscorides and eek Rufus;
> Old Ypocras, Haly, and Galien;
> Serapyon, Razis, and Avycen;
> Averrois, Damascien, and Constantyn;
> Bernard, and *Gatesden*, and Gilbertyn."

The work by which Gaddesden became celebrated is the "Rosa Anglica," in which he treats of fevers and injuries of all parts of the body, of hygiene, dietetics, and materia medica. It became a text-book of medicine throughout Europe. Although drawn largely from previous writers, it is illustrated by the author's own practice, and is interesting as a register of the medical and surgical knowledge of the time at which it was written. Such knowledge however, had not advanced far if we may judge by the author's well-known recommendation that a patient suffering from small-pox should be wrapped in scarlet or some other red cloth, "as," says he, "when the son of the illustrious king of England (Edward II.) had the small-pox, for I took care that everything about his couch should be red, and his cure was perfectly effected, for he was restored to health without a trace of the disease." The

[1] Prologue, line 429. Ed. R. Morris, Oxford, 1874. Clarendon Press.

style of the work is amusing, and the author quotes verses on almost every page.

John of Arderne.

John of Arderne, or Arden[1] (Plate II.), was born in 1307. He practised in Newark from 1349 to 1370, and being then sixty-seven years of age, and having acquired a large practice and a high reputation, he proceeded to London, whence, in 1377, he dates his book "De Curâ Oculi." Arderne describes himself as "chirurgus inter medicos" (a surgeon amongst physicians), but he nowhere affords us any information as to his place of study, or the means whereby he attained his knowledge. He was a specialist, devoting his attention to the treatment and cure of the various fistulæ which occur in the different parts of the body, though he by no means confined himself to this particular branch of his profession, and his chief work treats of a great variety of surgical subjects. He shall, however, tell his own story, and show how modestly he could advertise his merits. His Latin work on fistula was translated in the early part of the fifteenth century. It is from this translation that I extract the following :[2]

"OF YE PLOGE OF FISTULA IN ANO AND OF YE MANERE OF YE LECHE AND OF INSTRUMENTIS NECESSARY FOR YE FISTULE.

"John Arderne fro the first pestilence that was in

[1] Further particulars about John of Gaddesden may be obtained from Freind's "History of Physic," ed. j. part ii. pp. 277—293; and of John Arderne in the same work, pp. 325—332. The date of the latter's birth is fixed by the Sloane MS., No. 75, fol. 146.

[2] Sloane MS., No. 6, fol. 142 *et seqq.*

JOHN OF ARDERNE.

(From his "Praxis Medica": Sloane Collection.)

the ȝere[1] of oure Lord 1349[2] duellid in Newerk in Notyngham shire vnto[3] the ȝere of oure lord 1370 and ther I helid many men of *fistula in ano,* of Whiche the first was sire Adam E*uer*yngham of Laxton in the clay by side Tukkesford Whiche sire Adam forsoth Was in Gascone with sir henry that tyme named Erle of Derby and after was made Duke of lancastre a noble and worthi lord. The forsaid sir Adam forsoth suffrand[4] *fistulam in ano* made for to aske counsel at all the lechez and cirurgions that he myȝt fynde in Gascone at Burdeux at Bressac Tolows and Neyrbron[5] and Peyters[6] and many other places And all forsoke hym for vncurable whiche yse[7] and yherde[8] ye forsaid Adam

[1] The character ȝ has various forms. At the *beginning* of a word it is to be sounded as *y*, so that *ȝard* is our modern yard; in the middle of a word it had a guttural sound, still represented in our spelling by *gh*, as in *liȝt* for light; at the end of a word it either had the same sound or stood for *z*. In fact, the character for *z* was commonly made precisely like it, although sparingly employed; yet we find *marchauntz* for marchaunts, where the *z*, by the way, must necessarily have been sounded as *s*. This use of the character is French, and appears chiefly in French words. In early French MSS. it is very common, and denotes *z* only. ("Specimens of Early English," by Morris and Skeat. Clarendon Press, 1879, part ii. p. xvi.)

[2] 24 Edw. III.

[3] The characters *v* and *u* require particular attention. The latter is freely used to denote both the modern sounds, and the reader must be prepared at any moment to treat it as a consonant. Thus the words *haue, liue, diuerse* are to be read *have, live, diverse;* where it will be observed that the symbol appears between two vowels. The former is used sparingly (except when written in place of *f* in Southern MSS.), but sometimes denotes the modern *u*, chiefly at the beginning of a word. The following are some of the common examples of it . . . viz. *vce* or *vse* (use), *vtter* (utter), *vp* (up), etc. (Morris and Skeat; *op. cit.* p. xiv.)

[4] suffering from a. [5] Narbonne. [6] Poictiers.

[7] *Y*, prefix answering to the German and Anglo-Saxon *ge*, is usually prefixed to past participles, but also to past tenses, present tenses, adjectives and adverbs. (Morris and Skeat; *op. cit.* p. 483.)

[8] seen and heard.

hastied for to torne hom to his contre. And when he come hom he did of[1] al his knyȝtly clothinges and cladde mornyng clothes in *pur*pose of abydyng dissoluyng or lesyng of his body beyng niȝ to hym. At last I forseid John Arderne ysouȝt[2] and couenant ymade come to hym and did my cure to hym and oure lord beyng mene[3] I helid hym *per*fitely within half a ȝere. And aftirward hole and sounde he ledde a glad lif by 30 ȝere and more, ffor whiche cure I gatte myche honour and louyng[4] *th*urȝ[5] al ynglond. And the forsaid Duke of Lancast*re* and many other gentilez wondred therof.

"Afterward I cured Hugon Derlyng of fowick of Balne by Snaype. Afterward I cured John Schefeld of Briȝtwell aside Tekyll. Afterward I cured Sir Reynald Grey lord of Wilton in Waleȝ and lord of Schulond biside Chesterfelde Whiche asked counsel at the most famose leches of yngland and none availed hym. Afterward I cured sir henry Blakborne clerk Tresorer of the lord prince of Walez. Afterward I cured Adam Gumfray of Shelforde byside Notyngh*am* and sir John preste of the same toune and John of holle of Shirlande and sir Thomas Hamelden p*ar*sone of langare in the Vale of Beuare. Afterward I helid Sir John Masty, p*ar*sone of Stopporte in Chestre shire. Afterward I cured frere Thomas Gun*n*y custode of the frere myno*rs*[6] of ȝorke. Afterward in the ȝere of oure lord 1370 I come to london and ther I cured John Colyn maire of Northampton that asked counsel at many lechez. Afterward I

[1] put aside.
[2] sought.
[3] the instrument.
[4] praise.
[5] throughout.
[6] minors.

helid or cured Hew denny ffisshmonger of london in Briggestrete and William Polle and Raufe Double And one that was called Thomas Broune that had iij holes by whiche went out wynde with egestious odour that is to sey 3 holez of the tone *par*ty of the ersse and 7 on the tother side. Of whiche some holez was distant fro the towell[1] by the space of the handbrede of a man so that bothe his buttok*es* was so vlcerate and putrefied within that the quitour[2] and filthe went out ich day als mych as an egg shel miȝt take. Afterward I cured 4 frereȝ *pre*chours[3] that is to sey ffrere John B*r*itell, ffrere John Haket, ffrere petre Browne, ffrere Thomas Apperlay and a ȝong man called Thomas Vske of whiche forseid som had only on hol y distau*n*te from the towel by oon ynch or by two or by thre. And other had 8 or 9 holez procedyng to the codde of the testiclez And many other man*ners* of which the tellyng war ful hard.

"All these forseid cured I afore the makyng of this boke oure lord Ihsu y blessid God knoweth that I lye noȝt. And therefore no man dout of this *th*of al[4] old famo*us* men and ful clere[5] in studie haue confessed tham that thei fande nat the wey of curation in this case; ffor god that is deler or rewarder of Wisdom hathe hid many thingis fro wise men and sliȝe[6] whiche he vouchesaf*eth* afterward for to shewe to symple men. Therfore al men *th*at are to come afterward witte[7] thai that old maistereȝ war noȝt bisie ne p*er*tinaceȝ in

[1] anus.
[2] pus.
[3] preachers.
[4] although.
[5] renowned.
[6] cunning.
[7] understand.

sekyng and serchyng of the forseid cure. But for thai mi3t no3t take the hardness of it at the first frount, thei kept it vtterly byhinde *th*air bak. Of whiche forso*th* som demed it holy[1] for to be incurable o*thers* applied doutful opinions. Therfore for alsmyche in hard thingis it spedith to studiers for to *per*seuere and abide and for to turne subtily thair wittes ffor it is opned not to *th*am that are passand[2] but to tham *th*at ar *per*seuerand.[3] Therfore to the honour of god almy3ti that hath opned witte to me that I should fynde tresour hidde in the felde of studiers that[4] long tyme with pantyng breest I have swette and trauailed ful bisily and p*er*tinacely.

"As my faculte suffice*th* without fair spekyng of endityng I haue brou3t for to shew it openly to tham that cometh aftur, our lord beyng [nigh] me and this boke, no3t that I shewe my self more worthi of louying of suche a gifte than other, but that I greue not god and for the dragme[5] that he hath giffen to me that I be not constreyned for treson. Therfore I pray that the grace of the holy gost be to this werke that he vouchesaf for to spede it: that the thingis whiche in wirking trewly I am ofte tymes exp*er*te I may plenevly[6] explane tham in this litel boke. It is lefull forso*th* for to sey that[7] is knowen and for to witnes that is seene.

"And this I sey that I know no3t in al my tyme ne heard no3t in al my tyme of any man no*th*er[8] in yngland ne in p*ar*tie3 bi3onde *th*e see, that kouthe cure *fistula in*

[1] wholly. [2] passive. [3] persevering. [4] in which. [5] drachma or talent. [6] fully. [7] what. [8] neither.

ano outake[1] a frere minor that was wi*the* the prince of Waleȝ in gascon *and* gyan[2] whiche rosed *and* bosted hym that he had cured the forseid sekenes. And at london he deceyued many men. And when he miȝt noȝt cure som man he made suggestion to tham that no man miȝt cure tham and that affermed he with swering that ȝif the fistule was dried that the pacient at the next schuld noȝt eschape dethe whiche forso*th* ylefte *and* forsake of hym I cured p*er*fitely. And to remoue false opinions of ignorant men for witnes I putte experience. Avicen forso*th* sei*th* exp*er*ience ouercome*th* reson and Galien *in pantegniis*[3] sei*th* no man ow for to trust in reson aloon but ȝif[4] it be proued of experience. And he seith in ano*ther* place Exp*er*ience withoute reson is feble *and* so is reson withoute exp*er*ience fest vnto hym.[5] Ne*th*erlesse I afferme noȝt that I miȝt hele al *ffistulæ in ano* ffor som ben vncurable as it shal be seid within when I shal trete of tham."

The second extract from the same author is still more interesting, and although somewhat lengthy, it will well repay perusal. It shows what a thorough man of the world John of Arderne must have been : the quaint touches here and there are most humorous and graphic. The fees which he charged for making his cures are rather appalling when it is taken into consideration that money at the time in which he wrote was at least twelve times more valuable than it is at present, and even for the nineteenth century the sums he mentions are

[1] except.
[2] Guienne.
[3] A well-known treatise by Galen.
[4] except.
[5] to back it.

considerable. The principles laid down by the author for the guidance of the leech in regard to his patients as well as in his dealings with his colleagues cannot be held in too high estimation. It would surely be well if we could at all times act in the same gentle manner towards our brethren in the profession, and answer courteously when asked of their practice, "I have nought hearde of hym but gode and honest."

A fair estimate of a well-conducted medical practitioner in John of Arderne's time, may be obtained from his description

"Of ye manere[1] of ye Leche.

"Ffirst it behoueth hym that wil p*ro*fite in this crafte, that he sette God afore eu*er*more in alle his werkis and euermore calle mekely with hert and mouth his help. And somtyme visite of his wynnyngis poure men aft*er* his myȝt, that thai by thair prayers may gete hym grace of the holy goste. And that he be noȝt yfounden temerarie[2] or bosteful in his seyingis or in his dedes. And abstene he hym fro moche speche, and most among grete men. And answere he sleiȝly[3] to thing*es* yasked, that he be noȝt ytake in his wordes. Fforsoth ȝif his Werkes be oft tyme knowen for to discord fro his Wordes and his byhestis;[4] he shal be halden more vnworthi and he shal blemyssh his owne gode fame: wherfore seith a versifio*ur* *Vincat opus verbum, minuit iactantia famam.* Werke o*uer* come thi Worde, for boste lesseneth gode lose.[5] Also be a leche

[1] behaviour.
[2] rash.
[3] warily.
[4] promises.
[5] report.

noȝt mich laughyng ne mich playing. And als moche as he may withoute harm, fle he ye felowshippe of knafes and of vnaniste[1] persones. And be he *euer*more occupied in thingis that biholdith to his crafte, outhir[2] rede he, or studie he, or Write or pray he, for the exercyse of bokes Worshippe*th* a leche; ffor Why, he shal bo*th* be holden and he shal be more Wise. And aboue al *th*ise, it profiteth to hym that he be founden euermore sobre, ffor dronkenneȝ destroyeth al vertu and bringith it to not,[3] as seith a Wise man. *Ebrietas frangit, quicquid sapientia tangit.* Dronkenes breketh what so Wisdom toucheth. Be he content in strange places of metes and drinkes *tha*t yfounden, vsyng mesure in al thing*es;* ffor the wise man seith *Sicut ad om*ne *quo*d *est mensuram pon*ere pro*dest: sic sine mensura, de*per*it om*ne *quo*d *est.* As it p*ro*fiteth to putte mesure to al thing*es* that is, So without mesure p*er*isshe*th* alle *th*ing*is* that is. Skorne he no man ffor of that it is seid *Deridens alios non inderisus abibit:* He that skorneth other men shal not go away vnskorned. ȝif ther be made speche to hym of any leche nouther sette he hym at nouȝt, ne preise hym to mich or com*m*ende hym but thus may he curteysly answere, I haue noȝt eny knowleche of hym, but I lerned noȝt ne I have not herd of hym but gode and honeste; and of this shal honour and thankyngis of eche p*ar*ty encresse and multiplie to hym; aftur this, Honour is in the honorant *and* noȝt in the honored. Consider he noȝt *o*u*e*r openly the lady or the douȝters or o*th*er fair wym*m*en in gret men*nes*

[1] disreputable. [2] either. [3] nought.

houses ne profre tham noȝt to kisse, ne touche noȝt priuely ne apertely [1] thair pappes ne thair handes ne thair share, that he renne noȝt in to the indignacion of the lord ne of noon of his. Inasmoche as he may, greue he no seruant, but gete he thair loue and thair gode Wille. Abstene he hym fro harlotrie als wele in wordes as in dedes in euery place, for ȝif he vse him to harlotery in priue places, som tyme in opene place ther may falle to him vnworship of yuel [2] vsage aftir that it is seyde *Pede super colles, pedes vbi pedere nolles.* And it is seid in another place, Shrewed speche corruptith gode manners.

"When seke men forsoth or any of thair bysyde [3] cometh to the leche to aske help or counsel of hym be he noȝt to tham ouer fewe ne ouer homely, but mene in beryng after the askyngis [4] of the personeȝ, to some reuerently, to some comonly, ffor after [5] Wise men, ouer moche homelynes bredeth dispisyng. Also it spedeth that he haue semying excusacions [6] that he may noȝt incline to thair askyng without harmyng, or without indignacion of som gret man or frende, or for necessarie occupacion: or feyne [7] he hym hurt or for to be seke or som other couenable [8] cause, by whiche he may likly be excused. Therfor ȝif he will fauvour to any mannes askyng, make he couenant for his trauaile and take it byfore handeȝ. But avise the leche hym selfe wele, that he giffe no certayn answer in any cause but he se

[1] openly.
[2] evil
[3] relatives.
[4] requirements.
[5] according to.
[6] proper excuses.
[7] feign.
[8] convenient.

fust *th*e sikenes and *th*e maner of it; and whan he ha*th* seen and assaied it, *th*ofal[1] hym seme that the seke may be heled ne*th*lesse he shal make *pro*nosticacion to *th*e pacient *th*e *per*ile3 to come 3if the cure be differred. And 3if he se *th*e pacient perseiue bisily the cure, *th*an after *th*at the state of *th*e pacient asketh, aske he boldly more or lesse but eu*er* be he warre of scarse askyng*is* ffor ou*er* scarse askyng*is* sette*th* at not both *th*e markette and *th*e thing. Therfore for the cure of *fistula in ano,* when it is curable aske he co*m*petently of a worthi man and a gret an hundred marke or fourty pounde, wi*th* robez and feez of an hundred shillyns, terme of life, by 3ere. Of lesse men fourty pounde, or fourty mark*es* aske he, without feez. And take he no3t lesse *th*an an hundred shillyn*s*; ffor neuer, in alle my lyf, toke I lesse than an hundred shillyns for cure of that sekeness. Ne*therl*esse, do another man as hym *th*ink bett*er* and more spedefulle.

"And 3if the pacientes or thair frendez or seruantz aske by how moche tyme he hopeth to hele it, eu*er* more lat the leche byhete[2] *th*e double, *th*at he supposeth to spede, by half, that is 3if the leche hope to hele ye pacient by twenty wekes, that is the comon course of curing, adde he so many euer; ffor it is better that the terme be lengthed *th*an the cure; for *pro*longacion of the cure giffe*th* cause of dispairyng to the pacientes; when triste[3] to the leche is moste hope of helthe. And 3if the pacient considere or wondre or aske, Why that he putte hym so long a tyme of curyng, si*the th*at he

[1] although. [2] name. [3] trust.

heled hym by the halfe? Answere he, that it was for that the pacient was strong herted and suffred wele sharp *th*ingis, and that he was of gode complexion, and hadde able fleshe to hele, *and* feyne he other causes pleseable to the pacient for pacientez of syche wordez are proude and delited.

"Also dispose a leche him, that in clothes and other apparalyngis he be honeste noȝt likkenyng him self in apparalyng or beryng to mynstrallez; but in clothing and beryng sewe[1] he the maner of clerkes; ffor why it seemeth any discrete man ycladde with clerkes clothing for to occupie gentil menez bordes. Haue the leche also clene handes, and well shapen nailez *and* clensed fro all blaknes and filthe. And be he curtaise[2] at lordez bordez, and displese he noȝt in wordes or dedes to the gestes syttyng by. Here he many *th*ingis, but speke he but fewe, ffor a wise man seith, It semeth more to vse the eres than *th*e tunge. And in ano*th*er place, ȝif thou had bene stille thou had bene holden a philosophre. And whan he shal speke, be the wordze short, and als mich as he may, faire and resonable and withoute sweryng. Beware that ther be ne*uer* founden double worde in his mouthe; ffor ȝif he be founden trew in his wordes, fewe or noon shall doute in his dedez.

"Heve also a ȝong leche gode *pro*uerbez p*er*tenyng to his crafte in coumforty*n*g of pacientez. Or ȝif pacientes pleyne that ther medicynes bene bitter or sharp, or sich other; than shal the leche sey to the pacient thus: It is redde in the last lesson of matyns of the natiuitie of oure

[1] follow. [2] courteous.

lord, that oure lord Jhe*su* wiste come into this world for the helthe of mannes kynd, to the maner of a gode leche and wise. And when he cometh to the seke man he sheweth him medicynes some li3t and som hard. And he sei*th* to the seke man, 3if you wilt be made hole, *th*ise and *th*ise shal thou take. Also in another place, in an omely[1] vpon the gospel of the sonez of Zebedee, wher *th*er moder askid seying, Lord sey *th*att my two sones sitte in thy kyngdome *th*e tone on *th*i ri3t hand and the to*ther* on the lefte. And Jhe*su* answeryng seid, 3e wote[2] neuer what 3e aske. *Th*an seid he to the sonez of Zebedee, may 3e drinke the chalice that I am to drink, *th*ai seid to hym We may. As 3if he seid to *th*am, 3if 3oure soule or mynd couate *th*at delite drinke ye first *th*at sorrowe*th*[3] or ake*th*: And so by bitter drinke of confeccion[4] it is come to the ioyes of hel*th*e.

"O*uer* that hym ow to comforte the pacient in monysshyng hym, that in anguisshes he be of gret hert; ffor gret hert make*th* a man hardy and strong to suffre sharp *th*ingis and greuous. And it is a gret vertue and an happy; ffor BOECIUS sei*th*, *De disciplina scolarum*, he is no3t wor*th*ie of *th*e poynt of swetnes that kan no3t be lyrned with greuyng of bitternes; ffor why, a strong medicyne answereth to a strong sekenes. And *ther*eon sei*th* a wise man, Be no cure sene heuy or greuous to the, to whiche folowe*th* ane helefull effecte. And in ano*ther* place it is seid, Happy or blessid be *th*at day *th*at ordeyne*th* mery 3eres. And ano*ther* seith, He may neuer be in reste of body *th*at is oute of reste of soule. I will suffre

[1] homily. [2] know. [3] *i.e.* what causeth sorrow. [4] drug.

lesse *th*ingis *th*at I suffre noȝt more greuous. It seme*th* a gret herted man for to suffre sharp *th*ingis! he forso*th* *th*at is wayke of hert is noȝt in way of curacion: ffor why, forso*th*e in al my lyf I haue sene but fewe laborante in *th*is vice heled in any sikenes. *Ther*for it is to bewar to wise men *th*at *th*ei entremette[1] noȝt with sich: ffor whi, *th*e wise man sei*th*, Alle *th*ingis ar hard to a waik hert man, for *th*ai trow euer more yuellez[2] to be nyȝe to *th*aim; *th*ei drede euermore, *th*ai suffre no *th*ingis, *th*ai are u*er*more vnstable and vnwise *th*erfor a versifi*ou*r seiyth of them, *Quamvis nil pacior, paciendi me tenet horror*, *th*at is, *Th*of al I suffre no*th*ing vgglynes of suffryng holdeth me.

"Also it spedeth *th*at a leche kanne talke of gode talez and of honest that may make *th*e pacientes to laugh, as wele of the biblee as of other tragediez *and* any other *th*ingis of whiche it is noȝt to charge whilez that they make or induce a liȝt hert to *th*e pacient or *th*e sike man.

"Discouer neuer the leche vnwarly[3] the cou*n*sellez of his pacientez als wele of men as of wymmen ne sette noȝt oon to ano*ther* at noȝt, thof al he haue cause *th*at he be noȝt gilty of cou*n*sell, ffor ȝif a man se ye hede wele a no*ther* man*n*es cou*n*sel he will wist better in *y*e.

"Many *th*ingis forso*th*e bene to be kept of a leche withoute[4] *th*ese that ar seid afore *th*att may noȝt be noted here for ouer moche occupying. But it is noȝt to dout *th*at if *th*e forseid be well kepte *th*at ne *th*ai shal giffe a gracious going to *th*e vser to *th*e hiȝte of Worship and of Wyn*n*yng, for Caton sei*th* *Virtutem primam imputa*

[1] intermeddle. [2] evils. [3] unwarily. [4] in addition to.

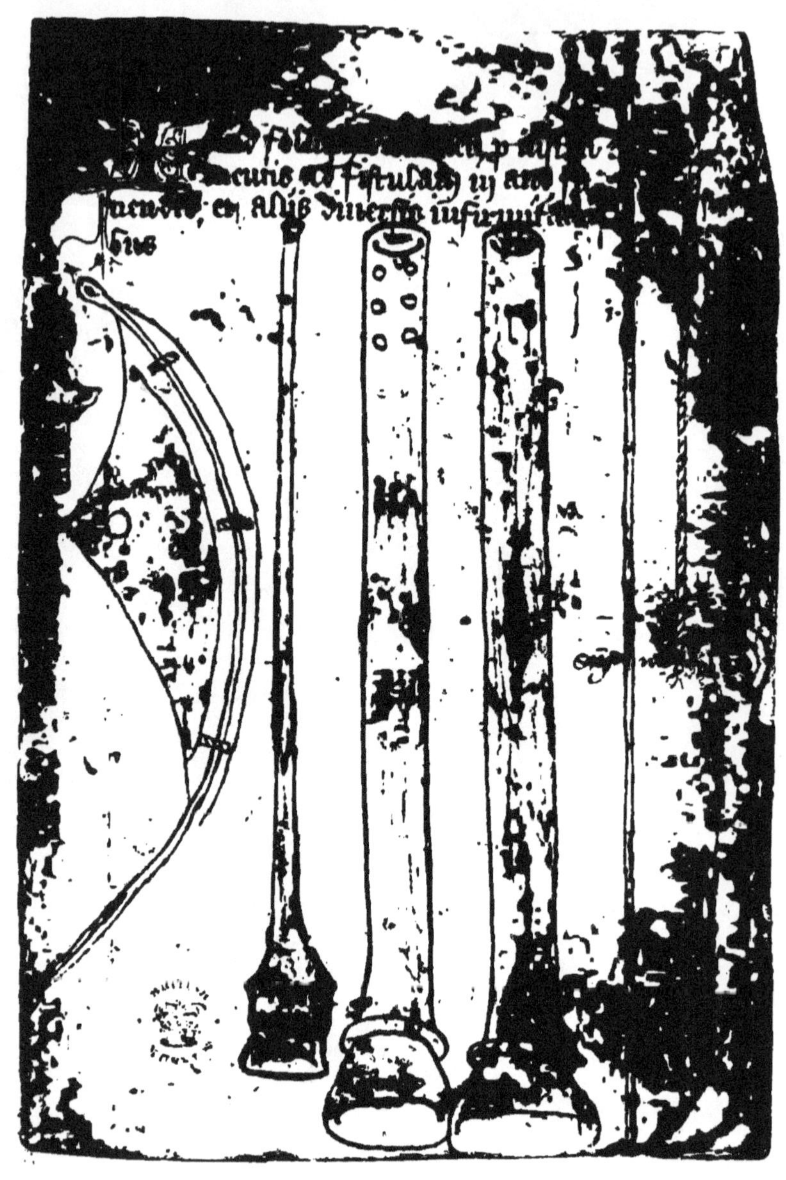

John of Arderne's Instruments for the cure of Fistula in ano.

(*Sloane MSS.*)

compescere linguam. The first virtu trow you to be to refrayne *th*e tong."

After this account of the line of conduct to be adopted by the fourteenth century surgeons, there follows a short account "of instrumentis necessary for the fistle," with the rude figures appended, which are represented in Plate III.[1] "Aftur al *th*ise it houeth that he knowe *th*e names of *th*e instrumentis *th*at perteneth to *th*e cure of *th*e fistule withoute whiche a leche may noȝt wele spede hym. Of which *th*e first is called *Sequere me* folowe me, whose shap is showed where *th*e instrumentez are paynted [Pl. III., Fig. 1]. And it is called sequere me for it is *th*e first instrument pertenyng to *th*at work, for a lech ow for to serche *th*erwith *th*e way of *th*e fistule Whider it go*eth*, Whe*ther* by *th*e middez of longanon[2] or noȝt. And it ow[3] to be made on *th*e same man*ner* as Wy*m*men us*eth* in *th*air heuedez[4] and of *th*e same metal, and it ow to be smal *th*at it may liȝtly be plied[5] and replied.[6] And be *th*e heuedez als little as thai may wele be ellez *th*ai miȝt noȝt wele entre *th*e mou*th* of *th*e fistule for *th*e streitnes of it. Ffor why oftymez fistulæ in ano hath riȝt smale holez. . . .

"Aftward is *th*er ano*ther* instrument *th*at is called *Acus rostrata* a snowted nedle [Pl. III., Fig. 2] for it hath *th*e tone heued like a snowte and in *th*e t'o*ther* an yȝe[7] like a nedel by which *th*redes ow to be drawen agayn by middez of *th*e fistule as it shal be seid agayn in his place. And it ow to be of siluer as it is paynted,

[1] Sloane MS., No. 2002, fol. 24, in the British Museum Library.
[2] the rectum. [3] ought. [4] heads. [5] bent. [6] rebent. [7] eye.

and it ow to be no gretter ne longer in *th*e snowte *th*an as it is paynted but it ow to be longer atte *th*e left *th*at it contene in al 8 ynches in leng*the*.

"*The third* instrument is called *tendicula* and it ow to be made of boxe or of ano*th*ir competent tree nou*ther*[1] lenger ne greter *th*an his shap is paynted [Pl. III., Fig. 3]. And it ow to haue an hole *th*rugh[2] in *th*e side as it is peynted. In whiche hole be *th*ere putte in a wrayst[3] by middez of whiche wraiste in *th*e ouer[4] end shal be a litel hole *th*rugh whiche shal be putte *th*e two endes of grete *th*rede four folde goyng atte firste by *th*e towel[5] and *th*e hole of *th*e fistule whiche *th*rede is called *frenum cesaris* [Pl. III., Fig. 4] and the whiche also goyng atwyx *th*e wraiste in wraistyng *th*e skynne atwyx *th*e towel *and th*e fistule be faste constreyned aboue *th*e snowte of *th*e nedel unto *th*at kutyng[6] be done.

"*Siringa* is an holow instrument by *th*e middez, and it ow to be made of the shappe as it is peynted here [Pl. III., Fig. 5] nou*ther* greter ne longer but even aft*er th*e shappe as it is peynted here, ne haue it noȝt but oon hole in the ne*th*er ende or smaller ende as it is peynted here."

Towards the end of the manuscript English translation of John of Arderne's work in the British Museum[7] are the series of little drawings reproduced in Plate IV.

The explanatory lines in the plate run as follows:

Above the top row of figures the first line is lost, but the second remains, as:

"& resonable[8] gouer*n*an*ce* of law & of lywyng."[9]

[1] nothing.
[2] through.
[3] twist.
[4] other.
[5] anus.
[6] cutting.
[7] Sloane MS., No. 6, fol. 175*b*.
[8] reasonable.
[9] living.

ILLUSTRATIONS FROM JOHN OF ARDERNE'S MANUSCRIPT.

(Sloane MSS.)

Above the second row of figures to the left are the lines :

> " *Æ*sculapi*us* helyd me*nne* wi*th*
> fernice*s*[1] & medicin*es*."

Whilst to the right may be deciphered :

> " Aschepi*us*[2] taught to geder[3] rots[4]
> *And* herbez, flou*ris*, & fro*tez*."[5]

Whilst above the bottom line of figures on the left is written :

> " Aschepi*us* schewed me*s*ure*s*
> And qua*n*titi*es*, weghtez & wases."[6]

In the middle :

> " Asclepi*us* teche*th*
> to mak pulu*eres*[7]
> [*Con*feccio*nis* & electuaries."

And finally on the right-hand side of the page is

> " Ypocras & galen schewe*th* certeyne
> Qua*n*tities i*n* reseyuyng."[8]

[1] ferns.
[2] Æsculapius.
[3] gather.
[4] roots.
[5] fruits.
[6] bundles : tow.
[7] powders.
[8] receiving (?).

CHAPTER IV.

THE ARMY SURGEONS—THE CONJOINT COLLEGE—THE FRATERNITY OF SURGEONS.

The army surgeons during the hundred years' war.

In the spring of 1415, Henry V. crossed the Channel to engage in that campaign which terminated so successfully for England on Oct. 25th at the field of Agincourt. The army medical arrangements during this expedition are preserved in the indentures[1] made by the king with his physician Nicholas Colnet and with his surgeon Thomas Morstede. The indentures are dated April 29th, 1415, and they set forth that Nicholas Colnet was to accompany Henry for a year into Guienne and France. As physician to the forces he was to be attended by three archers for a guard, each archer receiving sixpence a day, whilst Colnet drew twelvepence for his own pay. Thomas Morstede like his colleague had three archers assigned to him. He too received twelvepence a day in addition to the usual allowance of one hundred marks a quarter—the pay, it is stated, for thirty men-at-arms. The surgical department, however, from the nature of the warfare, which produced many lacerated wounds, was placed upon the more extensive footing, for whilst a single physician was considered to be capable of attending to the wants of the army, the surgeon was directed to take

[1] Rymer, "Fœdera," p. 237, vol. iv. p. 117. Ed. 1740.

with him twelve of his own craft. Each subordinate surgeon was to receive the pay of an archer (sixpence a day), and as a pledge for the punctual payment of the daily and quarterly allowances, Colnet and Morstede were permitted to take certain jewels belonging to the king.

On May 26th, 1415, shortly after his appointment, Morstede [1] petitioned the king to command and assign a sum of money for the purchase of such things as were necessary for his office so long as the campaign should continue. He also desired that the king would command such persons to act under Morstede, and that he would pay them such wages as that surgeon should appoint; and that in his wise discretion he would assign all kinds of conveyance necessary for the service, viz. one chariot and two waggons. The petitioner further inquires in a discreet manner, what war wages he is to receive for himself and for the other persons whom he engages to serve under him, and how many attendants are to be allowed him during the campaign. The answer to the petition runs: "The king has granted twelve persons of the craft, one chariot and two waggons." This cannot be considered as an overwhelming surgical staff for an army which, at the outset, consisted of six thousand men-at-arms, and twenty-four thousand foot, mostly archers. In a second petition, which is undated, Morstede prays the king "to grant his letters of Privy Seal directed to your Chancellor of England to cause him to deliver to your suppliant, letters of commission under

[1] Rymer, "Fœdera," vol. iv. part 2, p. 123. Ed. 1740

your great seal, by force of which he should have power to press, as well within as without franchise, twelve persons of his craft such as he should choose to accompany him and serve your most sovereign lord during your campaign."

Impressment of surgeons for the army.

The impressment of surgeons for military service appears here for the first time, but it was probably not a novel idea, and was frequently employed in those times when wars were frequent and surgeons' assistants few. The practice was continued for many subsequent years, and not only were the Barbers and Surgeons' Company and the Corporation of Surgeons after them called upon by Government in time of war to choose surgeons out of their own body to serve in the army and navy, but the physicians were from time to time subjected to the same regulation.

In accordance with the petition of Morstede, the following writ[1] was issued in 1416.

"THE KING to our beloved Thomas Morstede and William Bredewardyne our surgeons, Health.

"Know ye, that we have appointed to you, conjointly and severally, surgeons and other workmen to take and provide without delay for the making of certain instruments necessary and fitting for your mystery such as may be required for our present campaign beyond the sea, wherever they can conveniently be found, as well within the City of London as elsewhere.

"And therefore we warn you that ye may diligently attend to and execute these premisses in manner

[1] Rymer, "Fœdera," vol. iv. part 2, p. 166. Ed. 1740.

aforesaid. But we grant to all and every sheriffs, mayors, bailiffs, constables, and other officers, our servants and lieges, as well within the liberty as without, that they should consult and assist you as is fitting, according to the tenor of these presents, effectually in the execution of these premisses, in the commands which are to you and each of you entrusted.

"In witness The KING at Westminster, the fourteenth day of June. "PER IPSUM REGEM."

Rise of the physicians.

The medical and surgical knowledge of this period was at a very low ebb. The physicians, as has already been shown, were mostly ecclesiastics, reading the Latin medical authors, and writing fluently. The universities of Italy taught both physic and surgery *more antiquorum*, but merely as copyists and commentators, adding nothing to the general stock of knowledge. The surgeons were in some respects rather worse, whilst in others they were much superior to their brother practitioners. Few of them knew any language save their mother tongue, whilst those who were more learned carried on the old surgical notions and practice, rejoicing in knowing somewhat of the professional secrets of the physicians. The majority of the unlettered surgeons who really became surgeons in the true sense of the word, or handicraftsmen, were empirics. As empirics, however, they thought and acted for themselves, and laid up much useful knowledge.

About 1421 the physicians began to claim a recognition of their social position, and evinced a desire to free themselves from these ignorant impostors who had for too long been classed with their more respectable members. For this purpose the following petition was presented. "Hey and most myghty Prince,[1] noble and worthy Lords Spirituelx and Temporelx, and Worshipfull Com*mones*: for so moche as a man hath thre things to governe, that is to say Soule, Body, and worldly Goudes, the whiche ought and shulde ben principaly reweled[2] by thre Sciences that ben Divinite, Fisyk and Lawe, the Soule by Divinitie, the Body by Fisyk, worldly Goudes by Lawe: and these conynges[3] sholde be used and practised principaly by the most connyng men in the same Sciences, and most approved in cases necessaries to encrese of Vertu, long lyf and Goudes of fortune to the worship of God, and common p*ro*fyt. But worthy So*uv*r*a*ines, as hit is knowen to youre hey discrecion, many unconnynge and unapproved in the forsayd Science practiseth and specialy in Fisyk, so that in this Roialme is ev*er*y man be he ne*uer* so lewed,[4] taking upon him practyse, ysuffred to use hit, to grete harm and slaughter of many men. Where if no man practised theryn,[5] but al only connynge men and approved sufficiently ylerned in art, filosofye and fisyk as hit is kept in other londes and roialmes, then shulde any man that dyeth for defaute of help lyve, and no man perish by unconnyng.[6] Wherfore pleseth to youre excellent wysdomes that ought aft*er* youre Soule have mo[7]

The physicians' petition.

[1] Henry V.
[2] ruled.
[3] learnings.
[4] ignorant.
[5] therein.
[6] want of skill.
[7] more.

entendance to your body, for the causes above sayd to ordeine and make in Statuit perpetualy to be straytly yused and kept, that no man of no manner of estate, degre or condicion practyse in Fisyk from this time forward, but he have long tyme yused the scoles of Fisyk withynne som Universitie, and be graduated in the same . . . . undur payne of long emprisonement and payinge xli li to the Kyng; and that no woman use the practyse of Fisyk undre the same payne. . . . Also, lest that they whiche ben able to practise in Fisyk ben excluded from practysing, the which be nought graduated.[1] Plesith to your hey[2] prudence to send warrant to all the Sherrefs of England that every practysor in Fisyk, nought graduated in the same Science, that will practyse forth, be withynne one of the Universities of this lond by a certeine day, that they ben able[3] and approved after trewe and streyte examinacion be receyved to theyr degree, and they that be nought able to cese from the practyse unto the tyme that they be able, or never more entremette[4] thereof, and that thereto also be iset a peyne[5] convenient."

The last clause is apparently a liberal one, and shows that the physicians were willing to admit into their ranks the more skilful of the unlicensed, although but few would care to accept the privilege thus extended to them. The reply to this petition[6] directs the Lords

[1] on account of their not being graduates.
[2] high.
[3] skilful.
[4] intermeddle.
[5] penalty.
[6] "Rot. Parl.," tome iv. p. 130.

of the Council to see that the various recommendations therein contained are duly executed, but there is no evidence that any further steps were taken by the authorities in reference to the unlicensed practice of physic.

Conjoint faculty of physicians and surgeons.

It may therefore be assumed that the physicians took the matter into their own hands, associating themselves for this purpose into a society which was to co-operate with the pre-existing fellowship of surgeons. The conjoint college thus consisted of physicians and surgeons, each to be independent of the other as to their rights and privileges. The physicians appointed for their government two surveyors of physic to correspond with the two masters of the surgeons. The entire college, however, was under the control of a common head, bearing the title of Rector of Medicines. This officer was to be president or ruler of each of the associated bodies. It seems, however, that he was only to have been appointed occasionally as a dictator when there was urgent need for a common government. To bring the association more prominently into notice it was proposed to acquire an authorised place of resort for its members. With this purpose in view the City authorities were asked to assign to the community three houses situated within the City of London: one, to be furnished and desked for readings and disputations in philosophy and medicine, and to serve as a common hall; a second, for the congregations, elections, and consultations of the physicians; and the third for similar use by the surgeons.

The exact date at which the conjoint scheme was established is unknown, but it was at some time between May, 1421, and May, 1423. When the physicians petitioned the king and parliament at the former date it is certain that neither surveyors of physic nor rector of medicines was in existence, and yet in May, 1423, these officers, in common with the masters of the surgeons, presented a joint petition to the Court of Aldermen. During the months which had elapsed between these two dates events of considerable importance had occurred in England. Henry the Fifth had died, and his infant son had succeeded to the throne. The political difficulties which thereupon ensued were such as to leave little hope that the regent or parliament would assist so trivial a scheme (for thus it would appear in those days) as an increased provision for public health. Moreover, there is no lack of proof that then, as now, all classes of society were ready and willing to trust their lives in the hands of ignorant and impudent pretenders. Under these circumstances the physicians and surgeons very wisely resolved to obtain the concurrence and authority of the Mayor and Aldermen of London for the furtherance of their proposed purpose of improving the professional acquirements and social position of themselves and their successors. It cannot be doubted that the desire of the physicians and surgeons to place themselves under the powerful authority and countenance of so important a body as the Mayor and Aldermen of London then were was most sagacious. The execution of such laws for the control of the medical and surgical professions as the joint college might resolve, and the Mayor and Aldermen approve, was thereby ensured. The establishment of a college of physicians and surgeons in London had also this further advantage, that the larger concourse of people of all classes, the great assemblage of handicrafts of all kinds, and the frequent street fights which occurred at this period, would afford much greater scope for practical experience, practical teaching, and practical improvement than can even now be afforded by some of our justly venerated Universities.

The fellowship of surgeons.

If it be granted that the master surgeons[1] previously alluded to were not aldermen of a surgeons' guild, Thomas Stodeley,[2] in the year 1392, is the only person admitted as master of the surgeons, until in 1422 Thomas Morstede and John Harwe were sworn as supervisors of surgery. The surgeons, therefore, appear to have been a society distinct not only from the barbers proper, whose masters had been sworn before the Mayor and Aldermen almost uninterruptedly from the year 1378; but also from those of the barbers practising the faculty of surgery, over whom Simon Rolf and Richard Wellys were selected to act as overseers. The surgeons must therefore have been a society distinct from either the barbers or barber-surgeons, and they most probably originated in the association of the military surgeons, of whom, from the warlike proceedings of the age, there must have been no inconsiderable number, and who, by the

[1] Page 16. [2] Letter Book H, fol. 276.

very nature of their service, must have been persons held in considerable repute. Thomas Morstede and John Harwe had been surgeons to Henry the Fifth, and, as has been already mentioned,[1] the former had been with him at Agincourt. They were doubtless, therefore, men of importance amongst their fellow citizens. Indeed, Morstede afterwards became surgeon to Henry the Sixth, and subsequently a sheriff of the City.

It is not improbable, therefore, that the petition of the Associated Physicians and Surgeons to obtain authority from the Corporation for the foundation of their joint college was backed by the interest of such influential persons as Morstede, and that the success of the petition was in great measure due to them. Be this as it may, however, in the year 1423 the surgeons were a distinct body worthy of being associated with the physicians in obtaining from the Lord Mayor and Aldermen of London a very remarkable ordinance, having for its object the foundation of the conjoint college. By the regulations of this body, physicians and surgeons practising in the City of London and its liberties were required, after due examination, to become members of the Commonalty of Physicians and Surgeons. The examination of the physicians was conducted by the rector and the two surveyors of physic, or by the surveyors and the majority of the physicians, whilst the rector and the masters, or the masters and the

Regulations of the conjoint faculty.

[1] Page 47.

majority of the surgeons, carried out the surgical examination. As in the Barbers' Guild so in the new college, both physicians and surgeons were bound to report their cases within three or four days, the former to the rector and surveyors, the latter to the rector and the masters. In neither case, however, could the rector come to any decision without the concurrence of the physicians or of the surgeons, nor could he make any ordinance or constitution affecting either without their consent. Any physician convicted of bad practice or of open fault was to be reported by the rector and surveyors to the Mayor, who awarded the punishment for the offence. The surgeons also, under like circumstances, were to be reported for punishment by the rector and master.

Poor people who could not afford to pay for medical assistance might have a physician or surgeon assigned to them, without incurring any expense, on application to the rector and surveyors or masters. Care also was to be taken that neither physician nor surgeon should receive more than the patient could afford.

The rector, surveyors, and masters with two apothecaries assigned to them, were to visit all apothecaries' shops, throw away bad medicines, and bring the apothecary who had kept them before the Mayor and Aldermen.

No person was to be admitted as a graduate in medicine into the commonalty of physicians without letters of record or other proof of graduation, and all admissions were to be reported to the Mayor.

The rector, surveyors, and masters were to swear to

observe all the constitutions of their offices, "all hate, favor, or negligence left." The physician was to swear to practise "well and truly;" not to give "wittingly noxious medicines," nor to assent to any giver of them; neither should he neglect any sickness, although unknown to him; nor employing any medicine should he resort to sophistication or untruth. If he should know any person so acting or not admitted to the practice of physic, he was to report him to the rector and surveyors.

The surveyors in like manner were to swear not to employ any noxious medicines nor any sophistication nor untruth, neither were they to neglect any sickness, sore, or hurt; and knowing any person so doing or not admitted to the craft, report was to be made of them to the rector and masters.

The physicians were also to appear at the call of the rector and surveyors, and the surgeons at that of the rector and masters, in all lawful and honest causes, saving always the privileges, statutes, and customs of London commendably used.

In reference to penalties received as forfeits, "made in the faculty of physic" and the like "in the craft of surgery," the one half in either case was to be paid into the Chamber of London, and the other to the faculty or the craft, "as it best seemeth to the rector, surveyors, masters and their commonalty to be done."

To this ordinance the Mayor and Aldermen gave their sanction, retaining, however, the power to add or take away any article or "all" the ordinance "to put away as it to them most needful and speedful seemeth." This document is so interesting and so important in the history of English medicine, that it has been added[1] entire as extracted from the City records. The ordinance was granted on 15th May, 1423, and the surgeons lost no time in acting upon it; for on the 23rd May, "Magister Gilbert Kymer, rector of the faculty of physicians, Thomas Morstede and John Harwe, the supervisors of surgery, were presented and sworn before the Lord Mayor." The physicians were more tardy, for it is not until the 27th of September, 1424, that Master John Sumbreshede and Master Thomas Suthwell were presented and sworn supervisors, Gilbert Kymer, doctor of physic and rector of medicines being again sworn on the same day.

The establishment of a college of medicine and surgery within the liberties of the City of London was thus confirmed by the Mayor and Aldermen. No notice has been found, however, of the assignment to them of the houses for which they asked. But whether this request was granted or not there is no doubt that the college speedily enforced the power which they possessed not only to govern their own members, but to interfere with all who appeared to them ignorant and unauthorised practitioners. Amongst the latter class the college chose to include those barbers who practised the faculty of surgery. The barber surgeons soon became aware of the danger which thus menaced their very existence, for within eighteen months of the establishment of the

[1] See Appendix B, page 299.

college they obtained a fresh confirmation[1] of the power to practise surgery which had been granted to them in 1415 during the mayoralty of Thomas Fauconer, "notwithstanding the false accusation of the rector and overseers of the physicians and the masters of surgery."

There is no evidence to show how long the joint college existed; it has not even been ascertained whether it continued up to or after the assumption by Dr. Kymer of holy orders and his appointment to the Deanery of Salisbury, which took place in 1449. After September 27, 1424, there is no further notice of the swearing-in of the Rector of Medicines, nor any record of the existence of the conjoint college. We can only conjecture that the scheme was not found to work in practice, and that the enmity which existed for many subsequent years between the two branches of the profession was sufficient to prevent the physicians working in harmony with the surgeons. It is probable, however, that the rupture was not a violent one, as a few years later we find the physicians aiding the barber-surgeons to obtain a charter.

Nothing has yet been met with in the City records to show that any action was taken by the physicians to establish their society. The surgeons do not appear to have been much troubled by the reconfirmation of the barber-surgeons' privileges, for they steadily pursued their plan of consolidating the craft; and in 1435 they appear as an established body with a code of laws for the government of their society. At this date they consisted

[1] Letter Book K, fol. 27b.

of seventeen members, a not inconsiderable number if the time and place be taken into consideration. Their laws and regulations, contained in a small quarto volume written on vellum, are now in possession of the Barbers' Company, who probably came by it on their incorporation with the surgeons in 1540. The same book also contains a record of the laws of the Barbers' Company as established by the charter of Edward IV. in 1461, by which barbers practising surgery were confirmed in the privileges previously conceded to them by the mayor and aldermen; the craft of surgeons proper being wholly ignored. It would be needless to make any apology for printing in full this very interesting document, as it is really the standpoint of English surgery, and proves that although its practitioners had not escaped the narrow notions of the age in which they lived, their purpose was to improve the social position of the profession, and at the same time to provide for the maintenance of their poorer brethren.[1]

Ordinances of the Fellowship of Surgeons.

This remarkable document states that: "In the tenth day of May, the year of our Lord a thousand four hundred and thirty-five, in the year of King Henry VI. 13. By the good advice of the worshipful men of the craft or science of surgery in the City of London and all the cominalty of the same craft, a composition or an ordinance in this matter is made and assented stably[2] to stand ever hereafter."

The craft were to meet yearly on St. Cosmo and

[1] Appendix C, page 307. [2] firmly.

St. Damien's Day (27th Sept.)[1] to choose four masters "as old custom was" to rule and govern the craft and to hold the treasure and common goods, to be handed on by them to the succeeding masters, at the same time giving also an account of the same; and each of the outgoing and incoming masters was to receive six shillings and eightpence at this audit. At every yearly election two of the former masters might be retained, but other two from them were to be elected to make up the full number. But no master after two years' service could be elected for the next coming year "against his will." Every master within ten days of his election was to "be presented and take his charge," under penalty "to the box of the craft of thirteen shillings and fourpence," without provable cause, and then another master to be elected by the four men chosen from the fellowship for the same year, who together with the old masters were to present him within ten days or each of them "to pay to the box of the craft *three shillings and fourpence.*"

Every surgeon of the fellowship was "to pay *yearly twopence* a quarter to the box, that is *eightpence* a year to the profit and worship of the craft and in helping and relieving the need of the poor men of the same fellowship, "to be collected by the masters with one of the four men with them;" and a book of account was to be kept in which also were to be entered all fines received.

[1] Two brothers, who practised as physicians in Cilicia, and were martyred in the early part of the fourth century. They are supposed to have been the first practitioners who refused fees. For this reason perhaps they were selected as the patron saints of the guild. At a later period St. Luke fulfilled this function. F. Bœrner, "De Cosma et Damiano." Helmstadt, 1751.

All the craft were to meet "once a quarter of duty," besides on election day, "to hear and learn the good ordinances, rules, and governaunce of the said craft, and as oft as it be needful at other times." For non-attendance at the quarterly meetings there was a fine to the box of *sixpence*, at other meetings of *fourpence*, unless reasonable excuse could be given. The masters were to call the duty meetings, and if they failed to do so, or were absent from the meeting, in either case each offender was fined *three shillings and fourpence*, but in regard to other meetings each time *twelvepence*, excepting on reasonable cause. In the event, however, of reasonable absence "of the masters or their deputies, or of any of the other four men chosen for the fellowship whether it be one or two of them of either party the other[1] to proceed with their business. And if any one of the eight shall be proved to have made false excuse for absence he is to be fined double of his penalty set afore."

None "of the four masters, neither any other person of the said fellowship of the craft of surgery," is to "put any man out of his cure[2] otherwise than the honesty[3] of the craft will,[4] but that each of them be ready if need be or by any of the parties called thereto then honestly[5] to help each other with counsel or deed, that worship, profit, and honesty of the craft and helping of the sick be done on all sides" (a noble maxim *occasionally* forgotten in the present times), "and if any of the said craft do the contrary, that each such

[1] the rest.
[2] filch his patient from him.
[3] honour.
[4] allows.
[5] honourably.

content[1] with the owner of the cure[2] to the value of all the cure and over to pay to the box *six shillings and eightpence* for his trespass."

If any of the fellowship of the craft "disclaunder or deprave[3] any of the fellowship unrighteously or unhonestly," on proof by witnesses he shall "pay to the box *three shillings and fourpence*, and over that make amends to the person he hath disclaundered after the judgment of the honest masters and their fellowship note fault herein."[4]

No freeman of the craft of surgery was permitted to employ "a foreigner[5] over a month" unless within that time he brought him before the masters and fellowship for examination, who being satisfied allowed him to covenant with his employer for three years subject to the control of the craft. If the freeman disobeyed he was fined "to the box *twenty shillings* and to discharge his servant."

Any one of the fellowship having a case "likely to result in death or maiming or which to him may be unknown,"[6] was bound to show it to the masters under penalty of "*thirteen shillings and fourpence.*" And if the master did not attend he was to be fined "as oft as he was herein faulty[7] *six shillings and eightpence.*"

Whatever "profit or advantage of gift come to any of the four masters" from being called in as above or from

[1] make pecuniary arrangement.
[2] case.
[3] slander or run down.
[4] think faulty.
[5] a non-freeman.
[6] that he did not understand.
[7] inattentive.

any other source was to be divided equally among the four masters or their deputies.

Every examination or adjudication appertaining to the craft was to be "done and performed evermore honestly[1] by the four masters and their deputies and freemen of the said fellowship."

Any foreigner was to be received into the craft and made free by redemption with "the assent of all the four masters and at the least two of the four men chosen for the fellowship," and he was to pay "to the four masters their fees and a certain [sum] to the box and a dinner to the craft."

No person was to be made a master within seven years of his entering the craft and unless he had been "proved good and honest of governance and secret"[2] during that time.

When apprentices were made free of the craft they were to be called before the four masters to receive the charge of being ruled and governed by the craft, and then "as be goodly[3] give *three shillings and fourpence* to the box. But no person shall become a master till six years after his admission, during which time he must be proved well governed and honest, wise and secret, else he [shall] not be received to the office of mastership within other six years. And if he be not found in the twelve years well ruled in the manner forsaid he [is] never to be chosen master."

No one was to go to law with another "for no cause

[1] honourably. [2] discreet. [3] proper.

longing[1] to the said craft, on penalty of *twenty shillings* to the box" if "he inform not·the four masters, who are to take it into their hands and duly and truly examine it [the matter] and redress it, righteously and conscientiously for both parties of them within forty days at the most or farthest." If the dispute, however, occurred between a master and one of the craft, then the master might appoint in his own place "one of the four chosen for the fellowship" as one of the judges in his room.

The masters were yearly to visit, "as oft as it is needful, the householders of the craft," to ascertain what apprentices or covenant men they have, whether they be ruled and governed after the franchise of the City and their oath, and if they be found disobedient to the ordinances, "to make it known to the Mayor or Chamberlain as custom and manner of the City wills."

Should it be advisable that any penalties, then or thereafter enjoined, should be moderated, this was to be done by the masters and four men "to the furthering of the peace, profit, health and welfare of God's people and the king's." But if they could not "accord within the said fellowship, the masters [were] then to have recourse to the Mayor or Chamberlain . . . to correct them that are misgoverned against the good ordinances of the craft, and also untrue[2] workers in the craft, of the same fellowship and so proved [to be]."

Great caution was shown as to any proposal "of

[1] belonging. [2] dishonest.

amending and addition of[1] the composition," as the following law fully shows.

"And if [at] any time to come hereafter it seem to the craft anything in this foresaid ordinance and composition to be too much or too little, that then the said craft by one assent, and after their good advice and discretion it is to be commouned[2] discreetly, that is to say, that the matter be duly examined by[3] good advice in a convocation of the fellowship [for] four or five days, and that by[3] a copy had out of the original of the matter, and to be answered by profitable reason and writing, and otherwise not to be received at any time to come hereafter. And if any ordinance in this mannerwise is thus approved, afterward be it not impugned."

Not unwisely for the social benefit of the craft, for meeting old friends, and rubbing off any little asperities which might arise among its members, it was "ordained that every freeman of the craft of surgery pay yearly to the dinner of the craft, that is to say, on the day of Saint Luke, each man [an equal sum], whether he be present or absent, except our own poor, and the overplus thereof not spent, if any such, is to be kept and spent on the next great day."

The masters were to be bound "in a plain obligation" for the property of the craft, and of the four men chosen for the fellowship, and "truly keep it for the use of all the craft" during their year of office, and at the end thereof "to yield their account of the property to the foresaid four men, and they [being] content

[1] to. [2] considered. [3] with.

. . . then the bonds of the masters to be broken or stand for nought. And in the same mannerwise be bounden the four men chosen for the fellowship to the said masters, in an obligation of two marks for none other intent but to hear and receive again the said goods of the craft, and hear their account and allow them that."[1] After which their bond to be discharged. And so on for ever at every change of the masters or of the four men.

Every person admitted into the craft was to be sworn in the following words: "Ye shall swear that ye shall well and truly behave you in the working[2] of the craft of surgery in saving of God's people and the king's. And all the good ordinances and rules and secrets of the said craft, ye shall well and truly keep within the said craft. And to all the lefull and lawful biddings of the masters of this said craft that now are and hereafter shall be, ye to be thereto ever continually obedient when ye be called, and never it to forsake but to fulfille. So God help you and all Saints." A holy and honourable engagement upon all, whether they think reverentially or lightly on their invocation of God.

In the ordinance. "Of penalties of misgovernment,"[3] many provisions excite a smile, as whoever has had the chance to be among the members of a like deliberative body must have seen a representation of precisely the same conditions, with the simple exchange of the president's knock of his hammer for the fine.

[1] it. [2] the practice. [3] penalties for disobedience.

"When the masters at any time sit in judgment, or in examination, or in communication of[1] the said craft, with the whole fellowship or parcel thereof as place cause and time requireth, that then every person of the said craft that time present *keep silence at the first bidding or commandment of the said Masters, and not without license of them had, again to speak*. And if any will not at the first bidding cease, for the second time (of them) bidden to cease to pay for the said fault *twelvepence*, and if he will not yet cease for the third warning *two shillings*, for the fourth time *a noble*, and if he will not then cease at the fifth time to be taken for a rebel."

Personal quarrels among the craft were punished. No one was to be malicious or excite malice, "which" might be cause of disturbance of the good peace among the fellowship of the said craft, upon pain to pay to the box *twelvepence*. If "any draw any weapon in violence or unlawfully menace any person of the said craft to pay therefor *a noble*. And if any of them smite another of the same craft to pay to the box *twenty shillings*, and over that the parties to be justified by law or by ordinance of the craft." If the masters or either of them offended as above, they were "to pay the double of the pain[2] set afore." And "if any of the said fellowship revile, or menace either falsely any of the masters, or upon them unskilfully complain, he [is] to pay *two nobles* at each time, and as oft as any of them so doeth."

"The intent of this ordinance is thus for to

[1] with. [2] penalty.

undirstonde that all the said craft and every person thereof, be well ruled and governed within themselves, that is to say, both the masters and their fellowship and all the things that shall among them be done or said, that is, the Masters patiently their matters to hear and wisely and truly there to see, and the said fellowship in time reasonable to ask,[1] and in their complaints and sayings honestly to be mesurable,[2] and to[3] silence meekly to obey after the discretion of the masters as is aforesaid, by virtue and ordinance of this City ordained to masters and wardens to[4] crafts."

"The Charge and the Oath that the Old Masters give to the New.

"Ye shall swear that ye shall well and truly govern the craft of surgery, and the fellowship of the same craft after your cunning[5] and your power as long as ye be master for this year. And also ye [are] to keep and see to be kept all the good rules and ordinances of this said craft now made, and that none of you any other ordinance to make, neither to your knowledge suffer to be made without the assent of all the fellowship, and that also with condition and manner as it is before ordained and written."

"The conclusion of this composition above written is this, that no person of the said craft presume in any wise to break this said ordinance, neither any other to break it on pain[6] of *an hundred shillings*, into the time

[1] to inquire of.
[2] not unreasonable.
[3] in.
[4] of.
[5] knowledge
[6] penalty

that a better ordinance be founded or made, and so of the craft received."

The City records do not show any further disputes between the surgeons and the barbers practising surgery. It would, therefore, seem probable from the lists of the masters of cirurgy of the conjoint college before alluded to, and of the master barbers exercising the faculty of surgery already given and their days of swearing in before the Mayor and Aldermen being separated by the interval of a few weeks, that there was no attempt at fusion between them, but most probably some sort of truce was agreed upon.

In 1450, as appears by the probate of his will, Thomas Morstede died, a rich and influential man, who had been surgeon to Henry VI., Henry V., and probably also to Henry IV. His death must have been a great loss to the surgeons, whether the college had been previously broken up or not. As regards Morstede's position, it may be noted here that in the first Patent Roll of the 16th of Henry VI., No. 22 is the patent appointing "Thos. Morstede Scrutator omnium navium ac batellorum in portibus London : ac omnibus aliis crykes ex vtraque parte vsque Gravesend." And it does not seem very improbable that this was our Thomas, for it was by no means uncommon in those times to fit a round peg into a square hole, when profit and place were assigned to a favourite whether he were fitted for it or not. But Morstede's public services had entitled him to be rewarded with an appointment the actual duties of which might be done by deputy, as allowed by the patent, whilst he reaped the profit of the office. Thomas Morstede was buried by his own order in the church of St. Olave Upwell in the Jewry, to which, as Stowe says, he had built a fair new aisle during his lifetime.

By his will, after leaving money for masses for the souls of himself and his wives, as well as for those of such as were connected to him by family ties or friendship, money for the poor of parishes in Surrey and in Essex, with sundry bequests of church plate and of money to various persons, he bequeaths the whole of the rest of his property to his second wife Elizabeth, the daughter of John Michel. Among the bequests is the following:

"Item: I leave to Roger Brynard, my apprentice, ten marks sterling (£6 13s. 4d.) meum librum Anglicanum ligatum cum duabus latitudinibus, omnia Instrumenta mea Cirurgie cum omnibus suis pertinentibus, meum cornu Argento ornatum et meum magnum pyxidem argenti."

Shortly after his death, in the year 1452 the Barbers' Guild obtained a grant of arms in the following terms:

Grant of Arms to the Guild of Barbers of the City of London, 30 Henry VI.

"Be it knowen to all men that y Clarensew Kyng of Armes of the South Marche of Englond Consideryng the noble estate of the Cite of London by the name of Erle & Barons as in their ffirst Charter by scripture appereth and as now called mayre and aldermen and by good avyce of all the aldermen and the noble citezenis of London that euery alderman shuld haue a ward by hymself to governe and rule to the Worship of the cite and the maires power to haue euery alderman in his Ward with correccion of the mair beyng for the tyme and so notablie ordeyned to be custumed euery Craft clothyng be hem self to know o[ne] Craft from another and also synes of Armes in baner wyse to beer conueniently for the worship of the realme and the noble cite and so now late the Maisters of Barbory and Surgery within the craft of Barbours John Strugge[1] Thomas Wyllote[2] Hugh Herte & Thomas Waleys[3] come & praying me Clarensewe Kyng of Armes to devise hem a conysauns & syne infourme of armes vnder my seall of myn Armes that might be conveniently to ther Craft And where y Clarensewe Kyng of Armes consideryng the gode disposicion of them y haue devysed a Conysaunce in fourme of Armes that is to sey A felde sabull a cheveron bytwene iij flemys of siluer the which syne of armes y Clarensew gyve the same conysaunce of Armes to the forsaid Crafte and none other Crafte in no wyse shall not bere the same. To the which witenesse of this wrytyng y sette my seall of myn armes & my syne manuall wreten atte London the xxix day of the monthe of September the xxxth yere of the regne of our soueraync lord Kyng Henry the Sixt

"By Claransew Kynge of armes."[4]

[1] John Struggo or Struge was Master of the Company in 1449 and again in 1452. [2] Thomas Willot, Master in 1458 and 1466.

[3] Wardens of the Company at the date of the grant.

[4] See also Appendices J and K and page 124. The arms portrayed in the frontispiece are those granted to the United Company of Barber-Surgeons, which consisted of the old Guild of Surgeons incorporated with the Company of Barbers. The first and fourth quarterings in that shield show the arms here granted to the older Barbers' Guild.

CHAPTER V.

THE BARBERS' CHARTER—PRIVILEGES OF SURGEONS—RISE OF THE COLLEGE OF PHYSICIANS.

The barber-surgeons.

WHATEVER may have been the condition of the college, or of the surgeons alone if the college had been dissolved, the barbers in their vocation had been growing into an important civic body. In 1450 they desired the sanction of the Mayor and Aldermen to a code of laws of great length and stringency for the government of their own members and the protection of the craft. They insisted on their right to practise surgery by the insertion of a clause ordaining "that no barber nor able person using barbery shall inform any foreyner nor him teche no wise in eny manner of point that belongeth to the crafts of barbery or surgery." The entry commences, as is usually the case, in Latin; but the petition, which is a curious production, and the laws themselves are in English, whilst the ratification of them by the Court of Aldermen is again in Latin. The ordinance runs as follows: "Be it remembered that on the xxvth day of February in the twenty-ninth year of the reign of King Henry the Sixth after the Conquest (1450), the masters and wardens and other honest folk of the craft of barbers came hither into the king's court of Guildhall, and laid before Nicholas Wifold, the Mayor, and the Aldermen of the City, a certain Bill or supplication."[1]

An important epoch in the history of the English barbers was now at hand. Those members of the guild to whom such frequent reference has been made as practisers of surgery, either improved their surgical attainments, or, as is the more likely, sided with the popular party according to the wont of the citizens of London. For in the first year of the reign of Edward IV., the people's favourite, they received a charter granted to them no doubt as one of the rewards which were showered on the City in return for the readiness with which the citizens had espoused his cause, and for the acclamations with which they had received him after the battle at Mortimer's Cross. The charter was granted on Feb. 24, 1462, before Edward was firmly established on his newly won throne, and whilst he was yet a friend of the Aldermen. Ostensibly the charter was granted to the barbers as a class, but ostensibly only, for the shavers and the trimmers of beards are passed over in silence as if the "Barbitonsor" was to exist in name only, whilst his place was taken by the modest barber-surgeon. Both surgeons and barbers had been licensed, as we have seen, by the City, though till this date they had neither of them obtained a royal charter. The barbers, as appears from their records and from a journal kept in the office of the Town Clerk of the City of London, were a guild in the second

[1] Letter Book K, fol. 250. See Appendix D, page 321.

year of King Edward II., 1308, whilst they were a livery guild in 1387.

The Barbers' Charter.

The charter[1] begins by reciting how our "beloved, honest, and free men of the trade of barbers of the City of London using the craft or faculty of surgeons have for a long while diligently and laboriously occupied themselves with the wounds, bruises, hurts, and other ailments of our lieges, in tending and curing our lieges, as well as in bleeding them and in drawing their teeth." Barbery, *i.e.* the shaving and trimming of beards and the cutting of hair, is not mentioned. On account of the ignorance of those who practise surgery, both surgeons and strangers who do not belong to the brotherhood or freemen of the City, "other of our lieges have gone the way of all flesh, whilst others from the same cause are sick and incurable." At the humble petition of the aforesaid beloved honest freemen, the members of the craft are constituted one body and community. The two principals of the community well skilled in the art of surgery, with the assent of twelve or at least of eight of its members were annually to choose and make two masters or governors most skilled in the art of surgery, "to oversee, rule, and govern the craft and community aforesaid, and all men of the said craft and their affairs for ever." It seems probable that by the words "all men of the said craft," these barbers who acted as surgeons, and were

[1] See the charter in full in Appendix E, page 326; with the confirmations by Henry VII. in Appendix G; and by Henry VIII. in Appendix H.

truly barber-surgeons, hereby obtained the power of being alone elected masters of the Company of barbers or shavers and trimmers of hair and beards. The whole community, however, obtained a corporate seal, power of holding property to the amount of five marks [66s. 8d.], ultra reprisas, of pleading and impleading, of holding courts and making statutes for the government of the Company without hindrance from the king, etc., so long as they were not contrary to the laws of the land. They also had the power to oversee, scrutinise, and govern all freemen surgeons who acted as barbers in the City of London, and all strangers; whilst they had the oversight of all instruments, plaisters, and medicines for curing wounds. The powers of fine and imprisonment were also conferred upon them, a privilege which was only possessed by a few of the other city companies, whilst they were relieved from the duties of serving on juries and inquests.

The charter was enrolled by the Court of Common Council shortly after it was granted by the following ordinance. "At a Common Council, held on Friday, the sixth of May, in the 3rd year of the reign of King Edward the Fourth after the Conquest (1463), it was agreed that certain letters patent lately granted by the king," etc., should be enrolled, with the provision that anything enacted contrary to the liberties of the City should be wholly null and void.

From the provisions of the charter it is quite evident that the barber-surgeons had made for themselves a position which they did not relinquish but continued to improve until they finally became surgeons, when the Barbers' Company and the Guild of Surgeons were united by Henry VIII.

Barbers' apprentices.

The regulations of the Company in regard to the binding of the apprentices in the reign of Edward IV. have been preserved. They are dated 1483, and ordain that "no fraunchesed barbour w*it*h*in the Citee [of London] shall take any man or child to be his apprentice before that he hath *pre*sented the same man or child unto the maister and wardeyns" of the craft that they may "duely examyne, ou*er*see, serche, and beholde by the colour and complexion of the said man or child if he be avexed or disposed to be lepur or gowty, maimed or disfigured in an*y* p*ar*ties of his body, Whereby he shall fall in disdeyn or lothefulnesse unto the sight of the King's liege people. And also to be examyned of his birth and of his kyn[d]rede, or if there be on hym any bonde claymed. And if he be founde defectif in any of thise poynts that than no fraunchesed Barbo*ur* of the saide Citee shall take hym to be his apprentice vppon payn*e* to pay v^{li}."[1]

The victory at Bosworth and the consequent union of the rival roses found the Barbers' Company in the same state of disorder and disagreement as other classes of society. A petition of 1486 declares that "they of longe tyme haue been in discorde and not of oon[2] conformitie but euery man in effect of the saide craft or science hathe taken and folowed his owne singuler way

[1] Letter Book L, fol. 174.

[2] one.

and apetite,"[1] and they desire of the Mayor and Court of Aldermen certain articles for the better governing of their Company. The articles were that none but enfranchised barbers should keep open shop within the City under a penalty of a forty shilling fine. Not more than two "stranger" servants were to be kept by each barber, and these only on condition that they were presented before the master and wardens of the Company within three days of their coming into service. The fine for acting contrary to this ordinance was £5, and such importance was attached to its enactment that those masters and wardens who did not see it duly enforced or connived at its disregard were themselves subjected to a fine of thirteen shillings and eightpence. Each enfranchised barber might instruct three apprentices, but no more, except "that it shal be lefull to every suche persone oon yeere before the tyme of the apprentishode of any of his apprentices be expired to take another apprentice in the stede of hym that is nygh commynge oute of his tymes of apprentishode to th' entent that the same newe apprentice may haue his erudicion and larnyng in the said crafte or science of barbours before the tymes of the other apprentice" be expired. "Also if any persone of the said craft or science selle away his apprentice to another manne . . . that then it shall not be lefulle to any suche persone so sillyng away his apprentice to take any newe in his stede duryng the tyme to come of apprentishode of that apprentice so sold. Yet neuertheles if it fortune any apprentice

[1] Letter Book L, fol. 235*b*.

to dye withinne the tyme of his apprentishode that then it shall be lawfull to the Maister of that apprentice so dying to take another in his stede whensoeuer it shall lyke hym."

Surgeons privileges.

The next notice of the surgeons in the City records is of interest, as it forms the beginning of a long-continued struggle between the authorities and the surgeons. At their enfranchisement the surgeons had obtained an exemption from the keeping of watch and the bearing of arms, a privilege which the City would not recognise without compulsion, and which required constant petitions and unabated energy to maintain. A shadow of the immunity remains in the exemption from serving on juries which the medical profession still enjoys. Even now the Chairman of the Court of Examiners at the Royal College of Surgeons in Lincoln's Inn Fields, when admitting new members, enumerates amongst the privileges conferred by the diploma of that body that its holders are exempt from service in the militia. The humble petition of these surgeons in 1491 is, "that whereas they and their predecessours from the tyme that no mynd is to the contrary as well in this noble Citie as in alle other Cities and Burghes within this Realme or ell*es*where for the contynuell s*er*uice and attendaunce that they daily and nyȝtly at alle houres and tymes geue to the kyng*es* liege people for the releue of the same Accordyng to their science hath ben exempt and Discharged from alle offices and besynesse Wherein they shuld use or bere any man*ner* of armure or Wepyn.[1] And in like

[1] weapon.

priueleage hath ben entreated as heraudes[1] of Armes as well in batailles and ffelds as other places therefor to stonde unharnessed and unwepened According to the lawe of Armes, because that they be *per*sons that neuer used feat*es* of werre[2] nor ought to use but only the besynesse and ex*er*cise of their said science to the helpe and comfort of the kynges liege people in the tyme of their nede. And in this noble Citie from the time of their first incorp*or*acion When they haue been many more in nomber than they nowe be were neu*er* called nor charged to be on quest, Watche, nor other office whereby they shuld use or occupie Any Armure or defensible geere of warre, where thrugh they shuld be unredy and letted[3] to practise the cure of menne beyng in p*er*ill. Tille nowe of late at the last eleccion of constables oon of theym hath ben called uppon and [is] likely to be compelled to be a Constable contrary to the priuelage of their science as is abouesaid. . . . . Please it, therefore, [to your lordship and maisterships] considerynge the smalle nombre of your said suppliant*es*[4] in Regarde of the greate multitude of pacients that be, and daily by infortune increseth in this Citie. And also that if your said oratours shuld be compelled suche offices to occupy, that other at many tymes the kynge's liege people sodenly wounded and hurt for Defaut of helpe in tyme to theym to be shewed

[1] heralds. [2] war. [3] hindered.

[4] Who were, as the earlier part of the petition recites, "*your* poure orat*ours* the Wardeyns and other gode ffolkes of the ffelisship of surgeons enfraunchised in this Citee not passyng in nombre of viij p*er*sones."

shuld *per*isshe as godde forbede;[1] to enacte and establisshe that from hensforth your said suppliaunt*es* may be Discharged of Constableshipp, Watche, and of alle other maner of offices beryng[2] any Armure, and also of alle enquestes jurys within this Citie. And to contynue as they haue done in tymes passed and also that they may haue the Serche of alle the fforeyns that within this Citie usen the feate of Surgery to thentent that no man shuld occupie but such as your said oratours shuld thinke to be able[3] and hauing cunnyng and exp*er*ience in the said science, and this to be auctorised by your noble Court and your said suppliaunts shalle dewly pray for the prese*r*uacion of your worshipfulle lordship and maisterships." The petition passed unanimously and became an ordinance.[4]

In 1492 arms were granted to the Guild of Surgeons. The original grant still remains at the Barbers' Hall beautifully engrossed on vellum. It commences: "The ȝere of owre lord, MCCCCLXXXXII att the goyng ovyr the see of oure soueyn lord kyng Harry the VIIth in to Fraunsse. Thes armys were geuen on to the crafte of surgeons of London the vii^th^ ȝere of his reyng in the tyme of Hewe[5] Clopton, May*or*."

Arms granted to the Surgeons' Guild.

From this time it appears certain that the barbers who practised surgery and the surgeons who were members of the guild lived amicably together. The former indeed had the superiority in position and

[1] This passage is involved, though the meaning is clear enough.
[2] bearing. [3] skilful. [4] Letter Book L, fol. 293. [5] Hugh.

power by virtue of the charter granted to them by Edward the Fourth. The surgeons, however, probably held a better social position, although they were only enfranchised by the City authorities. It can hardly be doubted that they were of superior professional attainments to their brethren of the Barbers' Company. The craft had probably originated in the association of military surgeons, and would thus be composed of men who had seen service in France and Italy, and who must thus have been brought into contact with the surgeons of these countries, in which medical knowledge was at its highest development. Proof of the superior position held by the members of the surgeons' craft is to be found in a document which will be quoted almost immediately, and in which precedence of the barbers practising surgery is given to the surgeons. This important document is the "Wrytyng of Composicyons"[1] which concedes to the surgeons all the chartered privileges of the barber-surgeons, with the exception of admission to the freedom of the Barbers' Company. No evidence remains to show how this combination between the two sets of practitioners was brought about, or who were the chief agents in effecting it. There is no doubt, however, that it benefited both parties. The craft of surgeons was a small but energetic body, whilst the barber-surgeons, though numerous, lacked that experience which could be obtained from their more practical brethren. The Writing is dated the twelfth of

The barber-surgeons and the fellowship of surgeons.

[1] Appendix F, page 331.

July in the year of our Lorde God 1493, and is "of Composicyons made betwixt the ffelishippis of surgeons and the ffelishippis of barbours surgeons and surgeons barbours."[1]

Civic position of the Barbers' Company.

Nothing further of importance seems to have occurred in the history of the barbers and surgeons until the fifth day of December, 1499, being the fifteenth year of the reign of King Henry the Seventh, when the charter of Edward the Fourth was confirmed at Westminster under the great seal to Richard Heyward, James Holand, John Robertson, and John Boteler, the masters and governors of the craft of barbers and surgeons. Only a portion of this charter now exists,[2] but it is recited by the first charter of Henry the Eighth; it appears to have confirmed the previous charter of Edward the Fourth in every respect, and to have increased the number of masters from two to four. At the coronation of Henry the Eighth and Catherine, his consort, the barbers stand twenty-fifth in "*th*e ordo*r* of crafts as they shall stond (when the kyng and the Queyn shall passe by towards their coronacion) in the Chepe a litell from the olde Chaunge ende."[3] On the same occasion they were rated to make "xiv yerds of rayles at their prime[4] costs for their stondyngs." This order of precedence was the source of many pretty quarrels, which it required all the authority of the

[1] See Appendix F.

[2] Letter Book M, fol. 216.

[3] Journals x, fol. 370*b*. See Editor's Preface in regard to Journals.

[4] personal.

Mayor and Court of Aldermen to adjust. In course of years the Company obtained a higher place, for in 1532 it is ordained that the barber-surgeons shall "go in all processions, goings, standyngs, rydyngs, and other assembles for the worship of this Cite," and that they should be ranked as the xviiith of the City companies.[1] Two years later the Court of Aldermen "ordered the wardeyns of the mystery of barber-surgeons that theyre company shall no more goe yn *pro*cessions, standyngs, etc.,"[2] and that "the barbours shall be the xviith Company immediately to goe afore the company of cutlers."[3] In 1535 they were still ranked as the xviith company.[4]

Licensing of surgeons.

No improvement seems to have taken place in the practice of surgery as a result of the new Charter, nor does the practice of medicine appear to have been in better plight. Accordingly, in the third year of the reign of Henry the Eighth (1511) an Act was passed forbidding any person "in the City of London or within seven miles of the same to take upon him to exercise or occupy as a physician or surgeon except he be first examined, approved, and admitted by the Bishop of London or by the Dean of St. Paul's for the time being." Each of these dignitaries was to associate with himself four doctors of physic before granting a licence in medicine; and for surgery, other expert persons in

[1] Repert. viii, fol. 287*b*. See Editor's Preface for an account of the City repertories.

[2] Repert. ix, fol. 79. [3] Repert. ix, fol. 99. [4] Repert. ix, fol. 145.

that faculty, who were to certify after due examination as to the fitness of the candidate to practise his art. In other parts of the country the bishop of the diocese or his vicar-general acted as licenser.

By the passing of this Act the surgeons raised up for themselves a host of enemies. A petition was at once got up against it, which stated that "the Company and Fellowship of Surgeons of London, minding only their own lucres and nothing the profit or ease of the diseased, have sued, troubled and vexed divers honest persons, as well men as women, whom God hath endued with that knowledge of the nature, kind, and operation of certain herbs, roots, and waters, and the using and ministering of them to such as be pained with customable diseases, etc. And yet the said persons have not taken any money for their pains or cunning, but have ministered the same to the poor people only, for neighbourhood and God's sake and charity. And it is now well known that the surgeons admitted will do no cure to any person but where they shall know to be rewarded with a greater sum or reward than the cure extendeth unto. For in case they would minister their cunning to sore people unrewarded, there should not so many rot and perish to death for lack of surgery as daily do." So powerful were the agitators that "it was ordained, established, and enacted of this present Parliament that at all times from henceforth it is lawful to any person being the king's subject, having knowledge or experience of the nature of herbs, etc., to minister in and to any outward sore or wound according to their cunning." This enactment was practically a repeal of the former statute, and must have given a great blow to the legitimate practice of surgery in this country.

In 1513 the surgeon's guild again applied to Parliament, but on this occasion with a different object, viz. to be "discharged of constableship, watch, and all manner of office bearing any armour, and also of all inquests and juries within the city of London." And as proof of their continued friendship with the barber-surgeons, they pray that this exemption may extend to all barber-surgeons, admitted and approved to exercise the mystery of the surgeons "according to the form of the statute lately made in that behalf, so that it exceed not, nor be at one time above, the number of twelve persons." They thus raised again the old question, and their petition was granted by Stat. 5, Henric. VIII., cap. 6, enacting that surgeons shall be exempt from attendance at inquests, assizes, etc. The bearing of arms and the keeping of watches, was no light matter as may be gathered from the following precept addressed by the Mayor to the wardens of each City company about this date: "We will and charge you that for the honour of this citie ye do ordeyn and prepare against the watches to be kept within this citie in the nights of the vigilles of Seint John Baptist and Seint Peter nowe next comynge . . . honest and comely persones suche as ye will answer for, with Bowes and Arrowes clenely harnyssed and arrayed yn jaketts of whytte having the armes of this citie, to wayte and attende upon vs in the sayde watches and to come to Blackewell hall and there to be sett forthe. Not fayllynge hereof as ye tendre the honor of this citie, and also will answere at your perylles." The barbers are rated at four men. The concession thus obtained by statute was, as we have seen, and as the records of the Guildhall testify, the outcome of many bitter civic quarrels, which even this statute did not completely settle. The barbers ultimately became importunate, for in October 8th, 1545, it is recorded "The barber-surgeons have day over until this day seven night for their olde matter of dyscharge from offices, etc."[1]

The College of Physicians.

In 1518 the College of Physicians was founded, fifty-seven years after the granting of the Charter to the barbers and barber-surgeons by Edward IV., and thirteen years after the acknowledgment of the wardens and fellowship of the craft and mystery of surgeons as a distinct body. In 1522 the physicians' charter was confirmed by Act of Parliament. Linacre, the first president, used his position to interest the Universities of Oxford and Cambridge in his scheme of medical education. So lasting was his influence that in 1674 Charles the Second was induced to send a mandate to the college, ordering that no person should be admitted as a fellow who had not graduated in one of these universities. This mandate was obeyed until the last few years, and to its observance is due in great measure the high social position which the College of Physicians

[1] The matter was temporarily settled in their favour by a special ordinance in the year 1545, as may be seen at large in Repertory xi, fol. 324.

has always held, and the learning for which its fellows have been famous.

It is painful to turn from the College of Physicians and to compare the esteem with which its members were, even at this period, regarded, to the surgeons, for whose benefit it was necessary to make the following enactment:

An Acte concernyng Bakers, Bruers, Surgens, and Seryveners. 22 Henry VIII.; 1530.

"Whereas dyvers Estatutes penall heretofore have been made ageyn straungers artyfycers for exercysyng of handcraftes wythyn this Realme and for kepyng of houses, apprentyses and servaunts estraungers as by the sayde severall Estatutes more playnly is rehersed. Sythen the makying whereof, bere[1] bruers and bakers whiche bene comon vitaylers and also surgens and seryveners beyng straungers inhabyted and dwellyng wythin this realme, hathe bene putte to trouble and great vexacion by occasion of informations brought ageyne them upon the sayde Estatutes, supposyng that Straungers usyng bakyng, bruyng, surgerye or wrytyng shulde be handcraftesmen, upon the which information great doubtes and ambiguytes have rysen, whether [by] straungers usyng any of the sayde mysteres or sciences shulde be understande such handcraftesmen as were entended by any of the sayde Estatutes: For playne declaracion whereof hit is enacted by the Kyng oure Sovereign Lorde, and the Lordes Spirituall and Temporall, and the Commons in this present parliament assembled, and by auctoyty of the same, that no person

[1] beer.

nor persones straungers beyng a comon baker, bruer, surgen, or scryvenour, shal be enterpret or expounded hande craftesmen, in for or by reason of usyng any of the sayde mysteryes or scyens of bakyng, bruying, surgery, or wrytyng. And that all informations, sutes, accions, and processe had taken, or hereafter to be taken upon eny of the sayde Estatutes agayn any suche straunger or straungers beyng bakers, bruyers, surgeons, or scryveners, shall be by auctoryte of the present acte voyde and of none effecte. Responsio Regis. Le Roi le voult."

In 1530 the Barber-Surgeons' Company obtained from Sir Thomas More, on behalf of the king, a ratification of certain ordinances which established their corporation upon a new and firm basis. These ordinances it has been deemed well to print in full;[1] they are still preserved fairly engrossed at the Barbers' Hall, and they have appended to them the autograph of Sir Thomas More.

[1] Appendix I, pago 339.

CHAPTER VI.

HOLBEIN'S PICTURE—UNION OF THE BARBER-SURGEONS AND SURGEONS—THOMAS VICARY.

Incorporation of the fellowship of Surgeons with the Barber-Surgeons' Company.

ALTHOUGH the barbers and the surgeons had been long united in practice, it was not until 1540, or the 32nd year of the reign of Henry VIII., that a formal Act of Parliament was passed to unite and incorporate the two bodies. The bill was passed on July 12th, and received the royal assent on Sunday, July 24th. The preamble states: "that the king and parliament duly pondering that it is very needful to provide men expert in the science of physic and chirurgery, for the health of men's bodies when infirmities and sickness shall happen, etc. . . . . Yet, forasmuch as within the City of London, where men of great experience, as well in speculation as in practice, of the science and faculty of chirurgery be abiding, there are now two distinct companies exercising chirurgery, one the barbers incorporated, the other the chirurgeons not incorporated, nor having any manner of corporation, which two are necessary to be united. To the intent that by their unity and often assembly together, the good and due order, exercise and knowledge of the said science, or ffaculty of chirurgery, shall, as well in speculation as in practice, by their learning and ripe information, be more perfect. Therefore it is enacted that the barbers and chirurgeons be henceforth united, and as such to enjoy all privileges at any time befere granted by the Charter, 1st K. Edw. IV." These privileges are then confirmed to the united Company, and several additional ones are conferred. Thus they are allowed to take the "Bodyes of ffoure condemned persons yerely for Anatomies," etc. No manner of person within the City of London, suburbs, and one mile therefrom using any barbery, shall occupy any surgery, letting of blood, or any other thing belonging to surgery except drawing of teeth, nor any practising of surgery shall use any shaving." Power is also given to the four masters to have "the punishment, correction, and management of all defaults and inconveniences amongst the company using barbery and surgery, their authority to extend over freemen and foreigners, aliens and strangers." It is impossible to ascertain the exact circumstances under which the union was formed; it appears certain, however, that the physicians were favourable to it, and that their college assisted in the passing of the Act. Be this as it may, its delivery by the king to the incorporated Company has given rise to a celebrated painting by Holbein, a copy of which is here presented to the reader [Plate V.]. Of this picture, Mr. Carwardine gives the following account.

The chief treasure of the Barbers' Company is their fine picture, which is considered to be the best of Holbein's undoubtedly English works. The barbers having had wit and taste enough to retain it in their possession when they were separated from the surgeons under the Act of George the Second in 1745. It now[1] adorns the council room of their hall in Monkwell Street. The bluff monarch sits in his pride of place sumptuously apparelled with jewelled rings; the sword of state in his right hand. Without condescending to look at his humble lieges who are kneeling on either side, with his left hand he "reaches" the statute to Thomas Vicary, his sergeant surgeon, who was master of the Company for the time being. Vicary, with fourteen of his brethren arrayed in two rows, kneels to the left of the king, whilst to the right are three other figures. Thirteen of these worthies in the original picture have their names legibly printed on them in goodly Roman letters of white, done by some barbarian of a later age. These thirteen are Thomas Vicary, of whom more anon, with John Aylyf,[2] who take precedence of the rest as being surgeons to the king; next to them are Nicholas Symson and Edmund Harman, the king's barbers. It is probable that these persons being much about the Court had sufficient influence to compel the other surgeons to yield them precedence. Monforde and Pen, Alcocke and Ferries, complete the lower row; whilst Salmon and Tylly are the only two persons with labels in the upper row. The three figures on the right of the king represent

Holbein's picture.

[1] Written about 1825. The picture has since been removed into the Hall itself.

[2] First Alderman of Bridge Ward Without in 1551. He was a knight.

HENRY VIII. PRESENTING A CHARTER TO THE BARBER SURGEONS OF LONDON.

(From the Painting by Holbein.)

Master J. Alsop, the royal apothecary,[1] and Dr. John Chambre, with Dr. (afterwards Sir William) Butts, who was at this time physician to the king. It was owing to the influence of the latter with his master and his fidelity to his friend, that Cranmer owed his safety, whilst it gained for himself an undying memorial, true almost to the letter, in the pages of Shakespeare.[2] Dr. John Chambre was also physician to the king. He is mentioned first of the five physicians to whom the Charter of the College of Physicians was granted in 1518; and he was in attendance on Queen Anne Boleyn during her confinement of Elizabeth. He was in orders and held several church preferments.

The picture, which is an oak panel measuring within the frame ten feet two inches in width by about six feet in height, is finely coloured and elaborately finished. It is painted with the grave, unimaginative truthfulness so peculiar to the artist. The likeness of the king is regarded as the most perfect ever produced of him, and there is an autograph letter of James I. preserved in the Company's archives to the following purport:

"James R.

"Trusty and well-beloved, we greet you well. Whereas we are informed of a table of painting in your hall, whereon is the picture of our predecessor of famous

[1] In the Privy Council Register, vol. ij. p. 583, there appears under date ix September, 1549, a "warrant to Mr. Peckham for xiij li. to Alsope, poticary, for poticary stuff." In November of the same year Mr. Aylyf and Mr. Fferris are mentioned, a clear proof that the medical staff about the Court had not undergone many changes in the intervening nine years.

[2] King Henry VIII., act v. sc. 2.

memory King Henry the VIII., which being very like him and well done we are desirous to have it copied. Wherefore our pleasure is that you presently deliver it unto this bearer, our well-beloved servant Sir Lionel Cranfield, knight, one of our masters of requests, whom we have commanded to receive it of you, and to see it with all expedition copied and redelivered safely, and so we bid you farewell.

"Given at our court at Newmarket the
13th day of January, 1617."

There is a tradition that the Company had refused to take for their picture as many broad pieces as would cover its surface, and knowing that the memories of kings are apt to be treacherous when they borrow of their subjects, their worships declined to comply with his Majesty's requisition until he had deposited a pledge somewhat approaching to the value of the work. In late years Prince Albert visited the picture more than once, and at his desire (the Company proving more compliant than in days of yore) it was sent to Buckingham Palace, where it remained for more than a month. The following extract from Pepys' diary is well known, but will bear repeating here: "Aug. 29th, 1668. After dinner Harris and I to Chyrurgeons'-Hall, where they are building it new, very fine; and there to see their theatre which stood all the fire, and, which was our business, their great picture of Holbein's, thinking to have bought it by the help of Mr Pierce for a little money: I did think to give £200 for it, it being said to be worth £1,000." And then his opinion of it: "But it is so spoiled that I have no mind to it, and it is not a pleasant though a good picture." Baron was employed by the Company to make an engraving from this picture, and the following entry occurs in the minute-books of the Company in regard to this transaction: "27th Aug., 1734: Copper-plate of Holbein's picture ordered of Mr. Baron for 150 guineas; 50 guineas on finishing the drawing, 50 guineas on delivery of the plate, and 50 guineas on 100 prints." The engraving is well and boldly executed in line, and the likenesses are faithfully preserved. The face of the copper plate, however, has been made to correspond with the picture, and the figures are consequently reversed in the impressions when struck off, so that the king is handing the Act to his sergeant surgeon with his right instead of his left hand. This plate is stated to be the property of the masters or governors of the mystery and community of barbers and surgeons of London. It was published on the 15th day of October, 1736. In the usual place in the corner of the print is written, "B. Baron, del. et sculp., 1736."

The council room of the Company contains a beautiful drawing in red chalk, corresponding in size with the print, about two feet six inches by about seventeen inches; it is probably the drawing on a reduced scale made by Baron from the picture before he engraved it. The delicacy of the touch recalls that of Holbein himself. It seems in place here to notice also the curious preliminary design for the original picture which is now in the possession of the Royal College of Surgeons in Lincoln's Inn Fields. The cartoon hangs against the library wall on the staircase of the college. It is placed at a considerable height, and the light falls upon it in such a manner as to render a close scrutiny of it almost impossible. I have, however, recently made a somewhat careful examination of it. The principal portraits bear testimony to the masterly hand of Holbein. The back row of figures on the left of the king are evidently by an inferior artist, as they bear no resemblance to those of the finished picture in Monkwell Street, some of the heads being quite grotesque, and more like caricatures than the faithful portraits of Holbein. Van Mander, and more recently Mr. Cunlett, state that Holbein expired without finishing the panel in the possession of the Barbers' Company, and Mr. Black has shown that he died somewhat suddenly of the plague in the parish of St. Andrew Undershaft in the year 1543. The cartoon, moreover, is shorn of its fair proportions, being curtailed at each end so that it only measures nine feet two inches across, thereby docking off Master Alsop on the right, and Richard Ferris with the nameless one above him towards the extreme left of the king. It differs also from the panel in the colour of the drapery. The portraits are taken upon four pieces of paper which are joined together upon a canvas to form the picture. Saunders,[1] in confirmation of this view, states that the "portraits are separate pieces of paper pasted in their proper places and are evidently the original studies

[1] Knight's "London," vol. iii. p. 182.

made by Holbein from the life."[1] In the panel there is a piece of suspended drapery occupying a certain space at its upper part, and containing a long inscription. Some years ago Mr. Clift, who was then Conservator at the college, found, on cleaning the cartoon, in the corresponding space a window in place of the drapery; through the window could be seen the old church of St. Bride's. This appears to show that the event commemorated by the painting took place in the palace of Bridewell.

The following entry in the minute book of the Surgeons' Company explains how the cartoon was obtained: "At a Quarterly Court of Assistants on the 6th of July, 1786. The master (Mr. Watson) informed the court that having been informed that in a late sale of pictures of M. Desenfans a large and capital picture was exposed to sale, being a cartoon painted by Hans Holbein representing King Henry VIII. delivering the Charter to the barber-surgeons in the year 1535, and having examined the same, and being satisfied of its undoubted authenticity, and that it was the original picture of that subject, and Mr. Grindall having also examined it, and being of the same opinion, and thinking such an opportunity of procuring the possession of such a picture should not be missed, and being of opinion that the same might be procured for a reasonable price, they had treated for the purchase, and having reduced the terms to fifty guineas, they had purchased the same on the account and for the use of

[1] Mr. Wornum thinks that this picture is the copy made by order of King James I., which was mentioned above.

the Company, and the Court of Examiners had issued the price of it out of the Company's cash.

"Resolved that this court doth highly approve of the conduct of the Master and Mr. Grindall upon this occasion, and return their best thanks for their attention to the concerns of the company." So far the matter seems to have been discreetly managed, but three years later for want of a little business-like caution on the part of the court, there appears to have been a determined attack upon the finances of the Company, which was met by a remedy worthy of the drastic measures then in vogue. The circumstances of the case are reported in the minute books of the Company under the date July 2nd, 1789, Mr. Watson being still master.

"Mr. Lloyd, who had been employed to clean and repair the picture of King Henry VIII. presented a bill of his demand upon that occasion amounting to £400, which demand the court considering as highly enormous, came to a resolution offering him fifty guineas and no more. And he being called in was acquainted with such resolution, when he said he begged some time to consider of it and would call again.

"Ordered that if Mr. Lloyd shall signify his acceptance of such offer, the Clerk be authorised to pay him the said sum of £52 10s." Three months later, at the next quarterly meeting, "The Clerk reported that Mr. Lloyd had at length accepted the £52 10s. offered him by the court, and that the Clerk had paid him the same, and taken his receipt in full of all demands."

The exact date of the painting is not known, nor

can it be decided on what grounds the Surgeons' Company speak of Henry presenting a charter to their predecessors in 1535. The Act of Parliament was passed in 1540, but there is no allusion in the preamble to the Charter having been granted so long before. The king, too, did not often visit his palace of Bridewell after the trial of Queen Catherine and his marriage with Anne Boleyn; both of which events took place in 1533. If Sanders' conjecture be correct, therefore, and the Charter was really granted in Bridewell, the date of its presentation must have been previous to 1535, unless, indeed, this palace was used as a place for state pageants. The position of Thomas Vicary, too, is suggestive that the actual date of the event is not the one usually assigned. He was master of the Barbers' Company in 1531, and of the united crafts in 1542, as well as on several subsequent occasions. From his actually receiving the document from the hands of the king it is most probable that he was master for the year, and the date must therefore be assigned to one of these two years, and I am inclined to believe that it was in 1542, and that the event commemorated was not the granting of a charter, but the final delivery of the Act of Parliament into the hands of the new Company. It is possible, however, that the painting is merely commemorative, the separate studies being incorrectly pieced together after Holbein's death.

To return to the Act, however, the intention of the union was "that by the often assembling together, the good and due order exercise and knowledge of the said science or faculty of

The Barber-Surgeons' incorporation.

surgery should be rendered more perfect as well in speculation as in practice, both to (the members) themselves, and all their servants and apprentices brought up under them . . . than it hath been or should be if the two companies of barbers and surgeons should continue severed asunder . . . as they before this time have been, and used themselves not meddling together." How the due order, exercise, and knowledge of surgery was to become more perfect by the union and often assembling together of barbers and surgeons is a puzzling matter to explain, especially as, by a subsequent clause, each party was bound to the practice of its own profession. "No manner of person within the City of London," runs the Act, "or the suburbs thereof, or within one mile compass of the same, after the feast of Nativity of our Lord God next coming, using barbery or shaving or that, shall hereafter use barbery or shaving . . . neither he nor they nor none other for them to his or their use shall occupy any surgery, letting of blood, or any other thing belonging to surgery, drawing of teeth only except. And furthermore, in like manner, whosoever useth the mystery or craft of surgery, shall in no wise occupy nor exercise the feat or craft of barbery or shaving, neither by himself nor by none other for him to his or their use."

Although the improvement of those practising surgery "in speculation and practice," was not likely to be very great by "their often assembling" with those who were forbidden "to occupy any surgery," yet the surgeons were gainers by the union, inasmuch as they now shared by right in all those municipal and corporate advantages which they had hitherto possessed only in part and on sufferance. It was further ordained "there shall be chosen . . . . four masters or governors of the same company, two of them shall be expert in surgery, and the other two in barbery." The study of anatomy was also encouraged by the clause, enacting that "the said masters or governors of the mystery and commonalty of barbers and surgeons of London and their successors yearly for ever, after their said discretions at their free liberty and pleasure, shall and may have and take without contradiction, four persons condemned, adjudged, and put to death for felony, by the due order of the king's laws of this realm, for anatomies, without any further suit or labour to be made to the king's highness, his heirs and successors for the same. And to make incision of the same dead bodies, or otherwise to order the same after their said discretions at their pleasures; for their further and better knowledge, instruction, insight learning and experience in the said science or faculty of surgery."

The first master of the United Company was Thomas Vicary,[1] sergeant chyrurgeon to King Henry the Eighth, and overseer[2] of all the

Thomas Vicary.

[1] For many interesting particulars of Thomas Vicary's relations with St. Bartholomew's Hospital, see an excellent paper by Dr. Norman Moore, entitled, "The Physicians and Surgeons of St. Bartholomew's Hospital, before the time of Harvey;" published in the St. Bartholomew's Hospital Reports, vol. xviii. pp. 333—358.

[2] In this position Dr. Norman Moore points out Vicary was intermediate in authority between the Master of olden times, and surgeons who were subsequently appointed.

officers within the newly restored hospital of St. Bartholomew, in Smithfield. He was the author in "The English-mans treasvre with the true Anatomie of Mans body," of the "Profitable Treatise of Anatomy." This work was published in 1548. "A little treatise," he calls it in his preface, "for all suche young Brethren of his felowship practising chirurgerie. Not for them that be expertly seene in the Anatomie; for to them Galen the Lanterne of all Chirurgions hath set it foorth in his canons to the high glory of God, and too the erudition and knowledge of al those that be expertly seene and learned in the noble Science of Chirurgerie. And because al the noble Philosophers wryting vpon Chirurgerie doo condemne al such persons as practise in Chirurgerie not knowing the Anatomie." In the first chapter he declares "three poyntes very expedient for al men to knowe, that intend to vse or exercise the mysterie or Art of Chirurgerie. The first is, to knowe what thing Chirurgerie is; The second is how that a Chirurgion should be chosen; And the thirde is, with what properties a Surgion should be indued."[1]

This description of "what properties and conditions a man must have before he be a perfect Chirurgion" clearly proves that had surgical education been carried on as he proposed, the standing of English surgeons in society would have been much higher than for many years after it had attained, and if, as is not improbable, Mr. Thomas Vicary was not a very learned person, he had

[1] "A Profitable Treatise of the Anatomie of Mans body." Lond. 1577.

sufficient common sense to perceive that only by general education was his profession to be raised. Of the "foure thinges moste specially that euery Chirurgion ought to haue" . . . "The first (I sayde) he ought to be learned and that he knowe his principles, not onely in Chirurgerie but also in Phisicke, that he may the better defende his Surgery; Also he ought to be seene in natural Philosophie, and in Gra*m*mer, that he speake congruitie in Logike, that teacheth him to proue his proportions with good reason. In Rethorike, that teacheth him to speak seemely and eloquently: also in Theorike, that teacheth him to know thinges naturall, and not naturall, and thinges agaynst Nature. Also he must know the Anatomie, for al Authors write against those Surgions that worke in mans body not knowing the Anatomie, for they be likened to a blind man that cutteth in a vine tree, for he taketh more or lesse than he ought to doo. And here note wel the saying of Galen, the Prince of Philosophers in his Estoris!

"That it is as possible for a Surgion (not knowing the Anatomie) to work in man's body without error, as it is for a blind man to carue an image & make it perfyt. The ijd, I said, he must be expert: for Rasus sayth he oughte to knowe and to see other men work and after to haue vse and exercise. The thirde, that he be ingenious or witty: for al things belonging to chirurgerie may not be written nor with letters set foorth. The fourth (I sayde), that he must be wel manered, and that he haue al these good conditions here folowing," among which are some it were well always to follow, viz. "That a Chirurgeon must take heed he deceiue no man, with his vayne promises, nor to make of a smal matter a great, because he woulde be accounted the more famous. . . . Likewise, they shal geue no counsayle except they be asked, and then say their aduise by good deliberation, and that they be wel aduised afore they speake, chefly in the presence of wise men. Likewise they must be as priuie and as secrete as any Confessour of al thingis that they shal eyther heare or see in the house of their pacient. . . . And see they neuer prayse them selues for that redoundeth more to their shame and discredite than to their fame and worship: For a cunning and skilfull Chirurgion neede neuer vau*n*t of his dooings, for his works wyll euer get credite ynough. Likewise that they despise no other Chirurgion without a great cause: for it is meete that one Chirurgion should loue another, as Christe loueth vs al.

"And in thus dooing they shal increase both in vertue & cunning to the honour of god, and worldly fame, to whome he bring vs al. Amen."

Vicary, as we may see from these extracts, was, if not gentle by birth or by profession, one of nature's gentlemen, with the best possession, "an honest and true heart," overflowing with kindly feelings towards his fellow men, and earnestly desiring to lead his professional brethren along that path which could alone raise them to the station he was anxious they should attain. That Vicary's intentions were duly estimated, and that his "Profitable Treatise" became a great favourite is pretty clearly proved by its having attained a ninth edition in 1641 (ninety-three years after its original publication). The surgeons of St. Bartholomew's Hospital, William Bedon, Richard Story, Edward Baily, and William Clowes, thought it "so learned a worke of Anatomie," as worthy to be by them "students in Chyrurgerie, not without our great study, paines, and charges, newly revived, collected and published abroad to the commodity of others" long after the death of the author.

CHAPTER VII.

REGULATIONS OF THE UNITED BARBER-SURGEONS' COMPANY CONCERNING APPRENTICES AND THE EXAMINATIONS FOR LICENCES TO PRACTISE AS SURGEONS.

The barber-surgeons a livery company.

THE Company of barber-surgeons formed in 1540 was a livery company. The Mayor and Sheriffs as well as certain of the City companies wore liveries on days of festivity or public business. Each company might find its own livery, or if they wished to wear that of the Mayor, the clerk of the company was to collect the dues which "must amount to 20 shillings at the least, put into a purse with their name that gave it, and the wardens were to deliver it to the Mayor by the first of December in each year. For the which every man had then sent him four yards of broad cloth, rowed or striped athwart with a different colour to make him a gown, and these were called ray gowns, which was then the livery of the Mayor, and also of the Sheriffs, but each differing from the others in the colours." This is the account given by Stow,[1] who mentions several alterations which subsequently took place. The money paid to the Mayor for the livery he calls a "benevolence" and "this is got from 20s. to 40s. But Sir Thomas Lodge gave instead of four yards of

[1] "Survey of London," p. 196. Ed. J. Thoms, 1876.

broadcloth three yards of satin to make them doublets, and since that the three yards of satin is turned into a silver spoon, and so it holdeth." At the Mayor's feast, which, as may be supposed, was the chief public festival of the year, the representatives of each company were placed according to precedence and were grouped into "messes." The chirurgeons at the Mayor's feast in 1531, shortly before their union with the barbers, were reckoned the thirtieth of the City companies, the wardens and two persons to attend making one mess. The barbers on the same occasion ranking as the thirty-second company, the wardens and four persons being present and making two messes.

Early records of the united Company.

The records of the Barbers' Company do not commence until some time after the union had taken place. The earliest minute book at present existing is a thin folio measuring 15½ inches in height by 11½ inches in width. It is bound in a limp parchment cover bearing an almost effaced inscription. The writing is generally clear and distinct, but of different styles. The spelling is that of the period and is good; abbreviations are not very numerous, and the subject matter is well expressed; with marginal indices of the matters recorded. The volume commences on the "xxvith day of Auguste in the fowerthe yere of the reigne of our Souereigne Lorde Kynge Edward the Syxte," and "within the tyme of Maister George Geen, Thomas Jhonson, Thomas Stockdall, and Mathew Jhonson, Maister and Wardens of the Company and fellowship of

Barbers Surgeons of London." It contains too "Orders and Awardes." The earlier written pages have their several sentences widely crossed with erasure lines giving the idea of their having been transcribed into some other book, but the Barbers' Company have no such book at the present time. The greater number of pages are not crossed in this manner. The book is in a good state of preservation, though the lower corners of the leaves are worn as if frequent reference had been made to it. One of the first entries is in relation to the election of the master, who was to be chosen by ballot, "every man to prycke as his mynde dothe serve him wi*th*out any telling, and when every man hathe prycked about[1] the Courte, then the byll to be broughte to the M*aste*r and he that hathe the moste pryckes to have the roome of M*aste*r the yere ensuinge." The same proceeding was to be carried out for the election of the upper warden, the second warden, and the youngest warden. In 1551 the second warden was ordered to receive the "rente of the Landes and all maner of Receits" and he is henceforward known as the Renter Warden. In the same year a dentist was admitted into the Company, the entry being, "John Brysket toothe drawer" hath been admitted for a brother into this house.

On and after "the xxvith daye of November 1551 the Kinge Maijestye's Barber or Barbers and also his Ma*ijestye's* surgeons shall syt next to the Laste M*aste*r up on the benche where the M*aste*r nowe usually doo sitt." During the greater part of the reign

[1] in.

of Edward VI. and for the first two or three years of the reign of Philip and Mary the annals of the Company are a series of personal details such as the settlement of various demands, the appeasing of quarrels amongst the members, the regulation of the craft and the ordering of apprentices.

In 1554 a slight mark of the troublous times in which it was written occurs in the following minute :

An Order (for a Masse and other . . . .)[1] to be Sayd on the Ellection Daye and for the Lyverye to be thereat.

"Memorand: Yt is understanded and agreed the xijth day of August *which* was the ellection daye at the chosing of the newe *Master* and Wardeins . . . that there shall be a solemne (masse or)[1] other dyvyne and godly preyer sayd and songe, that the *Master* be chossen and the hole lyvery to be thereat in their best clothing and to meet at the hall at or before the hower of IV of the clocke, and he that faylleth his hower to paye xijd for his fyne *without* any redempcyon or gaynesaying. The sayed masse to be at the chardges of the Company."

The years 1555 and 1556 were actively occupied by the Company in the rearrangement of their guild, the results of which are formulated in a series of ordinances. Of these the first[2] relate to the establishment of a yeomanry within the Company, similar in its constitution to that of the corporate body itself, to wit, wardens, beadles, and clerk. The second series proscribed in the following words the shaving or trimming of any man upon Sundays:

Sunday trading interdicted.

[1] Erased. [2] See note, page 217.

"From this daye forwarde there shall no man, ffreeman, fforeiner, or Straunger of the clothinge or *with*out the clothing,[1] shave, wasshe a Bearde or tryme any man *w*ith any Instrument, as to make cleane teathe upon the Sondayes . . . in his owne house, or in any man's house or Chamber, or in any place els. He shall forfayte, at every tyme, beinge dewly proved, for a fyne to the hall[2] the sum of xl s. And further that no fforeyner, being no ffreman, shall carry out any Bathe, or clothe, or Instrument to make cleane teeth, to shave poll, or wasshe a bearde, or to tryme any man but *with*in the Lybertye where he dwelleth," under the same penalty. The remaining articles are of less interest, until we reach one which shows us that unrestricted freedom of speech was as little appreciated by the majority as it has been in the late Parliament. The minute is headed "An article for the paym*en*t of a ffyne if any man of the Assistance doo talke or interrupte any matter or questyon that shalbe moved in the sitting of the Assistaunce *wh*ich shal be touchinge the comodytie of the Company." It enforces under pain of fine the authority of the master, and lays down rules of precedence for the speeches.

On the 5th of March, 1555, an order is made "againste the letting out or lending of the hall to kepe bridalls or any other games in yt, whereby the sealinge or other things being broken in the hall or kitchen shall redowne to the Losse of the Company. Yf, therefore, the sayed M*aste*r and gouernors for the tyme being doo lett out the hall to any bodye to the intent aforesayed

[1] livery. [2] of the Company.

without the consent of the hole howse to be called for, the same shall forfayte and paye at every tyme so doying to the hall for a fyne xl s. And the sayed Clercke of the Companye for so letting the hall to any bodye as af*orsaid* shall lykewise forfayte and paye to the hall at every tyme so doyinge xx s." Yet as a special act of favour, in 1568, "Yt is . . . agreed by the authoritie of this courte that . . . the M*aste*rs mayde shalbe maryed in to the hall w*ith*out any penaltie to the M*aste*r and gou*er*no*r*s and clarke, any other acte notw*i*thstondynge." And again, upon the xxvijth day of January, 1567, "In this court Rich: Hughes is graunted to have the Hall to kepe a maryage in upon Sonday co*me* a sevenight next onsuinge the date above wrytten."

In the reign of King Philip and Queen Mary several important sets of resolutions were passed which are of sufficient interest to be transcribed at length from the records of the Company. The first series relates to the examination of persons for the Company's licence to practise. The more especially interesting articles run as follows:

Regulations concerning examinations.

"MEMORANDUM THE VTH DAYE OF MARCHE IN THE YERE OF OUR LORDE GOD 1555, AND IN THE SECONDE AND THIRDE YERE OF THE REIGNES OF OUR SOVEREIGNE LORDE AND LADY KING PHILIP AND QUENE MARY.

"Yt was fully undestanded and agreed by the M*aste*r and governors then being, Thomas Knot, Mr. Thomas Gayle, John Smith, and Thomas Fferrys, Governors, with th'assent of th'assistaunce of the same Companye being there present, Mr. Vycary, Mr. Hollande, Mr. Geen, Mr. Lambe, Mr. Sprignell, Robert Brommell, Thomas Stockdall, John Barker, Thomas Wittingham, William Green, James Wade, Richard Ellyot, Henry Lymocke, and John Bonar to these articles hereafter following, which be to the n*um*ber of XIII, as hereafter may and doth appear. . . . .

"The Seconde Artycle is that there shalbe chosen XIII examiners, whereof iiij to be alwayes present to examyn all suche as experte in Surgery, the M*aste*r and governors being present, whereupon the sayed examyners may sette their hands w*i*th the consent of the M*aste*r and governors hearing the mater. And that the sayed examyners shall not examyn nor give Lett*e*rs of lycense but that the M*aste*r and governors shalbe privy thereof.

An article for the election and chosinge of the XIII examyners.

"And that there maye be a booke made wherine every man's name that have Lycense to occupy Surgery being approved to be enrolled and what ys the yeare that ys to him or them appointed. And yf they take upon them to doo otherwise, then there ys given them the blame to redowne to him or them that so doo, and not to the examyners nor to the M*aste*r and governors and that there maye be alwayes at every courte daye two at the leaste of the sayed examyners during a month and so afterward monthely two of them to be there whose chance the monthe shall fall too yf there besiness be not the greater: because the M*aste*r and governors shoulde not be to seke if any bodye shoulde be examyned there. And for Default of men being there, having no reasonable excuse, to lose to the hall ij s. if he doo not send worde or come himself being in the Cytie of London, or desyre another examyner to be there for him when his course ys at every courte Daye, because we should not be without them who can answer the matter touching Surgery."

"The Nyenth Artycle ys that there be chosen every year two for the Anathomye, and other two also to be chosen for to be Stewards: so that two always shall stande for ij yeres, because they that doo not worke of the Anatomy the one yere being Stewards for the provysyon of the vyctualls thy shall worke the other yere following. And they that shalbe chosen shalbe alwayes for first yere Stewards, because that they shall see the makynge of them the yere before that they may be the more practysed in the doynge the next yere the sayed Anathomye, that after it may appeare by the workmanship that they be the dooers. Yf the M*aste*r and governors do goo about to breake the same acte they shall lose for a fyne to the hall xl s."

Regulations relating to anatomy.

An Article for the Eleccion of the Masters and Stewards of Anatomy yerely.

Of the further duties of the stewards of anatomy more will be said in a subsequent chapter.

The second series of ordinances relates chiefly to the apprentices. In 1556, regulations were made for the physical as well as for their moral welfare. Of these articles the first is "that all prentyces of the Companye as

Regulations relating to apprentices

I

well of the Clothinge as of the yeomanry shall weare no beardes during the tyme of his prentisshippe, the same to begynne at Whitsonday next. And if the sayd prentys doo weare any the *Master* to lose vj s. viij d. for every such defaulte. The sayed prentis not to weare a bearde past xv dayes growing upon payne of the forfayture aforesayed." Other articles relate to the actual granting of the Company's licence. In many points they bear a close resemblance to the conditions under which the Universities of Oxford and Cambridge confer their medical degrees at the present time :

"The Secònde Artycle is that from the feaste of Saint Michaell tharchaungell [29 Sept^r] next comynge no barber Surgeon that dothe occupye the mystery of Surgery in the Clothinge or out of the Clothinge shall take or have any prentys But that he can skyll of the Latin tonge and understand the same and can wryte and reade suffycientlye : and yf they or any of them doo take any that can not doo the same they that offende shall paye to the hall for a ffyne xl s.

An Article that every barber-surgeon occupying surgery shall take no maun*er* of prentice but suche as hathe some knowledge in the Lattin tonge.

"The Thurde Artycle is that prentisses that shalbe made ffree after Michelmas next comynge that doo occupye the mystery of Surgery and all other men that doo desyre to occupy the same and to be a brother w*i*th us, to be examyned and to passe according to the order of this house as a preferment of a yeare to him giuen as the order hereafter followeth as he shalbe demaunded and apposed.

An Article that all prentizes that hereafter shalbe made ffree and doo entend to occupye surgery shalbe examyned and to passe his first prefer-me*n*t of grace.

"The iiij[th] Artycle is that after Michelmas next comyng all p*re*ntices when they are made ffree must be demaunded by the M*aste*r and gov*er*nors and the iiij examyners what he intendeth to doo after he is made free, whether he will occupye the Mystery of Surgery or no w*i*thin the Cytie of London. yf he saye, ye, Then to be examyned what he can doo towards yt, how he knoweth what ys Surgery and also what an Anatomy ys and how many parts it ys; of what the iiij elements and the xij signes be, w*hi*ch ys the firste part of examynac*i*on for a prentyce and for other that wilbe brother with us as the examyners shall see cause for having of their preferment of their fyrste grace to them to be given.

An Article how the sayed prentizes shal- be examyned.

"The fyfte Artycle is that when he hathe answered to the firste article provinge that he hathe some Learninge or practyse Then to have his firste preferment of grace to occupye Surgery by the space of so many yeres or tyme as the M*aste*r and gov*er*nors and the examyners shall thinke meete and as his Industrye shall serve to receyve the grace of God and by his dilligent travell to study in the same and for an homage thereof if he be learned or can wryte to bring in an Epistell ev*er*y half yere: and to reade it himselfe openly at the day of Lecture before the hole house that they may see his furtheraunce how he hath profyted in his dilligent Labor and studye: and the unlearned that can not wryte nor reade to be examyned half yerely what they can doo in the practyse because they be unlettered

An Article that upon his examycon of Surgery the said prentis shall have his prefermen*t* of grace and if he or they can reade to bring q*u*arterly an Epistle.

by the M*aste*r and go*uer*nors and the examyners how they have taken payens in their studye to practyse because they be unlettered for the Savegarde of the k*in*g and quenes ma*je*sty'*s* people: (upon payen of a ffyne to the hall for neglecting to do their dewty a spone waying one ounce at every default.)[1]

An Article that any man desyring to have his firste preferment of grace shall paye to the Clercke etc.

"The Sixte Article is that any man occupyinge the mystery of Surgery being made free and desireth to have his firste preferment of yeare shall paye to the clercke of the Companye for the wrytynge Inrolling finding waxe to seale it and for the having of the Seale viij d.

An Article that no man occupying Surgery shall sue for the Bushoppes seale before he be admitted a Maister of Surgery by the M*aste*r and govnors and the iiij examyners.

"The Seventhe Artycle ys that no man of the Companye after the feaste of Saint Michell tharchaungell next coming shall call for the Bushoppes seale, which ys the confyrmacon of a Surgeon untill suche tyme as he hath passed his firste preferment of grace and the Seconde admyssyon to be admytted to be a Surgeon, and a Maister of Anathomye: and to paye for the having of the Seconde admissyon a spone of an ounce of Silver, and his name to be wrytten upon it to the hall, and the Clercke of the Company for the wryting and findinge waxe and enrolling of yt in the boke viij d., and if the *per*son doo not this passe orderly he to paye for a ffyne to the hall xl s.

[1] Erased.

"THE ORDER OF THE FFYRSTE PREFERMENT OF GRACE OF THE ADMISSYON OF PRACTYCYONERS THAT HAVE BEEN PRENTYCES AND BE MADE FREE WHAT THEY SHALL HAVE FYRSTE TOWARD THEIR PREFERMENT.

"Fforasmoche as yt is expedyent that no man occupye the worthye Scyence of Surgerye but suche as shalbe thoughte apte and industrious to execute the same truely and accordingly as well for the comodytie and proffyt of the comonwelth as also for the avoyding of the Inconvenyences and slaunder that otherwise mighte happen by the rashness and unconning of such lewde persons as taketh upon them to exercyse Surgery being neither expert nor of us admytted to the same. And forasmuche as it is not possyble that any shall attayne to the same wit*h*out instruc*i*ons firste learned of conning and well exercysed men of that facultie: being brought up therein as a practycyoner or otherwise under some well skylled M*aste*r for certayne yeres in whiche tyme he mighte applye his mynde to learne perfectly the rules and speculatyve part thereof. The Maister and gov*er*nors of the Barbors and Surgeons of London w*i*th the foure examyners and the rest of the hole assistaunce have thoughte it good after such tyme and terme of yeres expired every such prentis or otherwise Sarvaunt being made ffree of the sayed Companye and ffellowship shall also have a tyme appointed by us and the rest of the Companye to practise and put in use suche knowledge as he hathe that we in tyme afterwarde havinge intelligence

Temporary licence to practise as a surgeon.

of his connynge and well Doyings may constitute him a Maister of Surgery if his deserving so require. Wherefore we the sayed Maisters and Governors and the iiij Examyners *with* the rest of the hole assistaunce here at this instante doo admitt A. B. as a practycyoner who hathe served as aprentis *with* R. G. Maister authorissed of this Company the space of [1] yeares and now being a freeman of this Company to practise Surgery in all places according to his knowledge for the space of [1] yeares as A tryall and a proofe of his knowledge may be had. In witnesse whereof of the premisses we have caused this L*ett*re to be sealed *with* our seale of our hall touching the firste admission of his fyrste preferme*n*t of grace: the xxiiij Day of Julye Anno Dni 1556."

"THE ORDER OF THE ADMISSYON OF MAISTERS OF SURGERY AND OF THE ANATOMYE TO BE CONFYRMED FOR EVER BEFORE THEY HAVE THE SEALE OF THE BUSSHOP W*HI*CH MAKETH UP THE HOLE CONFYRMACION OF A MASTER OF SURGERY AND OF ANATHOMYE.

Final licence to practise as a surgeon.

The Dede of the admissyon of A M*aster* of Sugerye.

"We Thomas Knot M*aste*r Thomas Gayle John Smythe and Thomas ffishe Governors Thomas Vycary George Hollande George Geen and Richard fferres M*aste*rs and Examyners of the Company of Barbours and Surgeons of London *with* the rest of the whole assistaunce of the same Companye To all men to whome this wrytyng shall come greting We certifye youe by this Le*tte*r that whereas o*u*r welbeloued in Christ

[1] Left blank in the original.

T. A. ys not only a man of honest fame and good behaveor but is also expert connynge and well exercysed in the arte of Surgery as his well deserved Cures and prosperous successe w*h*ich cannot be dooen wit*h*out maturate iudgement and Learning Dothe make thereof moste certeyne truith and be a sure witness. ffurthermore we are assured by the experyence we have of the man that he is not only substancyally well exercysed in the curing of infyrmities belonging to Surgery of the p*ar*ts of man's bodye comonly called the Anathomye. Wherfor we as well in the behalf of Equyte reason and consyence as also for the prefermen*t* of Learninge knowledge and experyence Doo thynke yt mete convenyent and reasonable to constitute the same T. A. bothe a Maister of Surgery and also of the Anathomye and willeth him so to be taken for ever hereafter and to have authoritie to exercyse and occupye as well the one as the other wheresoever he shall come wit*h*in this Realme or ellswhere of the premisses. In witness whereof we have caused this L*ett*re to be sealed wit*h* the great co*m*mon seale of our hall the xxiij Daye of July An*n*o Do*m*ini 1556.

"IN WYTNES AND CONFYRMACION OF THE SAYDE ARTYCLES AFOREREHERSED AND OF THE ORDER OF THE PREFERMENT OF A PRACTYCYONER AND ADMISSION OF A MAISTER OF SURGERY WE THE SAID MA*STE*R AND GOVERNERS AND THE REST OF THE ASSISTAUNCE AFORE NOMYNATED HAVE PERTYCULERLY SUBSCRYBED OUR NAMES AND MERKS THE DAY AND YERES AFORE WRYTTEN.

"Thomas Knot Thomas Gayle John Smithe Thomas

ffyshe Thomas Vycary George Hollande Richard Bowle Robert Sprignoll Thomas Stockdale James Woode Richard Elliot and Hugh Lymeweke."

The ordinance requiring apprentices to be learned in the Latin tongue did not find favour with the majority of the Company, or else led to inconvenience, for it was repealed at the next court day, August 26th, 1557.

Before the end of the year the Company was in monetary difficulties, as is evidenced by the minute, "Mr. Thomas Vicary shall paye and dyscharge the debts of the house . . . and shall have the plate of the crafte in pawne or pledge, untyll such tyme as the sayde sumes of mony be unto hym payde agyne."

CHAPTER VIII.

REGULATIONS CONCERNING ANATOMY AND APPRENTICES IN THE UNITED COMPANY—THOMAS GALE—WILLIAM BULLEIN.

The Barber-Surgeon's Company during the reign of Elizabeth.

THE proceedings of the Barbers' and Surgeons' Company contain but little of interest as regards the progress of surgery during the reigns of Edward VI. and Queen Mary. From the latter sovereign, as well as from her sister Elizabeth, the Company obtained charters. The headings of their charters are reproduced on the next page.

The court minute books in the years immediately preceding the reign of Elizabeth are full of remarkable passages illustrating by quaint details the times in which they were written, and which carry us back to the days when the City was so small that each member of the Company residing within it knew every detail of his brother craftsman's life and practice; to a time when the guilds actually possessed absolute sway over their members, and when the Lord Mayor and his brethren the Aldermen formed the final court of appeal, and had the sole right of taxation within the liberties of London; to a time when the rules and ordinances of the Company could be enforced by fines under pain of arrest and subsequent detention in the Compter. And yet, again, to times when life was easier and merrier, since, in spite of

HEADINGS OF CHARTERS GRANTED TO THE COMPANY OF BARBERS AND SURGEONS BY PHILIP AND MARY AND ELIZABETH.

modern cavillers, England was "merry England" in those days, for over and over again may be read the records of pageants to be held at "any noble pere his comyng throughe the cytie oute of aney fforeyne contrey or lande," or a precept of the Lord Mayor directing "the master and wardens of the company to provide eight of the most grave persons apparelled and horsed to attend upon her highness' [Elizabeth's] royal person from the towne of Chelsey unto her princely palace at Westminster. And after that her Majestie shall be entered into her palace they doe attend with lightes and keepe such like goode order homewarde in quiet manner in their severall degrees and places." Frequent entries occur during the reign of Elizabeth about the barge; amongst others that the Company shall use "the maydens of hon*o*rs barge or the lyke in length and bredth." Enactments, too, had to be repeated against using the "hall of the Mysterie to kepe weddinge sportes or games therein, or playes or dauncinge, or for any other like entente" without due licence. "Provided alwaies . . . . that every one who hath borne [the] offyce of a M*aste*r shal be and may for the marriage of his child or kinsman or woman use the same with the consent of the master and governors onlie." There is an ordinance also that "whereas the whole body of the yeomanrye of this Mystery were compelled under a certeine fyne and penaltye to meet theire wardens at a certeine place by them appointed, beynge by the beadle warned, to go to offire at the weddings of their Brethren at all times when any of them weare married, w*hi*ch was to their great trouble and Divers inconvenyence and grieve thereof as absence from service and sermons one sonday mornyngs and other unmeete and Inconvenient metinges." Wherefore it is decided that there shall be no more any such meetings.

In 1561 during the mastership of Thomas Gale, and again in 1562 grants of arms were made to the Company by the Clarencieux herald, but as the heraldry was

INITIAL LETTER FROM GRANT OF ARMS TO THE UNITED COMPANY OF BARBER-SURGEONS BY SIR GILBERT DETHICK, GARTER, JUNE 2ND, 1569.

deemed bad it was modified in 1569 by Sir Gilbert Dethick, here represented in his Tabard, who made a gift of the arms portrayed in the frontispiece.[1]

The literature of surgery during this period is chiefly associated with the name of Thomas Gale, who was born in 1507 and studied under Richard Ferris sometime sergeant surgeon to Queen Elizabeth. Like all the best English surgeons of his day he served in the army.

Thomas Gale.

[1] See also page 71, and Appendices J and K, pages 351—360.

He was with the troops of Henry VIII. at Montreuil in 1544, and with those of King Philip at St. Quintin in 1557. During the latter part of his life he settled in London, where he was living in 1586. In 1563 he published "An institution of a chirurgian conteyning the sure Groundes and Principles of Chirurgiry moste necessary and mete for all those that will attayne the arte perfectly:" a well digested work intended to ground the young student in the principles of his profession and to expose that empirical kind of practice which busied itself with the mere collection of remedies without any adequate knowledge of their properties and application. It is interesting as being the first book on surgery published in the English tongue. It seems that things went but sadly with surgery when Gale wrote, for he observed in his introduction that "The greatest number of Chirurgians (I meane those that are ther*e* unto lawfully called) are so rude and unskilfull in their art because they have no methode, no exact waie, or yet order in learnynge the grounds and principles of Chirurgery. . . . So that they are confused in their studies and make rather a rude chaos than a perfecte arte of Chirurgerye. For nowe it is come to thys pointe that if they can get thys recepte and that recepte [prescription] as they terme it, they thynk they nede no farther stodye. So that the more receptes he hath the greater Chirurgian he thinketh hymselfe to be: such Ignorance now raygnes that recepts beareth the bell: theorike of Chirurgerye is quit forsaken or not regarded. . . . Wherefore we nowe accordynge to oure smalle poure and symple knowledge, wayinge the promisses and hartelye wisshynge the redresse of the same, haue sette oute the Theorike part of Chirurgerye in thys presente volume conteynynge the principles and sure groundes of the arte, in such order and methode as shal be most conuenient for the yonge student." The Institute then proceeds by way of dialogue between the author and his fellow pupil, John Feilde, who are required "to enter into some talke of chirurgerye" with one John Yates. Gale observes[1] that "chirurgerye is free for all persons, the more is the pitye, when as so muche hurte and damage doe daylye spryng, through the abuse of so noble an arte, so that I am more than halfe perswaded that Chirurgerye wyll hardlye, or neuer gather strength agayne, and florishe: excepte the prynce hauinge compassion of poore deplorated Chirurgerye, doth by vertue of lawes driue awaye from her theis her forenamed enemyes." To this answers John Yates, "But yet I doe meruayle at those, whyche are as it were Chirurgians by profession, knowynge themselues voyde of a number of the poyntes required in a chirurgian that for loue to their contrye, and conscience sake: they wil not yet at the leaste chose suche seruantes, as had some of the fornamed vertues. Whyche, yf they dyd there were yet some hope that Chirurgirie shoulde hereafter floryshe."

Feilde replies: "Their aunswere is, that when the skey fall we shall haue larkes. They thynke that they shoulde not haue servantes to serue in that vocation. For few, say they, that haue well brought up their sonne, will put hym

[1] "An Institution of a Chirurgian," fol. 10. Lond. 1563.

to the arte, because it is accounted so beggerly and vile." In another place Gale complains, after witnessing the surgical practice at the Royal hospitals of St. Bartholomew and St. Thomas in 1562, that "it was saide that Carpinters, women, weuvers, coblers and tinkers, did cure more people than the chirurgians." The first English mention of syphilis occurs in the Institution of Thomas Gale, and in referring to it he gives an account of the unlettered condition of his brethren in the following words, after discussing the various classes of tumours: "Yet amonge all theis I here you make no mention of *morbus*, whiche I thyncke is to be numbred amonge tumours against nature." To which John Feilde replies, "I do not well understande your meanynge, for manye a tumour is called *morbus*, for *morbus* is a Latine worde, and all sycknes and infirmyties be called *morbi*." "Yea," says Yates, "but I meane that scabbe whiche the common Chirurgians calleth the *morbus* and the *morbus* cause, and the better lerned nameth it *morbus Gallici*."[1] "It is a great thynge to be trayned up in ignoraunce or barbarous doctrine," replies Thomas Gale, "one rude Empirike called it the *morbus* and a nombre foloweth hym in his folly, and another because he wolde seme connynger then the rest, wyth as muche ignorance, nameth it *morbus Gallici*, brekynge the olde Pricians head, and yet wil not or cannot gyue hym a plaster. But if you will vse that usuall name you might haue named it *morbus Gallicus* or *Lues venerea*, and so have obserued co*n*gruitie in spekyng."[2]

[1] "Office of a Chirurgian," fol. 32*b*; 1586. [2] Gale, "Inst. Chirug.," fol. 30

The surgeons of his time must of necessity have been ignorant, for even the few and instructed surgeons can only have been taught by the very small number of those who were acquainted with Latin, and who were content to lecture and comment upon the classic writers; or of the still smaller number who had improved their book learning by the surgical experience which they had gained in the wars. In the Institution already quoted, Gale gives the following account of the qualifications which he considers necessary for a surgeon. The extract is of interest, as illustrative of the views of a well educated surgeon of the sixteenth century in regard to the position his profession should hold.

"Guido requyreth iiij thinges in hym that is or shoulde be a Chirugian. First, that he shoulde be lettered, secondly, that he shuld be expert, thyrdlye, that he be ingenious, and last, that he be vertuous and well maneryd. . . . And as touchyng theis iiij pointes, note first wher Guido sayth he must be lettered, he doth meane he shoulde be lerned, and that chefelye in suche doctrine as is in his art requisite. That is at the lest, that he can wright, red, and understande the mynde of latyne authour. If he had knowledge in geometrie for makynge his incision, for curynge fractures and luxations, and also in curinge malignant ulcers; knowynge what figure would moste spedely unite, and Ioyne the lyppes of the ulcere; it ware very commendable. Naturall Philosophie is a goodlye ornature to the Chirurgian, for it showeth hym what resulteth whan as there is diuers and sondry simples mixed to gether.

But I will go to the seconde pointe, he must be experte, that is, he muste be garnished wyth muche and longe experience, whiche is excogitated be firme and certayne reasons, and by them also confirmed, otherwise he is to be accompted rude, and an empericke if he hath not reason annexed and ioyned to his experience.

"Thirdlye, sayth GUIDO, He must be ingenious, unto whiche there are fyue thynges principally required. Fyrst is the redye and good conceyuynge; then a firme and sure memorye, nexte a sounde and ryght judgement, after a easey callynge thinges to mynde whyche he have harde or sene, and laste a lyuelye and sharpe redynes in findynge and inuentynge remedyes. The maners whyche GUIDO woulde haue in a Chirurgian are reconed of HIPPOCRATES and CELSUS, whiche briefelye I wyll numbre; they muste be bolde and wythout feare in suche cures as are without peryll, and whereas necessitye requireth. Also in cures that be doutful, not to be raishe and hastie, to be gentle and courtyous towarde the sicke pacient, to be frendlye and louyne towarde those of hys profession. Also wyse and circumspecte in Prognostications, last of all, he muste be chaste and temperate of body, mercefull towarde the pore, and not to gredy of mony, and this is sufficient touchynge the description of hym that must be admitted in Chirurgerye."[1]

In relation to his patient Gale observes that "The Chirurgian must also in theis his operations obserue six thynges principally; First that he doeth it safelye, and that wythout hurte and damage to the pacient; secondly,

[1] Gale, "Institution of a Chirurgian," fol. 11.

that he do not detracte tyme or let slepe good occasions offered in workyng, but with suche spede as arte wylle soffer, let hym finishe his cure; Therdly, that he worke iently, courtyously, and wyth so lytle payne [to] the pacient as conueniently you may, and not roughly, butcherly, rudlye and wythoute a comblenes.

"Forthly, that he be as free from crafte and deceyte in all his workynges, as the East is from the Weast. Fiftly, that he taketh no cure in the hande for lucre or gaynes sake only, but rather for an honest and competent rewarde with a godly affection to doe his diligence. Laste of all, that he maketh no warrantyse of suche sickness as are incurable, as to cure a Cancer not vlcerate, or elephantiasis confirmyde: but circumspectly to consider what the effecte is, and promyse no more than arte can performe; and you shall doe theis thynges muche the better (yea withoute theis, you can not any thynge profit your pacient) yf you understande the maner and exacte wayes of stichynge woundes, of makyng tentes, splanes, stuphes, bolsters and conuenient rollynges."[1]

William Bulleine.

After reading these accounts of Gale's ideal of a surgeon, it will not be uninteresting to compare them with that formed by his contemporary, William Bulleine.[2] The rules laid down by

[1] Gale, *op. cit.* fol. 46*b*.

[2] Educated at Cambridge; after travelling in Germany and Scotland, he practised in Norwich, and afterwards in Durham, where he was tried for murder, but acquitted. He finally moved to London, where he obtained a large practice. He died Jan. 7, 1576, and was buried in St. Giles', Cripplegate.

these two surgeons may be compared with those of John of Arderne, living two hundred and fifty years previously, whom we have already quoted in a previous chapter. The first extract is from a work by William Bulleine, entitled "A little dialogue betwene twoo men, the one called Sorenes and the other Chirurgi: concernyng Apostumacions and Woundes their causes and also their cures (1562)." His notion of a surgeon is, that

"He must begin first in youth with good learning and exercise in thys noble arte, he also must be clenly, nimble handed, sharpe sighted, pregnant witted, bolde spirited, clenly apparailed, pitefull harted, but not womenly affeccionated to wepe or trimble, when he seeth broken bones or bloodie woundes, neither muste he geue place to the crie of his sore Paciente, for soft CHYRURGIAINS maketh fowle sores. Of the other syde, he maie not plaie the partes of a Butcher to cutte, rende or teare the bodie of manne kynde. For allthough it be fraile, sore, and weake, yet it is the pleasure of God, to cal it his Temple, his instrument, and dwelyng place, and the Philosopher[s] dooe call it ORBICULUS that is, a little world."[1]

Eight thinges or properties of a good chyrurgian.

The following rules for apothecaries are appended to the same author's "Booke of Compouds":

"i. Muste first serue God, forsee the ende, bee clenly, pitie the poore.

"ii. Must not be suborned for money, to hurt mankinde.

[1] "A Dialogue betwene Sornes and Chyrurgi," fol. viii.]

"iii. His place of dwelling, shop to be clenly to please the sences withal.

"iiii. His garden must be at hand, with plentie of herbes, sedes, and rootes.

"v. To sow, set, plant, gather, preserue and kepe them in due tyme.

"vi. To reade DIOSCORIDES, to know the natures of plantes and herbes, etc.

"vii. To inuente medicenes, to chose by colour, taste, odour, figure, etc.

"viii. To haue his morters, stilles, pottes, filters, glasses, boxes cleane and swete, etc.

"ix. To haue Charcoles at hande, to make decoctions, syrupes, etc.

"x. To kepe his cleane wares close, and cast awaie the baggage.

Thapoticary.

"xi. To haue two places in his Shoppe, one most cleane for the Phisicke and a baser place for Chirurgi stuffe.

"xii. That he neither increase, nor diminishe the Phisicians bille,[1] and kepe it for his owne discharge.

"xiii. That he neither buie, nor sell rotten drugges.

"xiiii. That he peruse often his wares, that thei corrupt not.

"xv. That he put not in *quid pro quo*, without aduysement.

"xvi. That he maie open well a vein for to helpe the pleurisie.

"xvii. That he medle onely in his vocacion.

[1] prescription.

"xviii. That he delite to reade NICOLAUS MYREPSI, VALERIUS CORDUS, JOHANNES PLACATON *th*e Lubik etc.

"xix. That he doe reme*m*ber his office, is onely to be the Physicia*n*s Coke.

"xx. That he vse true measure and waight.

"xxi. To remember his ende, and the iudgement of God; and thus I doe commende him to God, if he be not coueitous or craftie sekyng his owne lucre, before other mennes health succour & co*m*fort."

Balthrop's ordinances.

During the mastership of Robert Balthrop, who attended Henry VIII. to the Field of the Cloth of Gold, and was subsequently sergeant of surgeons to Queen Elizabeth in 1566, were enacted, constituted, and ordained the Laws, Acts, and Ordinances, under which, with little variation, the united Company of Barbers and Surgeons exercised their authority for many subsequent years. The Acts were ratified by the Lord Treasurer, Chancellor, and chief justices of either bench. The most interesting facts for the present purpose are those which more especially relate to the surgeons. Thus "yt is ordayned that there shalbe hereafter none examinacions had or made of any p*er*son to be admitted in Surgery except it be in the com*m*on hall of the saied Mysterye, and that there be then and there present three of the examyners w*i*th the M*aste*r and governoures for the time beinge, and some one or more of the saied M*aste*r and governoures at the Leaste." "Also yt is ordayned that none of the saied

Misterie shall at any time hereafter bring or procure any pacient to come into the saide hall of the said Mysterye uppon any courte Daye uppon payne to forfet and paye for euerye suche offence iijs. iiijd." From this it is evident that the personal examination of patients by the candidates for the licence was not considered a necessary or desirable feature at this period.

Anatomy regulations.

Public demonstrations and dissections were held in the hall, and to preside over them, as we have seen, two masters and two stewards of the anatomies were annually chosen from amongst the members of the Company. There was an important clause, however, which for years retarded the progress of anatomy in this country, and which ran as follows. "Priuate Anathomyes and other Anathomyes by any of the mysterye shall not be made or wroughte at any time hereafter in any place or places, but onely within the com*m*on Hall of the said Mysterye. And further that uppon speciall Lycens graunted by the M*aste*r and examin*er*s for the tyme beinge any p*er*son or persons of the fellowshippe exercising Surgery lawfully may . . . . Take forth and bring home into the said Hall the Ded bodyes for Anathomyes, and then and there to make and worke the same for their knowledge and more connynge. And that all priuate Anathomyes shall reverently henceforth be buryed as publick Anathomyes for the worshippe of the said mysterye. Any skelliton to be made onelye excepted upon payne of forfeture of ten poundes." This regulation was strictly enforced, and for acting contrary to it, "John Deane was appointed to bring in his fine of x^{li}, in that he had an Anathomye in his house." At a somewhat later period it was the constant practice for the masters to dissect a body in private before they performed the public anatomy.

In addition to the masters and stewards there was a reader in anatomy. For many years the reader was a physician. The exact date at which the office was founded is not known. Wadd, however, mentions briefly that Dr. Willm. Cunningham lectured in 1563. The same Cunningham who "put in the Greek and Latin words in such sort as he thought good" to Gale's Enchiridion, as the author confesses that he "himself had not perfect understanding of the tongues." In 1577 "Mr. Thomas Hall of this companie [who was also a physician] is graunted to Desect Thanatomies private or publick for the terme of Tenne yeres that shall happen to be *within* this howse, and that there shall be yerelie forewer one private Anathomye at the leaste and one publick, yf some cause reasonable be not to the contrarie. The same pryvate alwaies to be Desected before the publicke. And the same to be allwaies fetched from the place of execucion by the M*aste*r and Stewards for the tyme being. And that as well the same Stewardes as Masters shalbe attendant upon the Desector During the tyme of any Desection." Masters and stewards of the anatomy, however vigilant, could not repress certain irregularities which occurred during the dissections. And the following minute shows that even at this period there existed that morbid taste for curiosities which gave origin to the tanneries of Meudon[1] during the French Revolution. "Yt is agreed and condescended that no *per*son or *per*sons of this Companie do presume at anie tyme or tymes hereafter of Anathomies to take and carrie awaie or cause to be taken or carried or conveyed awaie any *par*te of the skynn of any bodie which shall at any tyme thereafter happen to be wrought uppon w*i*thin the hall of the misterie, and the same tann or cause to be tanned like lether. Upon the payne of v^{li}."

The reader of anatomy.

As it would seem from the following article great care was taken of the reader of anatomy, the Mr Doctor here mentioned. The article is one of the series already quoted at page 113.

An Article that the *Masters* and Stewards of the Anathomye shall provyde and se all things necessary for the Docter as is further conteyned.

"The Enleventh Article is that they which be appointed for the Anathomye for the yere next following and must sarve the Docter and be about the bodye. he shall se and provyde that there be every yere a matte about the harthe in the hall that Mr Docter [be] made not to take colde upon his feete nor other gentelmen that doo come and marke the Anatomye to learne knowledge. And further that there be ij fyne white rodds appointed for the Docter to touche the bodye when it shall please him and a waxe candell to loke into the bodye and that there be alwayes for the Docter two aprons to be from the sholder downwarde: and two payr of Sleaves for his hole Arme with tapes

[1] Carlyle, "French Revolution," vol. iii. bk. v. chap. vii.

for chaunge for the sayed Docter and not occupye one Aporne and one payr of Sleves every day which ys unseemly. And the M*aste*rs of the Anathomye that be about the bodye to have lyke aprons and sleves every daye bothe white and cleane, yf that the M*asters* of the Anathomye *tha*t be about the Docter doo not see these thinges ordered and that there knyves probes and other instruments be fayer and cleane accordingly with Aprons and sleves if they doo lacke any of the said things afore rehersed he shall forfayte for a fyne to the hall xl s."

The following interesting and quaint document tells its own story, it is entitled the "Form of the Business at the time of a Publick Demonstration of Anatomy." It is evidently the directions given to a new clerk of the Company, but at a date long subsequent to the period now under consideration. It runs as follows:

"So soon as the body is brought in deliver out your Ticketts which must be first filled up as followeth 4 Sorts. The first fforme to the Surgeons who have served the office of Master you must Say Be pleased to attend etc, with which Sumons you send another for the Demonstrations; to those below the Chaire you say Our Masters desire your Company in your Gown and flatt Cap etc with the like notice for the Demonstrations as you Send the Antient Master Surgeons. To the Barbers if Ancient Mast*ers* you say, Be Pleased to attend in your Gound only; and if belowe the Chaire then Our Masters desire etc as to the others above without the Tickett for the Demonstrations. The body being by the Mas*te*rs of Anatomy prepared for the Lecture (the Beadles having first given the Doc*to*r Notice who is to read and taken Orders from the M*aster* or Upper Warden of the Surgeons Side concerning the Same) you meet the whole Court of Assistance in the Hall Parlour where every Gentleman cloathes himself and then you proceed in form to the Theatre (Viz^t^) The Beadles going first next the Clerk then the Doc*to*r after him the Severall Gentlemen of the Court and having come therein the Doc*to*r and the rest of the Company being seated, the Clerk walks up to the Doc*to*r and presents him with a [wand?],[1] and retires without the body of the Court untill the Lecture is over, when he then goes up to the Doc*to*r and takes the [wand] from him with Directions when to give Notice for the reading in the afternoon which is usually at five Precisely and at One of the clock at Noon, which he Pronounces with a Distinct and Laudable Voice by Saying, This Lecture Gentlemen will be continued at 5 of the Clock Precisely, having so said he walks out before the Doc*to*r, the rest of the Company following down to the Hall Parlour, where they all Dine The Doc*to*r Pulling off his own Robes and putting on the Clerks gownd first, which has been always usuall for him to Dine in, and after being Plentifully regalled they proceed as before untill the end of the 3^d^ Day, which being over (the Clerk having first given notice in the fforenoon that the Lecture will be continued at 5 of the Clock Precisely at which time the

[1] This word is indistinct in the original; I have, therefore, supplied the hiatus here as well as in the next line.

same will be Ended) he attends the Doc*to*r in the Cloathing Room where he presents him folded up in a Piece of Paper the Sume of $10^{li}:0:0$, and where afterwards he waites on the Masters of Anatomy and presents each of them in the like manner with the sum of $3^{li}:0:0$ which Concludes the Duty of the Clerk on this Acc*oun*t.

"NB The Demonstrator by Order of the Court of Assistants is allowed to read to his Pupills after the Publick Lecture is over for 3 days and untill 6 o' the Clock on each day and no longer after which the remains of the body is decently interred at the Expense of the Masters of Anatomy which usually Amounts unto the Sum of 3 : 7 : 5 as by a Bill in the Govern*men*t Dotation book will more plainly Appeare."[1]

The difference between a surgeon and a barber, and between one who has passed the chair and one who has not yet attained so high a dignity, is exquisitely defined by the different wording of the summons in each case. The performance of the public anatomy was always kept as a high festival by the Company, and was celebrated by a dinner, which was second only to that held at the election of the master and wardens. It was the duty of the stewards of anatomy to provide this dinner after the lecture, and at first, as the tenth article shows, it was a perquisite of the clerk's office to provide the necessaries for it, but the privilege was abused. Therefore

"The Tenthe Article is that where at the Anatomye

[1] See also pages 227 and 228.

the clercke of the Company [Apulton] clamith for a Custom that he shoulde fynde of his own all maner of naperye and vessell and a coke to dress the meete which he is not able to doo but for a lytle lucre of monye and to the dishonestye of the Crafte. Therefore yt is ordered understanded and agreed that hereafter he shall not claym any such custome but that they which be appointed for the Anatomyes maye every yere bring all maner of vessell that they will, and also all maner of napery and plate, and to have the coke which pertayneth to the hall when they list which may bring his stuffe with him; and to have the occupying of the kitchen buttery and there to laye their wood and cole that they mygt dresse their meate clenly and honestlye because of worshipfull M*aste*rs comyng ther unto: yf therfor the Clercke will not observe this order he to paye to the hall for a fyne his half yere's wage w*hi*ch is xl[s].

An Article that the *Masters* and Stewards of the Anatomy may bring in their vessells plate of theyr owne and also a coke. And that Apulton shall clayme no more any Custom for the fyndinge Glas upon payne of *payment* of a ffyne.

"In 1596, M[r] Doct*ou*r Paddy ys chosen to be the desect*ou*r of *ou*r Anothomies yf yt shall pleas him to accept of the same. And also xx[s] ys geven yerely to the Anathomists more than they were accustomed to have, in regard that suche Docto*u*rs of Phisick as shall associate the said M[r] Docto*u*r shalbe invited to Dyner at the good liking of the masters or governo*u*rs from tyme to tyme."

The duties of the Anatomists, however, were not always easy. The masters and stewards had to bring the bodies from the place of execution, and from certain complaints made by Clowes,[1] it is clear that this was by no means a pleasant task. Indeed, Clowes enters a formal protest against the illegal action of the Sheriff and his servants, in putting up for public sale the bodies of malefactors, and actually procures the dismissal of one of the servants on this ground. Useful anatomies, too, might be lost in other ways than through the rascality of the sheriff's officers. For it was "agreed That yf any bodie w*hi*ch at any tyme hereafter happen to be brought to o*u*r Hall for the intent to be wrought uppon by th' anathomistes of o*u*r Companie shall revyve or come to lyfe agayne as of late hathe ben sene. The charges aboute the same bodie so revivinge, shal be borne levied and susteyned by suche *per*son or *per*sons who shall so happen to bringe home the Bodie. And further, shall abide suche order or ffyne as this House shall Award." Either the executions were not conducted with proper care, or else evil-minded resurrectionists endeavoured to make a profit out of the Company, by selling to them subjects which could not be wrought upon.

Great care was taken that all things should be done decently about the anatomy, for in addition to the orders about burial and against tanning, there is another bearing date 1567, "that there shal be buyldynge don and made about the Hall for seates for the Company that cometh unto every publyque anothomy, ffor by cause that every *per*son comynge to

[1] An eminent surgeon who practised in London and was on the staff of St. Bartholomew's Hospital. His son was sergeant-surgeon to James I., and Master of the Barber-Surgeons' Company in 1626 and 1633.

see the Same maye have good p*ro*spect over the Same; and that one sholde not cover the syghte thereof one frome another, as heretofore, the Company hath much co*m*playned on the same. . . . And also there shal be pyllers and rods of Iron made to beare and drawe Courteynes upon, and aboute the frame where w*i*thin the Anathomy doth lye, and is wrought upon for bycause that no p*er*son or p*er*sons shall beholde the desections of the body, but that all may be made cleane and covered w*it*h fayer clothes, untyll the doctor shall com and take his place to read and declare upon the parts desected. And also, yet fordermore, that there shal be a case of weynscot made w*i*th paynters worke upon, as semely as may be done ffor the skellyton to stand in, That for the worshyp of the Company all these to be made through and don at the charges of the mistery and Common boxe of the Hall."

The attendance at the Anatomies here spoken of was not voluntary. Since by an ordinance of 1572, "every man of the company usinge the mystery or facultye of surgery, be he frema*n*, fforeyn, or alian straunger, shall come Unto the Anathomye, being by the Beadle warned thereunto. And for not keepinge their houre both in the forenoone and also in the afternoone, and beynge a ffreeman shall forfayt and paye at euery tyme iiij[d]. The fforeyn in Like maner, and the Straunger e*uer*ye tyme vj[d]. The said fynes and forfaytes to be employed unto the Anathomysts for the tyme beynge, towards theire charges w*i*thin the tyme of the sayed Anathomye. And also for not comynge in all the tyme of the Anathomye (having

Lawful occasion of Absence), the Freeman shall pay vj^{d}, the fforreygne viij^{d}, and the Straunger xij^{d}, to be employed in manner and fourme aforesaid. And also iij^{s} and iiij^{d} to the Masters and Gouernoures of the said Mystery, for their sommons brakinge Notwithstanding. Provided also that they come well and Decently appareyled, for their own honestye, and also for the worshippe of the companye."

In the yeare 1579, "was a motion before the courte of the company concerninge a Lecture in Surgery, to be had and made in our Hall, and of an Anuytie of x^{li} to be geven for performance thereof yerelie by Master Doctour Caldwall,[1] Doctour in phisick; but yt was not concluded upon nether was any further speche at that tyme." Nor indeed at any other, as far as can now be understood, for no subsequent reference is made to the proposal. In 1582, however, Lord Lumley, at the instance of Dr. Caldwell, founded in the College of Physicians the lectureship which is still known by his name. The surgeons thus appear to have lost a noble benefaction which should of right have belonged to them.

"Upon the xxvij of June, 1568, in this Courte John ffrende is comytted to warde for a pacient dying under his hand and not presented." And on the seconde of December, 1572, "Here was John ffrend and was commanded to lay down his fyne for not presentinge Mr. Watson of the Tower, which dyed of Gangrene in his fote. And he paid xx^{s}." This wholesome rule of "presenting" patients in danger of death or maim was

Presentation of serious cases.

[1] See further upon this subject, pages 184 to 190, note.

therefore in full force. As we have seen,[1] it had been handed down to the united Company from an early period in the history of English surgery. The regulation was an eminently satisfactory one, and when we consider the small number of the surgeons at this period, it was one which could be readily carried into effect by a little judicious management. The practitioner was bound to call in and consult with the master and wardens of the Company in all serious cases which he attended. He was thus enabled to shift the responsibility, whilst he derived benefit from the experience and suggestions of his seniors in the craft; they in turn added to their stock of knowledge from the various cases which they would thus see during their term of office. The exact words of the ordinance are perhaps worth quoting. "Also yt is ordayned That hereafter no manner of person of the mysterye shall take any Sicke or hurte person to his cure w*hi*ch is in perill of maym or dethe, But that he shall shewe and present the same sicke or hurte person within three Dayes next after any by hym receyved unto the M*aste*r of the Companye for the tyme beinge. And the said M*aste*r withe his governoures shall then go in theire owne *per*sonages or appoynte suche

[1] Chapter ii. page 26. A somewhat similar plan is still pursued, or rather has been of late years revived, at St. Bartholomew's Hospital. In this charity all surgical cases of an interesting or doubtful nature, or which are likely to result in serious operation, are brought into the operating theatre every Thursday afternoon, and are there publicly examined by the surgical staff. The surgeon in charge of the patient first introduces the case with a brief account of the symptoms, his colleagues, after examination, then give their opinions as to the diagnosis, prognosis, and treatment, the senior surgeon speaking first. When all have spoken, the surgeon in charge sums up and declares the method of treatment he intends to pursue.

*per*sons as at theire Discressions shall seeme most convenyente and experte in the said facultie of Surgerye, to go with them, or deputies in *the* absens of any the said governours. . . . And if any profitt advantage gyfte or rewarde, come or be geven unto any the said Mr Governoures, or Deputies, be it throughe calling and visitinge any *per*son as is aforesaid, That *euer*more it be by the M*aste*r for the tyme beinge Departed and Devided by even porcions amonge the said M*aste*r governo*ur*s or Deputies. . . . . And whoe offendeth in any of these twoe poyntes or cases, shall, for euerye tyme paye and forfeit to the use of this said mysterye for a fyne xx^s." Then follows an ordinance against any person taking upon himself the office of visitor in any presentation, unless he be duly called thereunto by the master. And lastly there is a very proper rule, "That hereafter yf the said M*aste*r and govornoures at any time within theire said office, be desyred to goe and see any sicke or hurt person or *per*sons as aforesaid and Do denye or make refusall but that he be rewarded for his or theire paynes takinge, shall for e*u*ery such Default forfet and paye for a fyne xx^s. Provided alwais that the said M*aste*r and governoures for the tyme beinge . . . . shall at any tyme hereafter take paynes goe and see *per*formed and geve theire goode counsell for the helth, as well of the poore for charitie as to those that are of ability. And beinge by any of the said company usinge surgerye Desyred thereunto, bothe for the health and curacion of the greeved pacient, the worshippe of the said mysterye, and also for the Discharge of the surgeon."

Apprentices.

The apprentices appear to have been a fertile source of trouble both to the Company and to their masters, for numerous ordinances were at divers times enacted about them, and the court books are full of references, some of which are of great interest. The ordinances of 1566, when Robert Balthrop was master, enter very fully into questions of apprenticeship, although the subject had been discussed at large only ten years previously.[1] In this second series of regulations "yt is ordayned that none of the said mystery occupyinge surgery shall at any tyme hereafter teache any the said scyence of surgery but his ap*pre*ntice, nor shall take any to be ap*pre*ntice but such as at leaste can write and reade uppon payne of forfeiture of v^li^." "No *per*son beinge in the clothinge of Lyverye shall have any more apprentises or *ser*vauntes retayned by covenant or otherwise to serue hym at one tyme but only foure." Other members of the Company who were not of the livery to have only three apprentices. Care was taken that the apprentices should be at least physically good, for none of the mystery are hereafter to "take any apprentice excepte he fyrste doe present the same ap*pr*entice unto the M*aste*r and go*ue*r*n*oures . . . and that they Do uppon the syghte of hym allowe hymm to be clene in *per*son and Lymm and mete for the exercysinge of the same mysterye, science, or arte." Perhaps it was in consequence of this rule that more than one entry occurs in

[1] Chapter vii. page 113.

the books which reflects in very severe terms upon the wives of those members of the Company who possessed unusually attractive apprentices.

Apprentices were protected from unjust treatment at the hands of their masters by the enactment that "no *per*sonne of the said mysterye doe or shall at anye tyme hereafter for frendshippe, Lucre or any other respecte put any his ap*pre*ntice awaye or out of his service untill his terme of ap*pre*ntish*ip* be fullye ended without reasonable cause, to be tryed and approved before the M*aste*r and governoures . . . Uppon payne of forfeture of xl[s] for euery suche offence." A good instance of such reasonable cause appears shortly afterwards. "Here was Henry Lushe and wittnessed how that hys ap*pre*ntice Rafe Soda raune awaye ffrom hym, and did contract hymself to three wemen and was Asked at Westm*inster* in the church. And also had Delt unhonestly w*i*th his maydese*r*vaunt. Whereuppon he Delyv*ered* his Indenture to his father." As a corporate body the Company on occasion were constrained to act with apparent harshness towards their erring apprentices by following an ancient adage, as the following extract shows, and as Wm. Fyshe must have experienced only too sadly. "Upon the xxij daye of August, 1569, . . . here was Rich. Upton playntyf against his app*re*ntis Wm. fyshe for that he is runne awaye frome his said M*aste*r the xxi[st] of the same moneth, and tooke w*i*th hym *cer*ten instruments of surgery and other things more. W*hi*ch *par*tyculers were here *pre*sently sene and by the said Wm. fyshe confessed and that he had no cause to go from his said M*aste*r but that he wolde have gon to the sea. And according to his desert had correction and punyshment unto auncyent custom w*i*th roddes." The which we will hope did not cure him of his longing, but the rather confirmed him in it, so that he might become a worthy follower of the renowned captains of his day.

The Company appear to have acted with the utmost impartiality, however, for on another occasion "The apprentice of Thomas Insoll complayned on his master ffor mysusing his ap*pre*ntice in beatinge hym unreasonablie as he sayed, and not gevinge him sufficient meat and drynke. And yt was ordayned that the saied Thomas Insoll should use him as he oughte to be used, and not to geve him any correc*ti*on for this complaynt." And again soon afterwards, "John the apprentice of Thomas Wayte was brought before the court for abusing Roger Laborne and his wyf, and his master agreeing he had favorable correction for his offences." And yet again, "Here Markes Gresvile complayned of his M*aste*r for geving him a blowe, and he was wiled to go home and serve w*i*th his m*aste*r deligently, and so departed." Master Laborne, too, was not happy in his apprentices, for on "the Laste of June 1574 here was the ap*pre*ntice of Roger Laborne and complayned one his M*aste*r ffor beinge evill entreated by his M*aste*r requiring to be released from hym because he would not teache his scyence. And he was commaunded to serve his M*aste*r agayne and his M*aste*r to geve hym corection for his evill behaviour, and also not to allowe hym half his lettynge bloude and Drawinge teethe, but geve hym reasonable apparell as becometh an ap*pre*ntice. And he submytted hymself one his knees before his M*aste*r and M[rs.]" A later entry shows the strict rule under which the apprentices were kept. "This daye it is ordered that Willm: Webbe apprentice or s*er*vant to Henry Needham was appoynted to bee co*m*mitted to the Compter uppon my Lord Maiors commaundem*en*t for being collered as an apprentice and yet marryed and havinge children." The following entry, too, is of interest as it gives an inventory of the worldly goods of an apprentice in the year of grace 1600. As a result of some controversy, "Edward Want did turne ov*er* his apprentice to Peter Ledsam for the residue of his terme to come. And did also *pro*mise to deliver such apparrell as hee had of the sayd apprentice . . . viz. one *pair* of hose, one *pair* of linnen drawers, twoe shirts, certeyne bandes, a new plaister boxe or salvatory, a splatter,[1] a payre of mulletts,[2] a *pair* of *pin*cers, a punche, a *pair* of crosse billes, fower bokes *and* one fflute." A truly miscellaneous collection, and worthy of the renowed Master Simon Tappertit.

The Company and the Universities.

Throughout Elizabeth's reign the Company appear to have maintained one scholar in the Universities of Oxford or Cambridge at an expense to themselves of xl[s] a year. The exhibitioners thus provided for were generally the sons or other relatives of prominent members of the Company. As early as 1566 "Yt is Ordayned that Thomas Hall

[1] spatula (?) [2] Small pincers for curling the hair.

shall have an excibicyon of fortie shyllyngs by the yere and yerely Towardes his studye (in Mawdelyn Coledge) in the unyversytie for Surgery anexinge physycke therunto, and thereby hereafter to perfet his other brethren beynge of this mystery . . . by Readynge Lectures unto them in *th*e com*m*on hall, and otherwyse by his councell, conynge, and knowlege." And, as we have already seen, Dr. Hall held the post of examiner in 1576, and reader in anatomy in 1578. These grants of money, however, do not seem to have been given solely to promote the study of surgery, for in 1578 "Yt is agreed that Nicholas Straugnishe, the sonne of Henrie Straugnishe of *our* misterie shall haue ffourtie shillinges by the yeare so longe as he shall continue his studie in the vniversitie of Oxforde in the learninge of deuinitie Duringe the good pleasure of this Companie and his well Doinge. And the same payment to begin ymediatlie after his admittance into some Colledge there." The university career of Master Straugnishe can be traced throughout, for in 1584 "Nicholas Straugnishe, scoller in Oxforde, shall haue his penc*i*on of xl^s yerely paid unto him Duringe the good pleasure of this howse: and also xxx^s w*hich* is behind unpaid of his said penc*i*on." In 1582 "Yt was agreed that Whereas there was sent unto Henry Straugnishe for his sonnes *pre*ferment to *pro*cede Bachelo*u*r of Arte xl^s, that he shall have the same wi*th*oute any reclayme frielie geven him." In 1586 "Mr Straugnishe nowe M*aster* of Arte in the universitie of Oxforde, shall haue his yerely stipend no longer then untill michelstyde next and then Willm. Parys, the sonne of Humfry. Parys who is student in Cambridge, shall haue that pencion Duringe the good pleasure of this Companie." Finally, in 1587, as the quondam exhibitioner had made good use of his natural abilities, there was "a consent unto the gevinge of xx^s to Mr Straugnishe, Preacher and Student in χρists Churche in Oxford, and sometyme Exhibitioner of this Companie." "Thomas Yates, sonne of Mr Warden Yates shall haue yerelie duringe the well plesaure of the *Master* and governo*urs* of this misterye towardes his maintenance in studie in the universitie of Cambridge yerelie the some of xl^s and xx^s in hand to buy him bookes."[1]

Licences to practise.

By the regulations of the Company the licences to practise surgery were granted after examination for the space of so many years as the *Master* and governors with the examiners should think fit. The examination appears to have been conducted fairly, and upon the results of it the licence was granted permanently if the candidate did well, or for a longer or shorter term of years if the results were less satisfactory. On very rare occasions, indeed, the aspirant was "referred" for a further period of study. The fees, however, appear to have been calculated on a sliding scale adapted to the rank and position of the applicant. The following minutes will illustrate these points, and

[1] This custom of granting presents of books and money to students at the Universities is still kept up by the City companies, as the editor is able to testify, for whilst a student at Oxford he has more than once experienced at the hands of the Merchant Taylors' Company a kindness of this nature.

will show that even the permanent licence was liable on occasion to be revoked: "Here was one Johnsonne a straunger and required to be a brother of *our* company. And he was willed to come the next court Day." Accordingly at the next meeting "Here was Johnson the Surgeon in Crowchett ffryers to make answere accordynge to his promyse the last Tuesdaye. And yt was agreed that he shoulde paye xxx^{s} in hand and x^{s} a quarter, and uppon Thursdaye nexte to be admytted." Johnson being already in practice was probably admitted to practise under the seal of the Company after passing a modified examination. "This daye[1] Richard Banester of Slyford in the County of Lyncolne Surgeon was examyned and approved before the M*aster*s of this Company by Mr Wood, Mr Baker, Mr Thorney, and Mr Willm Martin concerninge his skill in the practize of surgery. And was found an hable[2] and fitt man to use the same." "This daye Raphe Barret at his humble suite was examined concerninge his skill in Surgery before the M*aster*s by . . . the examiners in that behalf appoynted. And was admitted *and* sworne." On another occasion: "At the Request of Mr Doctor Julio, Gabriell Petiolio is Lycenced to exercyse and use Surgery for one hole yeare payeing the quarterage as other brethren strayngers do." "This daye Willm Pilkington paid to the M*aste*r towards his examinac*i*on xl^{s} and was comaunded to geve his attendance at this hall on Thursday next to be examined." On the following Thursday: "Willm Pilkinton uppon his examinac*i*on

[1] July 8th, 1602. [2] skilful

was tolerated to practize surgery for five yeares next ensuinge, Provided that hee paie quarterly to this howse ij^{s} vi^{d}, and that hee the said Pilkinton doe joyne *w*ith him in euery cure[1] he shall have in dang*e*r of Death or mayme some expert surgeon of this Company. A very provisional licence indeed, and one that was scarcely worth having, and not far removed from the state of affairs noticed under the date 17th day of Jan., 157$\frac{6}{7}$, when "Gilbert Scofeld made his request to have his le*tte*rs under the seale of the house, but upon his examinac*i*on being founde unable[2] he was wil'd to staie a while longer untill he had red more." We gather from a previous minute that Gilbert's ignorance arose from lack of application rather than from want of brain power, for on the 27th day of May, 1575, "John Wheck*er* complayned at this Co*u*rte of Gilbert Scofeld, for that the said Gilbert refused to paie unto the said Whecker for heling of a broken hed Don by the saide Scofeld. So he was at this Co*u*rte awarded to paie to the saide Whecker w*i*thin one ffortnight iiij^{s}." The next extract, on the other hand, appears to point to want of ability rather than to want of will. It affords a curious instance of the multifarious duties of the coroner's quest. "This daye John Ffoster a poore and unskylful man of this Company made his appearance before the M*aste*rs of this Company, And was examined concerninge his skyll in the arte of surgery, and was found altogether unskilfull in all the *par*tes thereof. Whereuppon it is ordered that M^{r}

[1] case. [2] unskilful.

Wilbraham Coroner to this Cytie be warned to be here *with* the Coroners Inquest on Thursdaye next by tenne of the clock in the forenoon to be satisfied by their owne hearinge of the unskilfulnes of the said ffoster."

The following are instances of the partial and complete revocation of licences: "Here was a Complainte determyned upon *wh*ich was made against Thomas Hoole. And for that he was founde ignorant he is bounde in xl^li^ never to medle in any matter of S*u*rgery but suche as he shall call some better experienced surgeon than him unto at *th*e seconde dressinge. This daye Oli*ue*r Peacock brought in his fine for not pr*e*sentinge his Cure beinge nowe Dead. And it was mittigated to five shillinges. And it is furth*er* added that he practize surgery no more." "This daye John Smith appeared before the M*aste*rs of this Company who was discharged by them from the practise of surgery. This daye one Willm Cell practiconer in surgery lykewise appeared before the M*aste*rs and was forbidden to practize any more in surgery."

The Bishop's licence.

The licence of the Bishop of the diocese or in London of the Dean of St. Paul's was sufficient to qualify for the practise of surgery; this was found to interfere very materially with the efforts made by the Company to improve the status of the profession. In 1599 it was found expedient to come to an understanding with the Bishop upon this question. Accordingly, "The Maister of the Companye made his petic*i*on to the lorde Bishopp of London that noe person shoulde be admitted to practice Surgerie, but suche person as shoulde haue the seale of this house to testifie his examinac*i*on before the Maisters: which was graunted, and order sett downe for the same." The matter was not however settled so easily, as will be seen at a later period, for succeeding bishops did not so readily forego their privileges.

From the subjoined entry it would appear that the freedom of the Company carried with it until 1582 a right to practise surgery. For "it was agreed and thought necessary that no ffreeman of this company hereafter to be made shall be put into the lecture bill although he have been broughte up in surgerie, unless he first be examined and have the seal of *ou*r house for his credit, which he shall have gratis paying for the wrytyng of his letters to the clark iij^{s} iiij^{d}.

Fees for licences.

In regard to the various payments necessary for licences, there are numberless entries showing that not only did they vary for different grades in the Company, as might be expected, but even for different persons of the same grade. It is amusing to observe how a little present judiciously selected and offered at an appropriate opportunity to the Company as a corporate body, would facilitate the delivery of letters of admission, or would smooth over any little hitch or difficulty which might occur. A curious entry in regard to these money matters occurs Sept. 15, 1576. "Whosoever sholde request to be a brother of this howse, yt is agreed that they shall paie redy money, otherwise not to be admitted." The difference in the fees occasioned from time to time subject for scandal, thus it appears that "George Baker complayned one Arthure Welborn forasmoche as the said Arthure hath reported to ffraunces Rosier that where he payd v^{li} for his Admyttaunce he would have undertoke to have gotten the said Admyttaunce for xl^s, and also a licens from my Lord Keper to occupye surgery for vj pounds. And alsoe he reported that he was admytted onely uppon the report of George Baker *with*out examynacion, *which* is untrue to the slaunder of the M*aste*r and gou*er*nores and the hole howse. Not*with*standynge the matter was left in suspence tyll some other tyme that the M*aste*r and gouernoures think good to call on yt." As regards gifts in 1572, "Here was one Thomas Hall of Howsbourne Crawley in the County of Bedfourd a brother of this House Admytted in the tyme of Mr. Mason[1] Beinge master, and had his L*et*tres of Admyttaunce corrected and Amended and newe seled. . . . Ffor the *which* he hath *pro*mysed to send to the howse a pece of fflėsh againste the Dynner Daye." And in the same year, "Mr. Baunester of Nottingh*a*m,[2] gentleman, was sworne and Admitted a Brother of this mystery. Whereuppon hee hath graunted to the howse yerely xx^s so long as he Lyveth, and to be Liberall and co*m*modious to the house in what he may: and will send yearly a buck or twoe and hath payed all . . . in hand x^s, and shall have his L*ette*r of Licence." Also, "This daye Rog^r^. Jenkins, a freeman of the Company of Weavers and an admitted brother in the practize of Surgery, for and in respect of the love he beareth to

[1] A.D. 1567. [2] John Banester the celebrated surgeon.

this Company became an humble suitor to this Courte, that they would be pleased uppon his discharge from the said Company of Weavers to incorporate him amongst them: w*hi*ch his request they willingly graunted. Whereuppon he freely gave to the M*aste*r to the use of the said Company the somme of tenne pounds in gowld, w*hi*ch the said Courte did very kyndely receawe of. And in respect thereof and for that the said Roger Jenkins was of the Lyvery and one of the Assistants of his Company they ordered, that uppon his translac*i*on from his said Company to o*u*r hee shal be admitted into the Clothinge, and further ordered that for and in respect of his yeres hee shal be by theis *pre*sents dischardged of the office of M*aste*r and Steward of the Annothomy."

For the honour thus conferred upon him the worthy member of the Weavers' Company was mulcted according to custom, for "it is ordered that the Assistance w*hi*ch were p*re*sent at the last Court of Assistance shal be warned to be at dynner w*i*th Roger Jenkins on Thursdaye next beinge the appoynted tyme for his translac*i*on from the Company of Weavers to this Company." "Garrett Key[1] a straung*er* appeared before the M*aste*rs of this Company and in respect hee hath vndertaken the cure of his patient, one ffeake a gowldsmith beinge in danger of death, w*i*thout makinge p*re*sentac*i*on thereof to the M*aste*rs of this Company, did voluntaryly geve to the said M*aste*rs to the use of the poore of the same Company iij^{li}. And thereuppon they have

[1] Dr. Norman Moore suggests that the non-freeman was an Irishman, and that his real name was Gearod Mc Aoth.

acquited him of all former offences done to this Company."

Quacks in the sixteenth century.

The quacks[1] who abounded in the sixteenth and seventeenth centuries were treated by the Company with the utmost consideration. Each and all who offered themselves were examined, and those who possessed even a minimum of knowledge were granted a temporary licence, whilst those who knew absolutely nothing were alone cast into outer darkness. There was also an attempt made to classify by means of the licences, thus: "Here was a letter directed to the M*aste*r and gouernores ffrom the e*a*rle of Lyncolne w*i*th Divers other gentlemens hands thereat in the behalf of one Henry Esthorpe of Sambringh*a*m in the county of Lyncolne Surgeon towchinge his well Doynges in Surgery in those *par*ts as by the said letter uppon the fyle doeth appeare. Whereuppon he was Admytted a Brother and had a l*ette*r of Admittance as uppon theire reports able to Deale there in, but not allowed as examyned and aproved thereunto, as by the forme and coppye of his l*ett*er in the boke of recordes of those Letters Doth playnely appere." The greater number of the extra-professional practitioners of this period were the cutters for stone, the healers of ruptures, and the couchers of cataracts, who were tolerated and admitted to fellowship, though the Company could not away with clerks of parish churches

[1] For an interesting account of the examination of a quack, see John Hall's "An Historiall Expostulation against the Beastlye Abusers, bothe of Chyrurgerie and Physike, etc.," fol. bbb. iii. *a.* Lond. 1565.

and sextons, concerning whom there is the following ordinance: "At a Courte holden the Second Daie of Novembre 1570 . . . it was Decreed and fullie thereupon agreed . . . that from hensforth after the Date of thies *pre*sents, no officer of church namelie Clarke or Sexton serving any perish church shall beare any office in this mysterie ether in the clothing or in the yeomanrie;" and the ordinance is, "to remain in full strength and virtue," in 1573.

But to return to the quacks; on Dec. 2nd, 1567, "Here was one John . . . deucheman for settinge up bylls frome the Blacke fryars gate Loodgate unto Westmynster; and so he was examyned upon his byll but he colde not answer none, w*hi*ch bill is upon the fyle. And for *tha*t he is not found able[1] nor suffycient he is lynguyshed and dysmyssed frome usynge or occupyng any p*ar*te of Surgery w*i*thin the citie of London or Subourbes and one myle compasse upon payne of the forfeyture of the statute in *tha*t beholf ordayned." Surely, to examine a man upon his own bill, and to find that he could "answer none," must have been as diverting an entertainment as could be provided; and yet the test was a just one, and effectual withal, for the Dutchman, whose very name the clerk was unable to record, never appears again before the court. "Here was one Robert Scrottell a Straunger and cutter for the Stone admytted a Brother and paid iij^{li} in hande and standeth bound in x^{li} to pay the rest by x^{s} per quarter. And also it is to be remembered that he is appointed

[1] skilful.

to repayre hyther agayne at Ester to be examyned." "John Gardener of Ring*ham* in the countye of Sussex Surgeon, a healer of the rupture and stone was Examyned and had his le*tt*er of Admyttaunce and payed xl^{s}, and the other xl^{s} to be payed at the viiith Daye of September [1573] next following." The next extract affords an instance of a temporary licence granted for the performance of specified operations: "Whereas James Vanotten and Nycholas Bowlden are this daye become humble suiters to this Company to be tollerated and *per*mitted to practize as Surgeons w*i*thin this Cytie of London for and durynge the space of Three months next ensuinge onely for the couchinge of the catarack, cuttinge for the rupture, stone, and wenne. It is uppon consideraci*o*n of their severall suites ordered by consent of this courte That hee, the said James Vanotten, shal be *per*mitted to practize for the couchinge of the Catarack, cuttinge for the rupture, stone and wenne for the space of three monthes next ensuinge w*i*thout contradicci*o*n or denyall of the M*aste*rs or Gouernors of this Company." On payment, of course, of the usual fees. "Provided," the minute goes on to say, "that neyth*e*r they nor eyth*e*r of them shall p*re*sume to hange oute any banners or signe of Surgery in any place oth*e*r then where they shall lye and make theyre abode w*i*thin the tyme aforesaid, or practize in any oth*e*r poynte of surgery then before is specified w*i*thout furth*e*r lycence of the M*aste*rs or Governors of this Company for the tyme beinge firste had and obteyned." Mathias Jenkinson, however, was not sufficiently versed in the science of his profession, for he was "examyned concerninge his skyll in the arte of Surgery and was lycenced to cut for the hernia or Rupture to touch[1] the catarac, to cut for the wry neck [and] the harelip, provided that he call the *pre*sent M*aste*rs of this Company to every such cure, or such of the Assistants as are examined & approved as the said M*aste*rs in such case shall appoynt. And is to enter into bond in xl^{li} for *per*formance hereof. And paid to the *pre*sent M*aste*rs xl^{s}. And is to paye xl^{s} more at midsomer next." He failed, however, to act up to his promises, for on June 20th, 1609, a year later, "Mathias Jenkinson is dischardged from his practize in surgery for that he hath not observed the articles of his tolleraci*o*n and for his evell and unskilfull practize." So also "Edward Stutfeyld a practiconer in bone settinge" "Josper Johnson practiconer in the Cure of a fistula" and "John of Mounepilier in ffrance" were at divers times "tollerated to practize for three monethes." Provided, as in the first case, "that they hang not oute their banners or other shewes and signes of their profession in any other place then at the howse where they shall from tyme to tyme lye."

In some cases the Company appear to have been called upon to judge of monstrosities, as witness the following letter:

"This daye was *pre*sented to this Courte by Humfrey Bromley a letter from the Lord Mair of this Cittie of London the tenor whereof is as followeth.

Occasional duties of the Company.

[1] couch.

"To the M*aste*r and Wardeins of the Companie of Barber Surgions Whereas S*i*r Henry Herbert Knight M*aste*r of the Revells hath authorised the bearer hereof Humfrey Bromley to shew a child presented to be naturallie borne haveing Twoe heades ffower armes and three leggs w*hi*ch I suppose not to be borne of any woeman or to be the perfect substance of a child in respect whereof I forbeare to p*er*mitt the said Humfrey Bromeley to make shewe thereof within the lib*er*ties of this Cittye vntill such tyme as I maye be truele satisfied from you whether the same child be of the substance as is pretended. Therefore I desire you that upon advised view of the said child you truly certifie mee in writing vnder yo*u*r hand whether the same be really a child as is presented to thend I maye not inadvisedly suffer his ma*jes*ty's subjects to be deceyved thereby. This second of November Anno D*omi*ni 1627 Hugh Hamersley Maio*ur* Whereupon the vew of the supposed body as aforesaid it is ordered that this answere be returned to the Lord Maio*u*r as followeth viz. Right Hono*ra*ble According vnto yo*u*r L*ordshi*p's reference vnto us directed, dated the second of November 1627 wee have taken a deliberate vewe of the supposed monstrous birth presented vnto us to be vewed as from your hono*u*r by one Humfrey Bromley And although wee cannot possitively affirme it proceeded not from a woeman yet vnder favor, wee conceive and soe deliver our opinions that the said supposed monstrous shape hath beene, either by Arte soe compozed and put together, from unnaturall and untimely birthes of children, or from other animalls, as apes, munckeyes, or the like, w*hi*ch have a greate resemblance of manns bodye, in many of their partes, and soe, by the cunninge subtiletye of the composer made into a monster, thereby to delude the worlde, and haveing a Bodye of Antiquitie cannot safely receive a flatt and manifest contradiction; And wee are induced the rather to suspect it, for that the producer thereof hath noe testimonye from any learned or judicious men; neither from any magistrates of the partes where it is pretended to have bene borne, w*hi*ch such offendors use aboundantly to be furnished withall. And in conclusion compareing his printed demonstrac*i*on of his monster, with the Author he siteth, and others, that have written of such and the like monsters, Wee finde a great deale of addition, and a manifest disagreem*en*t w*hi*ch is a playne badge of fixion and falsehoode. All w*hi*ch our opinions wee humbly submitt to yo*u*r hono*u*rs grave wisdome, to be further considered of."

Disagreements of the barbers and surgeons.

From time to time the Company made strenuous efforts to keep the professions of barbers and surgeons distinct from each other. Thus in 1568 "None shall call or name the hall but the barbo*u*rs and Chirurgeons Hall," under a penalty of xx^{s}. On June 5th, 1583, stringent regulations were made by the Lord Mayor and Court of Aldermen, which enforced the separation of the two crafts, probably on account of the increase in deaths from infectious disease, and a fear lest the surgeons should spread the contagion. "Mr. Banckes, the master, w*i*th his gouerno*u*rs . . . . went into the

Guildhall and in the Counsell howse before the L*ord* maio*ur* and Aldermen did provide that all suche of oure Companie w*hi*ch Did Deale in Barberye should not medle or Deale w*i*th any sick of the plague or infected cum morbo Gallico, and that he wolde take obligac*i*ons of eu*ery* one to that end, w*hi*ch was *per*formed accordinglie, and eu*ery* one Did enter into bond to o*u*r saide M*as*t*e*r governo*u*rs and their successo*u*rs." Individual members of the Company, however, were inclined to side with those who practised both arts. In 1597 a court was held at the request of certain members of the Company "who were by one Holmes, an Informer, put into her ma*jes*ties Corte of the exchequour for using both Barbery and Chirurgery. Whereupon yt was motioned what co*u*rse was to be taken ether to *pro*secute to a tryall . . . . or ells to agree w*i*th the informer. This being *pro*pounded the said *par*ties were called into the *par*lor,[1] and their owne opinions being demaunded, they made their request to consider thereof in the hall, and they wold *pre*sently make their answere. Whereupon the said masters or governo*u*rs and whole Corte consented. And they being in the Hall agreed among themselves to

[1] A room of peculiar sanctity in which the Court of Assistants sate. In the ordinances dated 1566 may be read, "Yt is Ordayned That none of the ffelowshippe other than the M*aste*r and governoures Assistants and Clarke of the sayed companye be [he] within the clothinge or without the Clothinge shall or Do *pre*sume at anye time hereafter to come into the perloure of the said ffellowshippe at any such time as any courte of the said companye shall then and there be sett or kept except he or they be first called in by the Clarke of the said Mysterye upon payne of forfeture for euery such default ij^s."

An Order that none being out of th'assistants com*me* into the *par*ler except the*y* be called ij^s.

agree *with* the said informer, *which* they related to the whole Co*ur*rte. Whereupon the Co*ur*te rose, and the said *par*ties made their owne agreament privately *with* the said informer to his best likinge." At this period there was evidently a section of the Company who were quite willing that the two callings should be carried on by the same individual, and a few years later this party had gained the ascendency. In 1601 "uppon hereing of the Controversie betwixt John Howe *and* his apprentice. It is ordered that hee shall take home his said apprentice and vse him well hereafter. And whereas the said apprentice hath complayned for that the said Howe doth not exercyse the said apprentice in his trade of Barbery *and* Surgery. It is further ordered that if the said Howe shall not take a shoppe and vse his trade before Christmas next, that the said apprentice shalbe turned over to anoth*er* of the same arte." Thus the class of barber-surgeons is distinctly recognised by the Company.

Surgeons' difficulties.

The lot of the surgeon, however, was often no more to be envied in the days of Elizabeth than in those of Victoria. Trying and difficult cases would arise, and when appeal was made to the Company the surgeon was not always supported by his brethren. "Here came Willm. Goodnep and complayned of Willm. Clowes for not curing his wief de morbo gallico. And yt was awarded that the saide Clowes sholde either geve the said Goodness xx^{s} or elles cure his saied wief, *which* Clowes agreed to pay the xx^{s}. And so they were agreed and eche of them made acquittance to other." A wise man was Clowes, and one who knew the world as well as his profession when he agreed to forfeit twenty shillings of good English money rather than undertake to cure a woman who could not be cured, and who was disagreeable to boot. The principle of underselling, too, in its meanest forms was rife at this time, for "Mr Ffenton complayned of Robert Money for supplantinge him of diuers cures. And for slaunderinge him in his *pro*fession. And also for his evell practize, and was for his said abuses fined at vli w*hi*ch hee is to bringe in at the next Court or els to be committed to the Compter." And at the next meeting of the Courte "It is ordered that Robert Money shal be com*m*itted to the Compter for his Contempt." Cash, too, appears to have been scarce, as the payments for professional services were often made in kind. Thomas Adams once "complayned againste John Padice who had receyved certayn money in hand, and a gowne in pawne for the remainder, to cure the daughter of the sayed Thomas, which daughter died." The master and governors ordered that "the gown should be redelivered to the father," who should in return "geve unto Jo. Padice for hys boat [hire] which he spent in going to the mayde at Putney vs." And again "Peter Wallis complayned againste James Wanadge, who had taken to cure the wyef of the saide Peter, and had taken xxs in money and one pillowe of Arrace work worthe iiijli, and did her smal gode."

This chapter may be fitly closed by quoting a few extracts which, whilst they do not relate so nearly to the practice of surgery as those already given, are yet of interest from the insight they afford us into the times in which the minutes were written. "Yt is ordayned that no per*s*onne of the Mysterye exercysinge fleabothomye or bloud lettinge at any time hereafter shall sett his measures or vesselles w*i*th bloude out or within his shoppe windowe but to hange or set his measures or vesselles cleane on the outsyde of the shoppe wyndowe and whoso Dothe contrary to this acte shall for ew*er*ye suche Default forfett and paye iij^s iiij^d." This was a very old and nasty custom of the barbers, for in the Liber Albus,[1] under the date 1273 is an injunction "That Barbers shall not place blood in the windows." And again, "That no barbers shall be so bold or so daring as to put blood in their windows openly, or in view of folks, but let them have it carried privily into the Thames under pain of paying two shillings unto the use of the sheriffs."

An Order that none lett any bloud stand to the annoyance of the people. iij^s iiij^d.

"Also It ys ordayned That god callyng oute or frome this Trancytory vale or worlde any of this saide Company decessed and beynge w*ith*in the clothing or lyvery of the same his best hood shal be layed upon the hearse, and unto the churche and *there* upon yt shall remayne untyll the takynge off of the saide fine cloth[2] w*h*ich is used at the goyng forth of the M*aste*r goue*r*nours and company of the clothinge oute of the churche and the corps goynge to be buryed.

[1] Lib. iii. pt. 2.

[2] This is the state pall or hearse cloth kept by each of the City companies to celebrate the obsequies of its deceased members.

And then and *there* the clarke of the saide Company shall take the same hood and shalbe his *pro*per owne of dewtie. Be yt Provided allwayes That yf the wyf or executo*u*rs of any *per*son decessed as aforesaide will not de*par*te[1] w*i*th the same sayd best hood that then they or any of them shall paye unto the saide clarke in redy Inglishe money vj[s] viij[d] And for the buryall of any woman beyng or *tha*t hath been wyf unto any of the said clothing the company of the clothing beynge warned to the same the clarke shal have — xij[d] and also the bedell for the caryeng of the M*aste*rs clothe at e*uer*y buryall shall have iiij[d] "

The answer made by the Company to a royal commission in 1576 was laconic, and put an end to further inquiries in the same direction. "At this Co*u*rte yt was agreed that whereas there was a precept directed unto o*u*r Companie in the Queenes maiesties name com*m*aunding the M*aste*r and Wardens to send in unto guildhall a true note of the revenue of o*u*r Landes and goods whiche Do belong unto our Companie. And the Answere was that the true revenue of the Landes was xxij markes wherof the moste *par*te went forthe and is disbursed in pencions and that we had no goods." And, indeed, this statement was in great measure true, for the Company was abundant in charity to its poorer members. Thus, " W[m] Eden Clark of this misterie shall have yerelie the som*m*e of six poundes in benevolence over and aboue the ffee of iiij[li] in consideracion of the smalness of his hyringe." The following extract, too,

[1] part.

shows that they were not harsh creditors, for "Whereas Willm: Ben *our* Clark ys indebted unto the howse in the some x^{li} w*hi*ch he should pay this yere yet in regard of the hardnes of the tyme and his greate charge he shalbe forbo*ur*ne this yere (1596) and to be paid afterwards notw*i*thstanding this forbearance." Six months later "There was geven unto Will*iam* Ben Clark of the Companie in regarde of his greate charge and the scarcitie of the tyme the some of v^{li} And for the x^{li} w*hi*ch he oweth he ys to pay the same as god shall inhable him."

Election feasts.

The dinners at the annual election day and after the public anatomy lecture appear to have been a constant source of worry to the worthy brethren. For after complaints had been lodged and due investigation made of certain alleged abuses occurring on these occasions, the following remarkable ordinance was published in 1600: "Whereas on the fifteenth daie of August wee Richard Wood, John Leycocke, James Bates, and Lewis Atmer were by consent of the Courte of Assistaunce appoynted to sette downe some order for reformacion of Abuses co*m*mitted at the ffeastes holden at the Com*m*on hall of this mistery both by some of the lyvery of this Companye and by their servaunts and apprentices. Wee therefore having taken deliberate consideracion of the premisses and finding as well of our owne experience as otherwise that the bodye of this Company hath sustayned much disparagement by reason that some of the Livery and others noe white at all respectinge the worshipp of this Company haue not onely by themselves but alsoe by their servants and apprentices disfurnished the tables att ffeastes whereat they have sytten to pleasur their private frendes contrary to all modestie and good government, Doe therefore order for reformacion thereof by the aucthority aforesaid, That noe *per*son of the Lyvery of this Company beinge not of the Assistaunce of the same shall not att any tyme hereafter suffer any of his children, frendes, servants, or apprentices to staye or attende uppon him or his wiefe att any ffeastes to be keepte in the said Co*m*mon Hall of this mistery. Otherwise than to attend uppon him or his wiefe onely to the said Hall and soe to departe till dynner be ended." It is easy to realise the scene; the worthy citizen and his wife seated in the hall of their guild with their children, apprentices, and servants grouped behind them ready to scramble for the dishes as they were removed, and to finish them up to the very last mouthful.

There were brave doings, too, upon the election day, which took place once a year. Certain electors were first chosen from the livery of the Company, "which electors [Aug. 14, 1598] after they had their chardge given them by the Maisters or Governors of this Companye, and their severall rules for the eleccon delivered unto them, after longe and deliberate consideraci*o*n had, did electe for the Master John Leycocke, and for the Upper governor John Burgis and for the second Govornor John Pecke and for the youngeste governor Roberte Johnson. Whiche saide John

Leycocke beinge not there present the garlande[1] accordinge to the manner and custome of the house was by the Maister for the yeare paste placed uppon the heade of Mr Doctor Browne as deputye for the saide John Leycocke. After which another garlande was likewise placed uppon the heade of the saide John Burgis by the upper governor. And a like garlande tendered to the said John Pecke which he utterlye refused. And for the same was fined at ffortie shillings which fyne he paide Accordinglye. Likewise another garlande was placed by the youngeste Governor uppon the heade of the saide Roberte Johnson and by him gratefullye accepted. And the said Maister Burgis and Mr Johnson were sworne standinge for the due execucion of their offices. Afterwarde a new eleccion was made by the saide Electors of the seconde Governor and therein Lewis Atmer was chosen, and the garlande proffered

[1] Mr. Shoppee, in his Description of the pictures and other objects of interest in the hall and Court Room of the worshipful Company of Barbers, Lond., thus writes of the successors to the garlands here mentioned. There are "four [three of which are visible lying upon the table behind the fire-screen in Plate VIII.] very handsomely wrought and chased silver garlands or wreaths for crowning the Master and Wardens, which are still used and worn by them on Court days in receiving guests. . . . Each of these is silver with the Company's arms and badges (the rose and crown) and other devices well rendered, and all mounted with silk velvet. The Master's, Prime Warden's and Middle Warden's being red and the Third Warden's being green." In the court minutes 20th June, 1629, is the following entry: "It is alsoe ordered that there shalbe made fower Garlands of silver enamelled garnished and sett forth after the neatest manner according to the discretion of the present governors for the choice of new Maisters and the coste thereof to be borne of the stock of this house." The green colour of the third warden's garland perhaps signified his office of rent collector, at a time when the Company owned green fields.

unto him but refused, and therefore was fined at xl[s] which he willinglye paide Accordinglye. After which refusal a newe eleccion was made and therein was chosen Thomas Thorney whoe willinglye accepted of the garlande and was sworne standing for the due execution of the place."

The Company, however, like all the guilds of the time, had its trade secrets, which its members were bound to respect; thus, "At this Courte John Yates, Thomas Lamkin and Edward Parkes were Dismist from their places and vote of the Assistance for revelinge of secretes contrarie to a rule in that case *pro*vided." The last entry that shall be made in this chapter is curious. It occurs under the date March xvi[th] 1573. "Here was Will*ia*m Carrington and put in a complainte agaynste Edward Parke for that he warned hym to the Courte of conscience w*i*thout Lycens of the M*aste*r and Governo*u*rs, for that the wief of the said Park had taught his children to playe one the virginalles.[1] And order was taken that Will*ia*m Carrington should paye v[s] at Ester nexte, and so withdrawe his suite."[2] In modern language, Parke brought an action in the county court against Carrington for the amount due for teaching

[1] An oblong spinet.

[2] In 1518 the Common Council of the City of London passed an act for the recovery of debts under forty shillings due to citizens, by a court to be called a Court of Conscience, held in Guildhall; and the debtors who refused to obey the award of that court were to be imprisoned in one of the City compters until they complied, although it were durante vita. In 1605 the powers of that court were established by act of parliament." "The State of the Prisons in England and Wales," p. 239; by John Howard, F.R.S. Warrington, 1784.

music to the children of the latter. Both were members of the same company, and Carrington by a mean subterfuge sheltered himself behind the bye-law common to all companies at that date, that no member should prosecute another without first obtaining permission from the superiors of his company. The Company, however, decided that Carrington should pay, on condition that Park withdrew his suit.

CHAPTER IX.

THE BARBERS AND SURGEONS' COMPANY UNDER THE STUARTS —THEIR PAGEANTS—DOMESTIC REGULATIONS.

Barber-surgeons' Company under the Stuarts.

TIMES changed with the barbers and surgeons under the altered succession. They no longer rode with their sovereign or lived in brotherly love with each other. The committals to the compter became daily more numerous, until after the execution of Charles I. and in the gloom of the Puritan period all trace of the Company's proceedings is for ever lost. To inaugurate this epoch the craft in its innocence and gaiety of heart contributed in April 1603 the magnificent sum of £12 10s. "towardes the supportac*i*on of the Chardges to be disbursed by this Citie as well for the receiveinge of the Kinges most excellent Ma*jes*tie at his first and next repayreinge from his realme of Scotland to this his highnes Citie and chamb*ou*r of London his imperiall crowne as towards his honorable coronac*i*on and all such solempnities and disburseme*n*ts as are to be *per*formed." A flicker of the times which were fast passing away, to give place to the solemn league and covenant amongst the people, whilst the festivities were confined to the court, is to be found in the order of the Mayor and his brethren the Aldermen for a pageant in which "not onelie *our*selves but the full number of five hundred of the best and gravest cittizens should according to *our* dueties wayte and attend uppon [the king's] royall person at his approchment to this Cittie in greater number and more statelie and sumtious shewe then hath bene at any time heretofore within the memory of man in the like case *per*formed. Towards the accomplishm*en*t of w*hi*ch number youre Company is appoynted to *pro*vide the full number of twelve *per*sons of the most grave and Comlyest *per*sonages . . . everie one of them to be well horsed and apparrelled w*i*th velvet Coates and w*i*th sleaves of the same and chaynes of gold. And not onelie yourselves but every of the saide *per*sons to have one comely *per*son well apparelled in his dublet and hose to attend uppon him one foote. All which *per*sons to be in a rediness well and substanciallie horsed apparrelled and appoynted as aforesaid w*i*thin one daies warneing to be signified unto you to attend on mee and my Brethren the Alderm*en*." And the order is still further emphasised by the injunction that "you are to have in regard that noe man for insufficiency in any respect be turned back to the disgrace and discredit of youre company as a man vnfitt furnished and appointed for so honorable a service."[1] Truly the mantle of the

[1] This pageant was never carried out, however, for in the minutes of the Carpenters' Company (Jupp, p. 67) the following occurs under the date July, 1603 : "By means of God's visitation for our sins, the Plague of Pestilence then reigning in the City of London and suburbs (the Pageants and other shows of triumph, in most sumptuous manner prepared,

distressed Netherlanders who were in their palmy days such masters of the art of pageant making seems to have fallen upon our citizens at this time. And as if to heighten the gaiety of the scene, the next precept tells of death and disease, and that as in life so in the reigns of princes,

> "Optima quæque dies miseris mortalibus ævi
> Prima fugit; subeunt morbi tristisque senectus;
> Et labor et duræ rapit inclementia mortis."[1]

The Plague.

"Whereas I and my Brethren the alder*men* duly consideringe *our*selves the present infecc*i*on of the plagve amongst us so greatlie dis*per*sed in all partes of this Cittie liberties and Suburbs and the greate multitude of poore people w*h*ich by reason of the said infecc*i*on have theire howses shut vpp and restrayned as well from goinge abroad as theire daylie trade and labors wherew*ith* theie were accustomed to mayntaine themselves, theire wives and families, and doe at this *pre*sent by reason thereof endure greate wante and extremities, have thought fitt that all publique ffeasting and common dinners at every the severall Halles and Com*m*on meetings of corporac*i*ons and Companies w*i*thin thys Cittie shall duringe the tyme of god's visitac*i*on amo*n*ge us be

but not finished), the King roade not from the Tower through the Citie in Royall manner, as had been accustomed, neither were the Citizens permitted to come at Westminster, but forbidden by Proclamation, for feare of infection to be by that meanes increased, for there dyed that weeke in the City of London and suburbs of all diseases 1,103; of the plague, 857." This was the plague of 1603. It went on increasing year by year until its virulence culminated in 1609, a year in which 4,000 were said to have died in London alone. It broke out again in 1625, in 1636, and finally in the great plague of 1665. (See Loftie's "History of London," vol. i. p. 355; ed. 2.)

[1] Virg., Georg., iii. 66.

wholely forborne and left of. And that one third parte of the chardges and expences intended to be bestowed and spent uppon the said feastings and dinners shalbe wholelie bestowed and geven for and towards the reliefe of the most miserable poore and needie persons, whose house it shall please almighty god to visit." And again in October 1603, "Theis are to will and require you that you take speciall knowledge herby that for avoydinge of infecc*io*n by assemblie of people this tyme of god's visitac*io*n, It is thoughte meete ther be noe shewe made. The morrowe after Simons and Judes daie next it is intended that youre Companie be dischardged thereof for theire attendance for that tyme." Four years later the pestilence, still gathering force, manifested itself in the very Hall of the Company, since it is recorded that "fforasmuch as Francis Rowdon clerk of this Company hath lately buried his child of the plague, w*hi*ch [i.e. the child] was carried through the gate of the hall of this company; by reason whereof there is express commandm*en*t from the Lord Mayor that neyth*er* wee, the said M*aste*r nor assystants nor any of o*u*r officers should [meet therein] for the space of 28 days after the buryall of the said Child. Therefore it is ordered by this Court that the audit of the M*aste*r's account . . . shall be howlden and taken at the new dwellinge howse of Joseph Fenton scituat in St. Bartholomewes court."

The order of precedence was strictly observed by the companies, and any oversight in regard to this matter was bitterly resented by the offended guild. In 1604 "the barbers and surgeons complayn of having been through ignorance misplaced at the king's and queen's passage through the city on the 15th of March last. Ordered (by the Court of Aldermen before whom the complaint was laid) that as since their placing in the mayoralty of Sir Stephen Pecock as the seventeenth company, the stockfishmongers have been dissolved, the masters or governors of the Commonalty of Barbers and surgeons henceforth be reputed, taken, and placed as the sixteenth company in this city."

Pageant.

In 1606 another pageant was held, and this was the last for many subsequent years. The precept authorising it states that it is "ffor the bewtefieinge of the streets and lanes *w*ithin this Cittie against the passage of the Kinge[s] most excellent Ma*jes*tie and the Kinge of Denmarke their nobilitie and trayne from the Tower through this Cittie. Theis are in his Ma*jes*ties name straightlie to charge and command you that all delayes and excuses sett apart you have and provide yo*u*r rayles in a readiness for the livery of yo*u*r company to stand in. . . . And likewise that yo*u*r railes . . . be hanged *w*ith blew azure cloth and garnished *w*ith bann*e*rs and stream*e*rs in the most bewtifule manner that may be. . . . And that you likewise have and provide sixe whifflers[1] at the least to eu*er*y score of yo*u*r livery well apparelled *w*ith white staves

[1] Halliwell, in his "Dictionary of Archaic and Provincial Words," defines wifflers as pipers and hornblowers who headed a procession and cleared the way for it. Dr. Norman Moore tells the editor that in pageants at Norwich they flourished a sword in front of a procession and so cleared the way.

in their handes to stand with their backs to the common railes over against y*our* companies railes for the better and quieter orderinge of the streets through which his Ma*jes*tie shall passe." The Company's share of the expenses of this pageant was five pounds, so that it was not as gorgeous by a half as the earlier one with which they proposed to welcome the arrival of their king to his new throne. In regard to this procession there is another characteristic entry, "Memorand*um* that the king's ma*jes*tie w*i*th the King of Denmarke and the Prince of Wales came through this Cittie fromwardes the Tower of London, attended uppon with the Lordes and gentry of this Land on the last day of this instant moneth of July. Att which tymé Mr. Foxe beinge one of the Co*m*mittee for placeinge of the Companies standings would have displaced us; But by the Lord Maior's order we were placed in the Seventeenth place accordinge as we ought to be placed."

The growth of luxury keeping pace with the increase of wealth amongst the burgher class is shown in the entry dated May 22nd, 1604: "It is ordered that from henceforth the Clark of this Companie shall against eu*er*y Courte daie bestowe iiij[d] in hearbes and flowers," or more probably the herbs were used as a prophylactic measure to ward off the infection of the daily increasing plague. Under the new monarch and owing to the spread of the puritan doctrines amongst the people, the edicts against trading on the Sabbath were rigidly enforced. The barbers appear to have been great offenders in this respect, so that

Prohibition of Sunday trading.

M 2

an ordinance was issued upon the subject as early as 1413 by the Archbishop of Canterbury.[1] During the reign of Elizabeth penalties for Sunday trading were only occasionally enforced, whilst in the reigns of James and his son such entries as "This day, William Stanthrop was fined for workinge on the saboth day and it was mittigated to xij^{d}" occur on each page of the minute books. If the fine were not paid a worse evil befell the unhappy barber, for "it was ordered" upon the same day as the last entry, "that Widdow Evans' man shalbe co*m*mitted to the Compter for workinge on the Saboth unlessse his M^{rs} bringe in her fine at the next Courte."

The Company's esprit de corps.

The ordinance "none to supplant or take another's cure from him on pain of v^{li}" was carried out to the letter, during the whole existence of the Company, although the fine was frequently reduced according to circumstances. Thus, "It is ordered that one Palmer, a practisioner in Surgery, dwellinge in Holborne, shall pay unto Humfrey Gorston ii^{s} iiij^{d} for and in respect he hath defrauded Gorston of a patient, which money the said Palmer promised to satisfie accordinglie." Whilst the esprit de corps was kept up by such notices as "John Udall complayneth against Richard Gyle for takeing away his patient w*i*thout seeing him satisfied for his paynes about the said patient." For which offence Gyle at a subsequent period paid v^{s} to Udall.

On another occasion: "It appearing that William

[1] See also Appendix G, page 335.

Baker did take away Thomas Hart pacient to Nicolas Boorne and William Watson before they were satisfied for theire paynes is thereupon fined at xx^s." Presumably, therefore, this was a more flagrant case of filching than the one last recorded, and on that account deserving of a more severe punishment. Malpraxis, too, was treated somewhat summarily: "This daie in the Controversy between Willm ffisher and Stephen Browne, It is ordered for that it appeareth to this Court that the said Browne hath not behaved himself well in a cure w*hi*ch he undertooke of the weif of the said Will*ia*m ffisher, That therefore the said Browne shall repaie vnto the said Will*ia*m ffisher the some of Twentie shillings of a more some w*hi*ch tofore the said Browne hath receaved of the said ffisher. And by a later order the social position of Willm ffisher is revealed as "Stephen Browne is to pay at this hall xx^s w*hi*ch he is ordered to repay to the Parson of Hamm w*hi*ch tofore he had receaved of the said Parson or ells he is to be Committed to the Compter."[1]

The Company's power of summary committal.

The power of summary committal, as Mr Sidney Young points out, seems to have been a privilege enjoyed by very few of the City companies. It was possessed, however, by the Society of Apothecaries[2] as well as by the College of

[1] The Compter here referred to would be the one in Wood Street, Cheapside. Its predecessor in Bread Street was disused after 1555 on account of the difficulty the City authorities experienced in getting rid of Richard Husband, who had bought for himself the office of Master of the Compter, and who scandalously abused his authority.

[2] "The Apothecary, Ancient and Modern," by G. Corfe, M.D. Blackfriars, 1885.

Physicians. The authority was confined to committal for minor offences, and did not extend to imprisonment in Newgate, for which an order from the Mayor or Court of Aldermen was required. Thus, in the records of the Court of Aldermen,[1] Walter ffaxon Citizen and Barber Surgeon of London was . . . for his wilfull disobedience to the M*aste*rs and Governours of the Company of Barber Surgeons . . . and refusing to conforme himself to the orders of the same company . . . and also for his contempt shewed to this courte commytted to the Goale of Newgate there to remayne during the pleasure of this Courte." And "Hugh Ward Citizen and Barber Surgeon . . . for his obstinate carriage and misdemeanor towards [his company] and refusing to pay a fine of forty shillings ymposed upon him . . . was now by this court for that his offence committed to the Gaole of Newgate there to remayne untill he conforme himself, or other order bee taken for his inlardgment." [2] As a further instance of the troubles resulting from malpraxis: "whereas Edwarde Knighte hath by his ignoraunce in the Arte of Surgery maymed one Richard Robinson a poore man it is ordred and the said Knighte hath undertaken to procure Robert Money to finish the cure.[3] Whilst such unqualified practitioners are being considered, it is interesting to note that "Henry Goodwin a Sorcerer was by the M*aste*r forbidden to practize any more in the Arte of Surgery;" and that "William Wryghte as well for diuers contempts by him co*m*mitted and done contrary to the good orders

[1] Repert. xxix. fol. 177. [2] Repert. xlix. fol. 254. [3] case.

of this Company, as also for his abuses and ignoraunce in the practize of the arte of Surgery is discharged of his practize in the Arte of Surgery, and is crossed out of the Lecture bill of Surgery."

From the recurrence of such cases it was considered advisable in 1606 to reconstitute the surgeons' portion of the United Company. "Fforasmuch" therefore "as divers persons ffreemen of this company who have very litle or no skill at all in the Arte of Surgery do neverthelesse make a publique profession of the said Arte, and thereby comitt many errours to the great disparagment of the worthie and experienced professo*u*rs thereof and to the hurte of divers of hys Majesties loveinge subjects. It is ordered that from henceforth no man be permitted to have his name entered downe for a Surgeon into the lecture bill except by the consent of the M*aste*rs or governours of the mistery or Co*min*altie for the tyme beinge. And that it shalbe lawfull for the presente M*aste*r or governours to dismisse out of the lecture bill the names of such persons as they shall think fitt to be put out: which persons dismissed and put out shall liue out of the protexion of this Company for and in respect of their practize in the Arte of Surgery untill they shall by them be thought fitt to practice in that Arte and admitted into such bill upon their humble suite." But against "disobedient obstinate or stubborne" persons who persisted in practising in defiance of this ordinance "suite in law was to be *pre*sented."

Reconstruction of the surgeons' side of the Company.

The teaching of surgery was not quite neglected, for attendance on a surgery lecture was strictly enforced; and in 1604 "Mr ffenton presented to this Courte 500 bookes of Horatius Morus tables translated into Englishe and delivered them to the *Master* of this Company in the name and behalf of Mr Deputie Caldwell[1] whoe freely gave

Surgery lecture.

[1] Richard Caldwell was born in Staffordshire in 1513. He was educated at Oxford, and was a Fellow of Brasenose and Christ Church. He subsequently studied physic, and was chosen censor of the College of Physicians. In 1570 he became president. Lord Lumley was by his influence induced to found and endow with a liberal salary a lectureship on surgery, which is still known as the Lumleian bequest. Caldwell translated into English the "Tables of Surgery," by Horatius Morus, a Florentine physician, copies of which still exist in the Bodleian Library, in the British Museum, and in the Library of the Medico-Chirurgical Society. The work was published in the year after his death, 1585, and is dedicated posthumously by his son or nephew, the Mr Deputy Caldwell spoken of in the text, in the following words:

"TO THE COMPANIE OF SURGEANS WITHIN THE CITIE OF LONDON, MUCH HEALTH, WITH GOOD SUCCESSE IN THEIR PRACTISES AND CURES.

"He which translated these tables of surgerie into our vulgar toong, is entred (as you know) the waie of all flesh, and dooth now rest (I hope) with the faithfull. That he was one, who sought by sundrie meanes to promote learning, and to doo the countrie good, which bred him up, manie deedes of his doo clearlie witnesse, and are confirmed as vndoubted testimonies thereof to continue vnto the verie end of the world. Againe, if you knew the man well in his latter daies, you cannot be ignorant, how feeble and crazie he appeered in the state of his bodie, and how well neere spent his spirits were: and yet, euen in that extreme weaknes, he toiled himself both night and day, and was a most painfull student for your furtherance, not refusing, beyond the strength of his drie and withered carcase (numbering then almost fourescore yeeres) to giue himself up to the compiling and writing of diuerse commentaries and other woorks, all which concerned your art onelie and profession, not vnlike herein to the candle or lampe, that wasteth it selfe whilest it giueth light to others. Although (to confesse simplie what I thought) I haue oftentimes wished with all my hart, that either hee had not at all giuen himselfe to writing, namlie of such obscure and darke matters, so late: or them to this company to be distributed amongst the professors of Chirurgery freemen of this Company."

The Company exercised a censorship over the writings of its members, for as early as 1588, "yf any man of this misterie shall at anye tyme hereafter make any Booke or Bookes of Surgerie the same shall not be published unles the same booke or bookes be firste presented unto the masters governo*u*rs and examino*u*rs of this Companie for the tyme beinge upon payne of x^{li}."

The rights of the masters over their apprentices were upheld during the Stuart dynasty with much greater severity, even when the masters

Apprentices.

els, had begun it 20 yeeres sooner, what time, for freshnes of memorie, wit and reading, he was far more readie and pregnant for such an enterprise. Which thing was manifested in the sequele. For not long before his end, acknowledging his owne defects, feeling in himselfe the assured signs of an vtterlie decaied nature, and forseeing that his daies would be ended, before his bookes were finished, he gaue me verie streightlie in charge to gather his notes and writings diligentlie together: to reduce them into that method which he intended, and wherewith he made me throughlie acquainted: to commend them to some deere familiar friends of his, whome he named vnto me, to be perused, censured, and allowed, and so foorthwith to be committed to the printer's presse. Notwithstanding that I haue not published anie of his doings, before this time, diuers iust occasions haue restreined me, but especiallie two aboue others: whereof one is the great ingratitude and negligence you shew to be in your selues, in not frequenting the lecture which he founded onelie for your sakes: and so much the more is your fault therein aggrauated, sithens he hath procured so rare and excellent a learned man as M. D. Forster is, to be your reader; whose eloquence (in my opinion) is of sufficient force to allure you to his schoole, if men, of purpose, were not wilfullie bent to giue occasion to be thought more froward and obstinate than those craggie rocks and hard stones, which were mooued with the musicke of Orpheus harpe, to come and couch themselues in the building of the walls and towers of Thebes. Againe, his method, perspicuitie, and plainnesse in teaching is such, that there is no man so simple and grossewitted, but that he may seeme to be capable of the doctrine which he deliuereth. Seeing therefore the paines is wholie his, and the profit acted harshly and unjustly towards them, than it had been during the reign of Elizabeth. In the following case the Company clearly sided with the apprentice, and yet were constrained for the sake of example to uphold his master: "Whereas it pleased the right honorable the Lord Maior of this Cittie by his order bearinge date the seaventeneth daie of this instant December [1604] to refer the hearinge and endeinge of all controversies betwixt Will*ia*m Wrighte of thone *par*te and Thomas Marston his ap*pre*ntice of the other *par*te to us the M*aste*rs and governo*u*rs of the Misterie and Comminaltie of Barbers and Surgeons of London. We . . . doe finde that the said Marston bound himself app*re*ntice to his

wholie yours, if you will accept thereof, who can denie, but that you, in all equitie, are as well bound to heare, as he to read? The other cause is a reuerend feare and conscience which I haue, least happilie by my rashnes and ignorance I might some waies in the edition of the old mans woorks, discredit or disgrace the great learning, grauitie, and iudgement of him now dead vnto whom, by nature and dutie I was most bounden whilest he liued. Wherupon (to tell the truth) I durst neuer haue ventured to haue medled with the cumbersome and difficult knitting of Oribasius knots, nor with the cunning and comelie rolling of Galens bands: no nor so much as once to haue been tempering with Hippocrates and Heliodorus instruments most ingeniously deuised for the helpe of luxations of ioints, and fractures of bones (which the hand onelie cannot performe) had not a singular learned man incouraged, or rather drawen me on with a most friendlie promise, to lend me such a clew of thred, I meane his good helpe and counsell, as should be able to direct and lead me through all those labyrinths and mazes into which I might vnaduisedlie [unadvisedly] cast my selfe by vndertaking a charge of so great weight and so farre beyond my reach. Hauing therefore confidentlie entred into the matter by these meanes, I thought good first to begin with the printing of this short treatise, because it contoineth in a few leaues verie compendiouslie the summe of all surgerie: and furthermere I am minded to make triall, by this little tast, how the residue of his greater trauels [works] are like to be accepted when they come abroad. Which thing (God willing) shall be accomplished with such conuenient expedition as

Master by Indenture bearinge date the last of June anno *dom* : 1601 and gave to his *Master* the so*mm*e of eight pounds in money and three suits of apparrell, w*i*th whom hee contynued a diligent s*er*vant and was of his said *Master* soe accompted of, till the daie of the solempnisac*i*on of his highnes coronac*i*on last when the said Wrighte p*re*tendeinge That his said app*re*ntice had robbed him did bringe his said app*re*ntice before one of his Ma*jes*ties Justices of peace who committed the said app*re*ntice

may be vsed. And thus crauing earnestlie at your hands, that you would hencefoorth studie to deserue well of him, who was at such great cost, and tooke such intollerable paines for your benefit and advancement, I bid you hartilie farewell.

"ED. CALDWALL.

"*From my lodging at Ludgate Hill*
16 *Julie* 1585."

"Tables of Svrgerie by Horatius Morus, a Florentine Physician, and Faithfullie Translated out of Latine into our English toong, by Richard Caldwall doctor of physicke. At London 1585" (p. 32).

Holinshed, in his "Chronicles" (ed. J. Hooker, vol. iii. p. 1,349), gives the following interesting account of Dr. Caldwell, and the foundation of the Lumleian lectures on surgery:

"This yeare, 1582, was there instituted and first founded a publike Lecture or Lesson in Surgerie, to begin to be read in the College of Physicians, in London, in Anno 1584, the sixt day of Maie, against that time new reedified in a part of the house that doctor Linacre gaue by testament to them, by John Lumleie, Lord Lumleie, and Richard Caldwell, doctor in Physicke, to the honour of God, the common profit of hir Majesties subjects, and good fame, with increase of estimation and credit, of all the Surgians of this realme. The reader whereof to be a doctor of physicke, and of good practise and knowledge, and to haue an honest stipend no lesse than those of the vniversities erected by King Henrie the eight, namelie, of law, diuinitie, and physicke, and lands assured to the said college for the maintenance of the publike lesson; whereunto such statutes be annexed as be for the great commoditie of those which shall giue and incline themselues to be diligent hearers for the obteining of knowledge in surgerie, as whether he be learned or unlearned that shall

Publike Lecture of Surgerie founded in London and presentlie red (as also in the life of the Founder) by doctor Forster, to his high praise and credit.

close pr*i*soner for seaven daies. And findeing as yt shold seeme noe iust cause to detaine him hee was dischardged as wee understand. And afterwards his said M*aste*r arrested him uppon an acc*i*on of trespas for carryinge away w*i*th him (when his M*aste*r had geven him leave to

become an auditor or hearer of the lecture, he may find himselfe not to repent the time so imploied. First, twise a weeke through out the yeare, to wit, on Wednesdaies and Fridaies, at ten of the clocke till eleuen, shall the reader read three-quarters of an houre in Latine, and the other quarter in English, wherein that shall be plainlie declared for those that vnderstand not Latine what was said in Latine. And the first yeare to read Horatius Morus Tables, an Epitome or briefe handling of all the whole arte of surgerie, that is, of swellings or apostems, wounds, ulcers, bonesetting, and healing of broken bones, termed commonlie *fractions*: and to read Oribasius of knots, and Galen of bands, such workes as haue beene long hid, and are scarselie now a daies among the learned knowen, and yet (as the Anatomies) to the first enterers in surgerie and nouices in physicke; but amongst the ancient writers and Grecians well knowne. At the end of the yeare, in winter, to dissect openlie in the reading place, all the bodie of Man, especiallie the inward parts, for fiue daies togither, as well before as after Dinner; if the bodies may so last without annoie.

What exercises are to be followed in the said college by the Will of the Founder. The first yeares exercises.

"The second yeare to read Tagaultius Institutions of surgerie, and onelie of swellings or apostems, in the winter to dissecte the trunk onelie of the bodie, namelie, from the head to the lowest part where the members are, and to handle the muscles especiallie. The thirde yeare to read of wounds onelie of Tagaultius, and in winter to make publike dissection of the head onelie. The fourth yeare to read of vlcers onlie the same author, and to anatomize or dissect a leg and an arme for the knowledge of muscles, sinewes, arteries, veines, gristles, ligaments, and tendons. The fift year to read the sixt booke of Paulus Aegineta, and in winter to make anatomie of a skeleton, and therewithall to show and declare the vse of certeine instruments, as Scamnum Hippocratis, and other instruments for setting in of bones. The sixt yeare to read Holerius of the matter of surgerie, as of medicines for surgians to use. And the seventh yeare to begin again and continue still. A godlie and charitable erection doubtlesse, such as was the more needfull, as hitherto hath beene the wante and lacke so hurtfull: sith that onelie in ech

The second yeares exercises.

The Third yeares and fourth yeares exercises.

The fift and sixt yeares exercises, and so to continue with *Repetitis principii.*

goe into the Cuntry) a lytle unguent in his plaister Boxe worth some twoepence ffor w*h*ich hee recou*er*ed fortie shillings Costes and dammages, Whereupon hee detained him in p*ri*son by the space of fifteene weekes. . . And afterwards uppon the app*re*ntice Conplainte to the

vniversities by the foundation of the ordinarie and publike lessons, there is one of physicke, but none of surgerie, and this onelie of surgerie and not of physicke, I mean so as physicke is now taken separatilie from surgerie and that part which onelie vseth the hand as it is sorted by the apothecarie. So that now England may reioise for those happie benefactors and singular welwillers to their countrie who furnisheth hir so in all respects, that now she may as compare for the knowledge of physicke, so by means to come to it, with France, Italie, and Spaine, and in no case behind them, but for a lecture in simples, which God at his pleasure may procure, in mouing some hereafter in like motion and instinct to be as carefull and beneficiall as these were to the helpe and furtherance of their countrie. At the publication of this foundation, which was celebrated with a goodlie assemblie of doctors, collegiats, and licentiats, as also some masters of surgerie, with other students, some whereof had beene academicall; Doctor Caldwall, so aged, that his number of yeeres, with his white head adding double reuerence to his person (whereof I may well saie no lesse than is left written of a doctor of the same facultie verie famous while he liued

> Conspicienda ætas, sed et ars prouectior annis,
> Famaque Pœonio non renuenda choro).

Euen he, notwithstanding his age and impotencie, made an oration in Latine to the auditorie, the same by occasion of his manifold debilities unfinished at the direction speciallie of the president (Doctor Gifford) who (after a few words, shortlie and sweetlie, vttered) gaue occasion and opportunitie to D Forster, then and yet the appointed Lecturer, to deliuer his matter, which he discharged in such methodicall manner, that ech one present indued with iudgement, conceived such hope of the doctor, touching the performance of all actions incident unto him by that place, as some of them continued his auditors in all weathers and still hold out; whose diligence he requiteth with the imparting of further knowledge than the said publike lecture doth afforde. When the assemblie was dissolued, and the founder accompanied home, diligent care was taken for the due preferring of this established exercise: insomuch that D. Caldwell and D. Forster, to furnish the auditors with such bookes as he was to read, caused to be printed the epitome of Horatius Morus, first in Latine: then in English, which was translated by the said doctor Caldwall. But before it was half perfected, the good old doctor fell sicke, and as a candle goeth out of it selfe,

Masters of this Companie of that and other wronges by the said Wrighte done to the said apprentice It was thoughte fitt by the *Masters* of this Companie to dischardge the apprentice from his *Master* unles hee wold

or a ripe apple falling from the tree, so departed he out of this world, at the doctors Commons where his vsuall lodging was; and was verie worshipfullie buried (but of his death hereafter in the yeare 1584, where the daie of his decease being mentioned, matter worth the reading shall be remembred)."

Page 1369: "1584, in this yeare and the twentieth daie of Maie, departed out of this life, that famous Father of physicke and surgery, the English Hippocrates and Galen, I mean doctor Caldwell, and was buried on the sixt of June, immediatelie following at S. Benet's Church by Paules Wharfe, at the vpper end of the chancell. His Bodie was verie solemnlie accompanied to the Church with a traine of learned and graue doctors, besides others of that facultie, the heralds of armes doing him such honor at his funeral as to him of dutie apperteined. Of this mans rare loue of his countrie hath beene spoken before, where mention is made of the institution to a surgerie lecture perpetuallie to be continued for the common benefit of London, and consequentlie of all England: the like whereof is not established nor vsed in anie vniversitie of christendome (Bononie and Padua excepted), and, therefore, the more to be esteemed. Indeed, the like Institution was in towardnesse whiles Francis the French of that name the first liued; but when he died, as the court that he kept in his time was counted an vniversitie, but after his deth made an exchange thereof with another name, so likewise discontinued or rather vtterly brake off that purposed institution of a surgerie lecture at Paris, so that in this point London hath a prerogatiue excelling the vniversities."

The Chronicle goes on to tell of numerous charitable bequests, etc., and among them "He left large sums of monie to be emploied by his executors at their discretion where charitie moued; as also to the publishing of such learned bookes of physicke and surgery (with sundrie chargeable forms grauen in copper and finished in his life), as he meant (if he had liued) to see extant."

In the margin it mentions: "His commentaries upon some part of Paulus Aegineta and other bookes." Holinshed further states that there is a monument erected to his memory in St. Benet's, with his arms, and the arms of the College of Physicians under it. "On either side of this latter scutchion are set certeine binding bands and other instruments of surgerie in their right formes, with their proper vse, also to be practised upon ech member; be the same head, leg, arm, hand, or foot, all workemanlie wrought."

use him after that tyme well and as an apprentice ought to be used: which he then promised to doe. And thereuppon the said apprentice wente hom with the said Wright. And yet neuerthelese sithens that tyme as wee vnderstan and finde hee hath keapt him in his house more lyke a prisoner then an apprentice manaceinge and threathing [1] him that hee wold hange him, and employeinge him for the most parte in the makeinge of wastcotes and Stokenges beinge not the trade wherevnto hee was bound, and lytle or noe whit at all in the arte of Barbery or Surgery, and useinge unfitteinge correction without any cause at all for ought wee can perceive. . . Ffor which causes and for that wee find the said Wright very maliciously bente against the apprentice and more lyke to doe him some mischief then to use him as becometh an apprentice . . . wee doe order . . . that the ffather of the said apprentice . . . paie to the said William Wrighte the some of fortie shillinges of lawful money of England. And in Consideracion thereof the said Wright is to deliuer uppe to the Masters of this Companie his said apprentice Indenture and all such thinges as hee hath of the said apprentice. . . And also wee further order and award that Wright shall not from henceforth for any cause or former matter troble or molest the said apprentice. And also wee doe injoyne . . . the said apprentice to Carry himself duetifully towards the said Wrighte."

The said Wrighte had been shortly before ordered to "bringe in his fyne at the next Courte for his

[1] threatening

absens from Lectures," perhaps because his time was fully occupied in setting his apprentice to work on the waistcoats. Be this as it may, however, at the next court day "The father of Thomas Marston the Apprentice of William Wrighte paid to the use of the said Wrighte the somme of ffortie shillinges which the said Wrighte then refused to receive, and also to stand to the order of the Masters . . . according to a referrence to them by the lord Maior referred. Wherefore it is ordered that for his said Contempt hee be committed to the Compter uppon the Lord Maior's command." This committal appears to have brought him to his senses, for at the next court, "according to a former order William Wrighte hath deliuered upp to our Master the Indenture of Thomas Marston his apprentice . . . whereuppon the Master paid unto the said Wrighte xl^{s} which he had receaved of the father of the apprentice." This would have settled the matter in an ordinary case, but Wrighte was litigious, and on the first opportunity he "appeared before the Masters of this Company and required a Coppie of the order betweene him and Thomas Marston his late apprentice. To which the Masters answered That they would be ready to showe that Order when and where they should be called by authoritie." And so the matter ended as far as the Company is concerned, since no further entry occurs in the minute books either about Wrighte or Marston, who, we will hope, went their respective ways in peace, and ceased from troubling each other.

An entry which is of interest as showing the ages of the apprentices at this period, occurs when "Peter Saunderson certiefied the Court that hee had offered to inroll his apprentice before the Chamberlen and it was Denyed him because the apprentice could not at the end of his terme accomplishe the age of xxiiij yeres." There was still the old objection to having any but comely apprentices, as "it is ordered for that John Knott hath taken John Doweinge a decrepitt boy to be his app*re*ntice for the terme of vij yeres . . . he shall kepe and maynetayne the said app*re*ntice with sufficient of meat drinck and apparell during the said terme although he shall contynew lame or decrepitt. And he is to be bound in obligacion for the performance hereof."

The bishop's licence.

The ecclesiastical authorities maintained their right of licensing, which brought them into collision with the Company. In 1607 the following entry occurs, which shows that the examiners were not so incorruptible as they had hitherto proved themselves, and that they had listened to the suggestions of the bishop's officers, and had examined persons who did not belong to their Company. "No examiner . . . shall p*re*sume to examine and approve any p*er*son in the Arte of Surgery w*i*thout the consent and orders of the M*aste*rs or governors of this Company". . . . nor shall they "pr*e*sent any p*er*son *pra*ctisinge surgery to the Bishop of London or to the Deane of Pawle's to the intent to get or procure such Surgeon Lycence or admission to practize

Surgery unlesse such Surgeon at such tyme shall have his letter of admittance from this company under the Common seale of the same testifieing his admision to practize Surgery, uppon payne to forfyt his [*i.e.* the examiner's] places and degree in this Company." The ceremony of presentation to the bishop appears to have borne the same relation to the Company's examination as the formal graduation in the older universities bears to the public examinations, and it was necessary to render valid the surgeon's licence to practise. Thus "Roger Jenkins heretofore examined and ap*pro*ved in the Arte of Surgery was p*re*sented before the Deane of Pawles and his letters of admittance from the said Deane" were read. And at the same time Abraham Allen was lykewyse admitted by the said Deane."

The Company a mystery.

The Company was a mystery,[1] in the more recent sense of the word, for "Mawris David appeared before the M*aste*rs and was rebuked for shewinge the copie of o*u*r Ch*arte*r to a scrivener." In 1606 a special ordinance was made, "ffor avoydinge of giui*n*g Controversy and debate in this Company. It is ordered by this Court that if any Assistant of this Company shall malitiously or otherwise at any tyme hereafter reveale report or publysh to any *per*son being not an Assistant of this Company any sensure, order or decree . . . . of this Company . . . . whereby any *per*son sensured or ordered may be ympeached or hurt in his name or creditt," he shall be fined.

[1] See Editor's Preface.

Partial licences were granted as in the reign of Elizabeth. "Garrett van Kettwick Stranger practitioner in Surgery accordinge to a former order was examyned concerninge his skill in the Art of surgery. . . . And upon his said examynacon he was tollerated by the M*aste*rs and Governo*u*rs of this Company as much as in them is to practize in Surgery untill he shalbe of riper experyence in the same Art: Provided that in all Difficult cases that he shall undertake he ioyne w*i*th him an approved Surgeon. And that he present to the M*aste*rs of this Company for the tyme beinge all such Cures w*hi*ch he shalbe possessed of w*hi*ch shalbe in danger of death or mayme. . . . And whereas in considerac*i*on of his said tolleracion he hath paid to the M*aste*rs and Governo*u*rs of this Company the some of ffortie shillings towards the mayntenance of the pore of the same, It is ordered that upon further tryall of his better practice and experyence in the saide Art hee shall upon his humble suite be examyned agayne. . . . And if then he shalbe thought fitt upon his examynac*i*on to be absolutely admitted he shall haue his letters of admittaunce under the seale of this howse, Payeing three pounds more to the M*aste*rs of the Company to the use of the pore of the same, and payeinge the Clarke of this Company his fee for the same." The inference to be drawn from this paragraph appears to be that Kettwick was a quack with an extensive and lucrative practice, from whom the various members of the Company hoped as advisers and consultants to derive a rich harvest of

Partial licences.

fees, whilst the adventurer derived benefit from the position given to him by the licence of the Company.

Midwifery.

The following is an interesting note of the early practice of midwifery as a speciality: "31st Jan. 1610. This daie James Blackborne was examined touchinge his skill in the generatyve *par*ts of women and bringenge of women to bedd in their dangerous and difficult Labors. And he the said Blackborne was found fitt and alloued to practize (in that chirurgicall *par*te of Surgery touching the generatyve *par*ts of women and bringinge them to bedd in their dangerous and difficult Labours) by letters under the seale of the house beinge the date above wrytten. And was at this court sworne and admitted a fforayne brother; and in consideracion thereof he paid to the *prese*nte Ma*ste*rs att this court x^{li}."

Charter granted by James I.

The first two years of James' reign in England were times of anxiety and expense for the united Company, for during this period they were struggling to obtain a fresh charter, and in the end they were successful. As early as 1588 it was felt that increased powers were required by the Company for the purpose of coping with that remarkable growth of quackery which occurred during the later years of Elizabeth's life. A Bill was therefore promoted in Parliament, as appears from the following entry, dated March 10th, 1588: "Also yt ys agreed that the Bill w*hi*ch ys to be exhibited into the parliament howse shall *pro*cede and shall be bo*r*ne at the charge of the howse, *pro*vided nevertheles that yf the said Bill happen to passe and be allowed of, and the charge growe unto some large somme of money then yt shall be borne upon and at the particuler charges of the Companie." Persons were also appointed to prefer and present this bill, but as no further entry occurs about it, we must suppose that it failed to become law. On the accession of James I., however, the agitation was recommenced in good earnest, and the first ordinance is dated Feb. 9th, 160¾, to the effect that "a bill be preferred into the parliament howse for reformacion of abuses committed against the weale of this Company according to that bill which was preferred in the tyme when Mr Baker [159⅞] was Maister. . . . And it is ordered that Mr Recorder of this cytie and Mr Wilbraham be reteyned for our councell." And "after they have finished and ended our present suite in the parliament howse, or when they shall think it fit, that they procede for the confirmacion of our Charter and ordinances." That the project was carried out with zeal may be concluded from a minute which occurs three or four months later, and which shows that then, as now, it was no inexpensive matter to obtain special legislation. "This daye it is thoughte fitt by this Courte that thear be no greate Election Dinner holden in the hall this yere for that the Company are lyke to be at great chardges in obteyneinge their ordinances and renewinge the Charter. Yet the ordinary allowance is to be spent uppon a small pittance uppon the Company that shall then be here."

In October of the same year (1604) the bill had passed

the Commons, for it is recorded, "that the *Masters* of this Companie shold *pro*cede w*i*th the reneweinge of the Charter of this Company accordinge as they shall thinke fyt, and shold contynnue the articles hereafter expressed or so many of them as it shold please the *Lords* to whom the same is referred to geve allowance of." The draught charter[1] is interesting, as it affords a picture of the customs of the profession at the beginning of the seventeenth century. It relates almost entirely to the surgical side of the Company, and gives to the surgeons a better position than that which they previously occupied. The clause[2] relating to the idea that "the openinge, searinge and embalminge of the dead corpes [was] properly belonginge to the science of barbery and surgery" is remarkable; whilst the conclusion of the paragraph gives us a curious idea of the undertakers of the period, for it says that "the same is intruded into by Butchers, Taylors, Smythes, chaundlers and others of divers trades unskillful in Barbery or surgery[3] and unseemely and unchristianlike defaceing, disfiguringe and dismemberinge the dead corpes so that by their unskillfull searinge and embalming the corpes corrupteth and groweth presently contagious and ofensive to the place and persons approachinge."

[1] Appendix L, page 361. [2] Clause xvi.

[3] The chandlers appear to have been especially busy in acting as embalmers, for, "Oct. 26, 1612. This daie it is ordered that at the chardges of the house the *presente* Masters w*i*th the Clark shall seek in the Rowles for the Charter of the wax chaundlers and to tak a coppie of that *parte* of the Charter touchinge the libertye gyven unto them for the embalmynge of dead bodyes. And as they shall finde the same soe to tak the advice of my lord chief Justice about the same at the chardge of the house."

The bill became law in the second year of James I. It enacts that there shall be four governors [of whom two are to be surgeons] and twenty-six assistants in the mystery of the barbitonsors or polers. The masters to be chosen annually, the assistants for life. The masters are chosen from among the assistants, and the assistants from amongst the most worthy members practising in London. The masters or governors to have the supervision and correction of all persons practising the craft of the barbitonsors in London or within three miles of it. They also were to possess authority to enter into any shop in London, or within the same radius of it, belonging to a surgeon, for the purpose of inquiring into the quality of the medicines, plasters, instruments, etc.; to examine all persons who practise in London, or to restrain them by virtue of the power granted by former charters. At the formal request of the master and governors, the Mayor, Sheriffs, and other authorities were to commit to ward persons who practised without licence until security was given for their appearance at the assizes. It will thus be seen that the bill had not undergone any very important modifications in its passage through the two Houses, with the single exception of the limits set to the authority of the Company. In the draft bill the radius over which they sought power was seven miles, whilst in the act it is reduced to three miles round London.

Within two years of the passing of the act the Company appear to have been assailed with doubts about their privileges, for, "it is ordered that advice be taken by Councell concerning the valedity of *ou*r Charter and Acts of Parliament." And as a result of this inspection "it is ordered that another Bill be preferred unto the Parliament Howse by advice of Councell to the same effect that the other Bill was. . . And that part of the Bill w*hi*ch concerneth the practize of Phisick is to be shewed to the Phisic*i*ons." No further action, however, appears to have been taken in the matter, but a general act of parliament was passed in 1607 "for the confirmac*i*on of the landes of this and the rest of the severall Companyes of this cytie," into which the special bill of the Company may have been absorbed.

Monetary troubles.

The Company, never very wealthy, and which once before had been compelled to apply for relief to one of its members (Mr. Thomas Vicary), appears to have again fallen on evil days. From the accession of James to the Restoration, the minute books, so far as they exist, bear tokens of the most desperate money straits taking the form of endless devices for procuring cash. As early as 1603 the King's demand for a loan was met by the answer "We have no money." This, however, was probably a mere subterfuge, for in spite of it the Company were assessed by the Privy Seal at £75, two-thirds of which they were constrained to borrow of Mr. Thomas Thorney. Three years later "Mr John Pecke M*aste*r of this Company for the better accomplishment of the necessarie business of this Company the same now beinge in want of money, did of his voluntarie good will, proffer to lend to this howse l^li^ gratis for a yeare . . . which was by this Court thankfullie accepted." On the same occasion "Mr Joseph ffenton fowerth or youngest M*aste*r or governo*u*r of this Company knowinge this howse to be indebted did of his owne free will proffer to lend . . . the somme of c^li gratis for one whole yeare . . . towards the payment of the debts of the same, if the howse shall please to accept thereof. Which his kind offer was kindlie accepted by this Courte." The scarcity of money experienced by the Company at this period appears to have resulted from expenses incurred in altering their hall and erecting additional buildings. They were harassed, too, by continual demands of money from the Parliament. So poor were they in 1606 that "for the avoydinge of charges It is ordered that no publique Anothomy shal be holden in the Com*m*on hall of this mistery for the space of theis three yeares now next ensuinge. Yett notw*i*thstandinge . . . the M*aster* and Stewards of the Anothomy . . . shall once in ev*e*rie yeare . . . dissect a private Anothomy in the Com*m*on hall of the said mistery for their better experience and cunninge." So that even in their utmost need the guild took care, so far as in them lay, that the poor little spark of anatomical lore which they kept alive should not suffer eclipse.

The Company's plate.

The following is a good instance of the way in which the City guilds became possessed of their renowned plate: "August xix. 1605 This day Mr Peter Proby a very lovinge member to this Company did out of his love and affection to this house of his owne voluntarie good will give unto this company a very fayre large and serviceable Standinge Cup of silver and double gilte wayghinge xxx ounces and a penny waight *w*ith a cover unto it whereupon are his armes placed. And hath this supscription ingraven upon it. (This cup is given to this hall by Peter Proby gentleman free of the same. A seruant to Queene Elizabeth and to Kinge James and for both armees keeper of the recordes in the Tower of London, postm*ast*er for the service of Ireland, and for speciall Service a pentioner for leife, who was twice of the highe courte of Parliament.) Which cup was kindly accepted by this Court. And in gratificacion thereof It is by this Court ordered that the said Peter Proby shalbe excused and discharged from all ffines within this company, except the office of M*aste*r or Governo*u*r of this company, And from all other attendances for the seruice or affaires of the said Company. And when it shall please him to come to the hall he is to be kindlie and friendlie intertayned."

"**Finis coronabit opus.**"

CHAPTER X.

THE COMPANY BECOMES BANKRUPT—ALDERMAN ARRIS' BEQUEST.

In the last chapter we saw that money was becoming so scarce with the Company that they were obliged to postpone the lectures on anatomy. But now in the later years of the reign of King James I., and in those which followed upon the accession of his son, demands for money came pouring in with the most alarming rapidity. Of these, one which bears the date January 30th, 1632, is interesting as being a letter from "your very loving ffriend Guil: London: of London house" written by William Laud, subsequently the famous Archbishop, and pointing out "the decayes of St. Pawles Church here in London, being the mother Church of this Citty and Diocess and the greate Cathedrall of this kingdome:" a church which he loved so well, and which he endeavoured so manfully to bring back to its historic state of magnificence. "A greate Dishonour," he proceeds, "it is not onely to this Citty but to the whole State to see that auncient and goodly Style of building soe decayed as it is. . . . Theis are therefore hartily to pray and desire you, the Maister, Wardens, and other Assistants of this worthy Company of the Barber Surgions to contribute out of your publicke Stocke to the worke

Repair of St. Paul's.

aforesaid, what you out of yo*u*r Charitye and devosion shall thinke fitt, and to pay the summe resolued on

BARBER SURGEON'S-HALL,

MONKWELL STREET.

Published Sepr 29th 1800, by John Manson, N° 6 Pall Mall.

by you into the Chamber of London at or before our Ladyday next, praying you that I may receave by any servant of yo*ur* Company a note what the Summe is which you resolve to give. And for that Charity of yo*ur*s whatsoever it shall prove to be I shall not onely give you harty thankes but be as ready to serve you and every of you as you are to serve God and his Church" "And thereupon this Court deepely considering the contents of that letter together with the p*re*sente ruines and dilapidac*i*ons of the said Church, and as faithfull and charitable members obliged largely to contribute to soe pious and religious a worke Doe now order that out of the Stock and revenue of this house there shal be paid into the Chamber of London towards the said repaire x^{li} p*re*sentely and x^{li} yearely for nine yeares following to make it upp compleate a c^{li} as of the free gift of this house. And if att any tyme hereafter the worke doe cease then our payments to cease likewise." The cost, too, of erecting their second hall, which was built in 1636 by Inigo Jones, must have been a serious drain upon the resources of the Company. The annexed representation of the appearance which its exterior presented towards the end of the last century, immediately before its fine theatre was pulled down, is taken from the interleaved copy of Pennant's "London,"[1] now in the print room of the British Museum.

The Company borrows money.

Ten years later, in 1642, "it is ordered that 400^{li} shalbe taken up upon the Seale of this howse and payd over for the reliefe of Ireland," *i.e.* for the relief of the people who had been

[1] Part viii. pl. 211.

ruined by the rising of October, 1641. This was the beginning of woes unutterable; from henceforward the minutes are filled with endless devices for raising money to pay the interest on this sum. The Company, however, were not responsible for the debt, as they were constrained to borrow, for "by order of the Lords and Co*m*mons in Parliament assembled Our company was proporc*i*oned to pay and lend 800^{li} at 8^{li} per cent towards and for nothing else then the releife of Ireland. Upon which our Company did humbly certifie the Lord Maior of the Citty of our inability to pay any part of the said 800^{li}, yet for that the present occasions forceing for present Releife, It was ordered that 400^{li} should be taken up at Interest and payd over into the Chamber of London for that and noe other purpose. And that nowe the Company would take up such a quantity of money but cannot obteyne soe much. It is now Ordered that the King's Cupp and Cover shalbe kept and that the M*aste*rs Wardens shall take out all our plate and view it and see which of it is fitt to be sold; and that [of] every parcell of that plate Which shall be sold a patterne or fashion of it shalbe drawne in paper and the perticuler Letters, Writeings, or Gravenings shall likewise be written on that paper. To the Intent that when the said ffoure Hundred pounds with the Interest at 8^{li} p. centum shalbe payd, the said Plate shalbe made agayne in the same ffashions and writeings upon them as now they are: Which this court doth Order shall fully and really be performed accordingly." What a picture this would form, the quaint figures in their old world habits looking with eager eyes at the ancient plate which they had so often used, and which they were now compelled to sell in order to assist in quelling the great Irish rebellion. The money, however, was urgently needed, as the reports from Ireland became daily worse, so that very shortly after these efforts on the part of the Company, "it is peremptorily Ordered that the 400^{li} shalbe payd into the Chamber of London on Munday next in the morneing."

Well might an old man seek rest in these troublous times, as did Mr Thomas Walton, who wrote: "To the worshipfull the M*aste*r Wardens and Assistants of the Company of Barber chirurgeons: I Thomas Walton a member of that Company haveing formerly desired at severall Courts and once by Peticion in writeing doe hereby once againe make it my request being now very sick and soe have continued by the space of six weekes and small likelyhood as yet of my recovery. Therefore heareinge by a messenger from the hono*ra*ble the Lord Maior that you were with him and did warne mee there to appeare, I thought good to lett you know my willingnes to give you contentm*en*t. And that is, that you wilbe pleased to shew mee now at the latter part of my dayes, being aged 68 yeares in this next moneth it cannot be likely my time can be long in this World, That I may obteyne to have my Writt of ease and not be any more warned unto more courts. And that you wilbe pleased to take in one other into your Assistants in my place fitter for it. And this my request I hope I shall obteyne before God doth take mee out of this World. And soe I do here now heartily pray to God for your health and Union together in love and ffriendship." The request of the old man was granted, and we may hope that his end was peace.

Mr. Arris. The Company, however, were spared the indignity of selling their plate, owing to the liberality of one of their number, Mr Alderman Arris. The name of this worthy member is still commemorated in the Arris lecture at the College of Surgeons, founded by him for the study of Anatomy, whilst his portrait adorns the walls of the Barbers' Hall. The Alderman advanced of his substance the sum of £214, and the plate remained intact, since the credit of the Company was sufficient to enable them to borrow the £186, which was required to make up the £400. At the following court, therefore, "The Orders made at the last Court were read and confirmed saveing the Order for selling the plate which is adnullated." The plate, however, was hardly yet secure, for the Company had no sooner paid over this large sum than they were again assessed. In May, 1643, "Whereas this Company is assessed at 8li per weeke for 3 moneths which they are noewayes able to pay In regard wee are so much in debt and that the Hall may be preserved from violence. . . . It is now Ordered that the plate in the Hall that is not guilt plate shalbe by the M*aster* and Wardens delivered over at the best rate to deliver soe much amounting to 96li." This order was actually executed, and the plate was pawned.

Bankruptcy of the Company.

In the autumn of 1643, the Company, apparently driven to desperation by repeated assessments, deliberately declared themselves bankrupt in the following terms: "This Court being sensible of those vast Debts they are involved in Doe desire that a Certificate be drawne up of our Debts and the House Seale passed beyond its ability and creditt and therefore not able to lend and soe to be delivered unto the Lord Maior." And yet, as the very next entry at the same court shows, they had money in hand, for, "Whereas there is 40^{li} per annum gathered of the ffreemen of this Company to be adventured into Ireland this Court is willing and doe Order that the said moneyes soe collected shal be made up [to] 50^{li} and by way of adventure that 50^{li} shalbe payd into the Com*m*ittee at Grocers Hall haveing formerly adventured 50^{li}." The speculation appears to have been a successful one, for the Company still possess and derive a small part of their revenue from estates in Ireland, and to this day a map of their possessions in that country is suspended in the anteroom to their hall. The seal, too, was not passed wholly beyond its credit, for in November, 1643, "the M*aste*r and Wardens may take up One hundred pounds at Interest upon the com*m*on seale of this Company for the paying of interest money and dischargeing other necessary payments." In this little matter the Company were accommodated, as appears from a subsequent entry, by "Robert Whitchurch Citizen and Butcher of London."

In 1644 a commission sate at Grocers' Hall, and to this commission the Company made application, showing the amount of their debts, which amounted to "Three Thousand pounds or thereabouts. And that one Thousand pounds borrowed of M[r] Watson and more vizt: In all Twelve hundred pounds and upwards have bin lent by the said Company (in expectation to have bin long since reimbursed of the same againe) unto the publique uses of the Kingdome Whereof not any part principall or Interest hath been repayed." Vain hope, when the whole nation was ablaze with civil war! Poor Mr. Watson ("auncient master of *our* Company," as he is styled) fared but badly in these times, for the M*aste*r and Wardens were required to pay to the Committee the money acknowledged at the last Court of Assistants to be due to M[r] Watson "being adjudged to be a Papist and a delinquent to the State." "This Court findeth that they are altogether unable to pay the same money unlesse the money by them disbursed to the publique be repayd." Yet the commission had its will, "since the Company was required forthwith to pay 400[li] and 100[li] more to be secured to them at 2 yeares by our Co*m*mon Seale in composic*i*on for the said Debt, or else doe threaten a most extreame Course for recovery of the whole. This Court doth consent That the 400[li] may with all possible speed be raised and that the Seale may passe for payment of the said 100[li] 2 yeares hence as is Demanded." To obtain this money, "the Court did pawne all the Companyes plate both guilt and white weighing 1120[oz] $\frac{1}{3}$[dwt] or thereabouts

The Company's plate pawned.

BARBER SURGEONS' HALL.

THE COMPANY ADMITTING A NEW MEMBER.

(From an Old Print.)

unto Mary Crosse of London widdow for the sum of 280^{li} . . . with a provisoe of redempcion." The sale of the plate thus enforced was much more extensive than the previous one had been, and it included the king's cup, which was specially exempted on the former occasion. The latter, however, was bought back by Mr. Arris, who was subsequently surgeon to St. Bartholomew's Hospital, who presented it to the Company in 1647, by which time the rest of the plate had been redeemed. In 1648 the plate was sold for the third time, the king's cup being again restored to the Company after the accession of Charles II., by John Knight, one of its members.

The King's cup.

The cup here mentioned as the "King's cupp" (seen in Plate VI. as the smallest of the three cups standing upon the table behind the fire-screen),[1] was presented to the Company by Henry VIII. to commemorate the union of the barbers with the surgeons. Mr Shoppee describes it as "a silver grace-cup and cover weighing 26½oz., elaborately chased with the Royal Badges of England and France, viz. the Tudor rose, portcullis and fleur de lys, and with the arms of France and England, quarterly, the lion and greyhound as supporters, and crown imperial on the cover."

"The design of this cup is quite in the style of Holbein, and in all probability was from his studio. There are four pendant bells, and the custom which is

[1] The larger cup, surmounted with the crown, is known as the Royal Oak Cup; it was presented to the Company by Charles II., in 1676. The third cup, with a Roman soldier standing upon the cover, was presented by Alderman Frederick in 1654.

referred to by Pepys, is that each person drinking from the cup empties its contents, and in handing it to his neighbour rings the bells. "Among other observables we drank the King's health out of a gilt cup, given by King Henry VIII. to this Company, with bells hanging at it, which every man is to ring by shaking after he hath drunk up the whole cup."[1]

As Mr Shoppee has pointed out,[2] "This cup has passed through many vicissitudes of fortune. The Company's minutes of the date of 11th November, 1615, give the following record. . . . 'At this Court our M^r^ acquaytinge them how unfortunatlie it hath happened that the Hall on Tewsdaie night last beinge 7 November was broken open and what losse the howse susteyned thereby. Whereupon it was presentlie considered and then ordered that a present Course be taken for the spedie repaieringe of the howse and tresory howse and that the same shalbe forthwith stronglie borded and made up at the charges of the howse . . . . . . Note that the xj^th^ daie of November Thomas Lyne confessed how he was the plotter for the Robbinge of o^r^ Hall and how o^r^ plate was carried to Westm' and our monie was devided amongst the theues who were these Thomas Jones, Nicholas Sames and Water ffoster wch did break open the hall. Whereupon the Clarke haveinge order from o^r^ M^r^ went to Westm' and upon search there made found our plate locked up in a trunke in the howse of

[1] "Pepys' Diary," 27th February, 166⅔.

[2] "Description of the Pictures and other Objects of Interest in the Hall and Court Room of the Worshipful Company of Barbers," by C. J. Shoppee, Master in 1878. Lond.

one ——[1] a shoemaker xj^{li} xviij^{s} of the monie Mr Warden Coop found the same daie in the howse of one ffulses in Fleete Stret. About the xvjth of Nov. then followinge Thomas Jones was taken who beinge brought to Newgate in December followinge Jones and Lyne were both executed for this fact.

"'In January followinge Sames was taken and executed. In April 1616 ffoster was taken and executed. Now letts pray God to blesse this howse ever from any more of these damigees. Amen.'"

Quarrel with the College of Physicians.

Throughout the subsequent years of its existence the United Company lived in a perpetual conflict with its neighbours the College of Physicians. Under the Tudor dynasty there had been occasional disagreements between the two branches of the profession, but they had on the whole lived amicably together. In 1616, however, in consequence of the incorporation of the apothecaries as a separate company, the College of Physicians obtained a new Charter from James I., which conferred several additional privileges upon them, and amongst others, that of proceeding against all persons who administered any "inward medicines." Under this clause the surgeons were constantly brought before the college and were as constantly fined. When the physicians endeavoured to get their Charter confirmed by parliament the barber-surgeons petitioned against the "grant whereby they did not only seeke to have a superintendancy over the Pettioners in theire owne

[1] Blank in the original.

profession, but also to abridge and restraine them from useing part of their art [viz. administering any internal remedies,] which they have served for and have done and doe lawfully use, and without which many times they cannot performe their cures, nor give such ease and remedy to their patients as is fitt. They therefore pray that such graunt may not be confirmed by Act before they be heard, or that a provisoe may be inserted that the petitioners may not be thereby restrained, but in all things use their profession as they have heretofore lawfully done. On which petition the king, the 4th of ffeb., 1620, ordered that the petitioners should be left to seeke any lawfull remedy either in Parliament or other wise."[1] The Company accordingly presented a petition to the House of Commons, in 1624, who ordered "that the physitian's patent should be brought into the committee of grievances, and that both parties should be heard by their counsel. The consequence of this was that the Physitians proceeded not with their bill."

Charter of 1629.

In 1629 the Company obtained a new charter by which, amongst many other provisions and declarations, it was enacted that no person, whether free-man, foreigner, native of England or alien, should exercise the science or art of surgery within the Cities of London or Westminster, or within seven miles of them, for private lucre or profit, without first undergoing an

[1] The account of the disputes between the physicians and surgeons is in great part taken from a manuscript apparently written during the reign of James II., and endorsed, "Observations on the Charters Concerning the Company of Barbers and Surgeons, and Remarks on the Legall State of the Practice of Surgery."

examination by four examiners in the presence of two or more masters of the mystery. Freemen and surgeons so examined and admitted might lawfully use the art of surgery in any city, town, borough, or place in the kingdom of England. By the same Charter it was enacted that no one should go out from the port of London or send out any apprentice, servant, or other person from the same port, to act as surgeon to any ship whether in the service of the Crown or of a merchant, unless they, their instruments, and their chests had first been examined and allowed by two of the governors of the mystery. No member of the Company was to receive any apprentice but such as upon trial could readily construe any Latin author proposed to him. A lecture on surgery was to be given every week to such persons as the masters should permit to attend. A penalty of forty shillings was incurred by every surgeon who within London or seven miles thereof should have a patient under his care so ill as to be in danger of death, and who yet failed to consult with the masters of the craft in regard to his case. These were ample powers, and the Company would have done well if it could have retained them. They roused the jealousy of the College of Physicians, however, who in June, 1632, procured an order of council with a clause to the effect that no chirurgeon "doe either dismember Trephan the head, open the chest or Belly, cut for the stone, or doe any great opperation with his hand uppon the body of any person to which they are usually tyed to call their Wardens or Assistants, but in the presence of a learned physitian one or more of the College or of his Maj[ties] physitians." To reinforce this iniquitous order the physicians afterwards exhibited a bill in the Star Chamber. The surgeons successfully petitioned against it, and Charles I. ordered the petition to be expunged in 1635.

The Barber-Surgeons' Company by this time hated the physicians most heartily, and their wrath was apt to vent itself in a variety of ways. A ludicrous instance of this kind appears in the following minute: "Whereas by Order of the honourable house of Co*m*mons assembled in Parliament of the 28th of June last (1644) the President of the Colledge of Physitians was appointed to call this Company before them and to tender the Coven*a*nt to them. This Court conceiveing their Priviledges to be thereby infringed . . . Doth Order [after advice with counsel] that a Petic*i*on be framed to be preferred by all the Assistants that are now present or the Maior part to the House of Com*m*ons to have the tend*er*ing of the Coven*a*nt themselves to theire owne Members, and the chardge to be allowed out of the Com*m*on Stock." The covenant here spoken of was of course that solemn league which, during the Commonwealth, was eagerly subscribed by persons of all ranks, and which in the reign of Charles II. was as solemnly burnt by the hangman in Westminster Hall.

The surgeons, as we have seen, considered themselves to be the superior members of the united Company, and they were bent from time to time to act as such, and to resent any slights which might be put upon them. Thus in 1625 Mr. William Clowes addressed the following letter to the authorities of his Company. "Right worthie Maister and Governors and Assistants of the Companie of Barbers and Surgions in my true love I wish all health and florishinge government of yo*u*r company to the glory of God, the honor of the kinge and the goode of God's people, Amen. Now whereas I have bin not only by many Brothers advertised, but also of yo*u*r officer legally by (letter and otherwise), given to understand that you had chosen me renter warden of the yeomanry[1] from the w*h*ich election I desired . . . I might be freed yett could not. I then knew well that in Duty I owed you an answere which might well beseeme my reverence to yo*u*r Authoritye, and my tender regarde of the Kinge my M*aiste*r's honor. . . And nowe not once questioninge the troublesomenes of the place nor other hindrances w*h*ich God Almighty did then sende; I this answer (because I will be free of Ambition or pride), that if you can make that appeare upon yo*u*r Records that any of my predecessors did beare the office of Warden after he was sworne Serieant Surgion to any of the K*ing*s or Queens of England I shall humbly serve it; if not, I must praye *your* pardon, for I must not soe poorly value the Kinge my M*aste*r as think him less mighty, less absolute a Prince than any kinge whichsoever hath raigned before him. And so as his servant

The position of the surgeons in the United Company.

[1] A comparatively humble position in the Company, since the yeomanry were freemen below the rank of liverymen; they consisted of the journeymen.

I expect from the Companye as good respect as any Sergeant Surgion heretofore hath had for my *Masters* honor. . . . And further because I am many tymes summoned to *your* Courts and other meetings whch service I am very willinge to performe when I shall knowe my place in the Company which I must leave to *your* grave consideracion only if you please to take notice how the College of Physicians and the Company of Apothecaryes of London have ranked the King's physitians and Apothecaryes, you may thereby guess what place I expect." From no one could such a request have come better than from William Clowes the younger, whose father had been sergeant surgeon to Elizabeth, and who would therefore be thoroughly imbued with court traditions, and with all points of the etiquette which pertained to his high office. The Court decided that "whereas he was chosen renter Warden of this Companie for this yere ensuinge *which* place by reason of this contagious tyme and other respects he is not able to execute. It is thereupon ordered by this courte that he shal be discharged from the said place of youngest warden and seconde warden of this Company. And it is further ordered that he shall take place next unto the youngest of our Assistants that hath served the place of Upper Warden."

Intestine disputes.

Disputes in the Company, however, were not always conducted in the same gentle spirit, and the plainness of speech which characterised Cromwell was too often heard at the Court of Assistants. Thus in 1642, "Mr. Lawrence Cotton summoned for his absence from 4 courts of Assistants . . . was fined at 5s.

a time for each. . . And being further questioned by the Court he gave ill language and wilfully left the Court, who being then but 15 in number, could dispatch noe busines till afterwards more Assistants appearing." For this insolent behaviour he was fined vj^s^ viij^d^. At a subsequent court "Mr. Cotton layed downe his ffine imposed on him at the last Court of Assistants, vizt xxvj^s^ viij^d^. The M*aste*r of the Company moveing by the consent of the last Court Mr. Cotton to withdraw himselfe according to Orders and Custom, he gave this court this peremptory answer. I will not goe out of Court nor the M*aste*r hath noe power to bidd one goe out, and that the court had noe power to fine him. Then he threatened Mr. Warden Arris as he sat in Court saying Winter will come. Alsoe he abused Mr. Dye in his delivery of speech to this court that his speeches were rotten speeches, and thwarted him to the generall disturbance of the Court. And to Mr. Martin Browne threatening him I will make you know it better in another place. Alsoe he sought to disable the M*aste*rs hand which was signed to Ticketts for his own and other appearances at the Hall for defaults saying they might choose to appeare or not. Alsoe he told the Court with high language, I will not be dismissed." For these misdemeanors "this Court doth dismisse Mr. Cotton out of and from his place of an Assistant and being an examiner."

Fifteen months and later Mr. Cotton's ardour had died away, and he writes: "To the Worshipful the M*aste*rs, Wardens and Assistants of the Worshipfull Company of

Barbers Surgeons theis, gentlemen my true respectes presented unto your wor*sh*ips. It is a trouble and greife to me to be deprived of the wonted Society I once had with you after soe long experience you had of my care and endeavoure to the utmost of my power for the generall Good I am imputed as a greate offender. I desire your wor*sh*ips to take mee into your serious Considerac*i*on. I was not conscious then of what I stood upon, My choler being such at that time That it transported mee beyond myselfe. And that which was imposed on mee I confess may be my just demeritt. Should I speak much for myselfe, yet could not I speake more than your wor*sh*ips know (though a cholerick yet an honest man). My request is that I may be by your good leaves and favours readmitted to my place, Which as heretofore soe shall it still oblige mee in an Obligac*i*on of care and thankfullness. Thus waiteing your wor*sh*ips answer to these my lines, I take my leave and rest your loveing Brother Lawrence Cotton, March 18th, $164\frac{3}{4}$." The Company bore no ill will, and doubtless knowing that Mr. Cotton was as he described himself an honest man, decided that "Mr. Cotton submitting himselfe to this Courte concerneing his ffines . . . and layeing downe the same in this Court and acknowledging here in open Court to have done Mr. Martin Browne wrong, This Co*u*rt doth thinke fit, and soe Order that Mr. Lawrence Cotton be restored into his place in the Assistants againe, and doe remitt unto him his said ffines except xx^{s}."

Another fertile source of those intestine dissensions which did so much to weaken the power of the

Company during this period is to be found in the following minute, dated Jan. 17th, 1643, which sums up in a few words the entire matter. "Whereas the Government of this Company doth consist of 4 M*aste*rs or Governours to be yearely eligible, whereof two to be expert in Surgery, and the other two in Barbary, other freemen of the said Company exercising any other Arts Sciences trades or misteryes than surgery have bin of ancient time past reputed and taken for Barbers, and elected into the said office of M*aste*rs or Governo*u*rs. Which usage in elecc*i*on of M*aste*rs or Governo*u*rs hath been continued and observed till of late our Soveraigne Lord the King that now is through his princely care . . . by his letters mandatory under his signett beareing date at his highness Pallace of Westminster the three and twentieth day of March in the ffourteenth yeare of his Ma*jes*ties Raigne [1638] . . . did streightly charge and co*m*mand them that from henceforth it be duely observed . . . that none be chosen into that Government but two expert in Chirurgery, and the other two in Barbary. . . And the same have been ever since observed and performed accordingly. But now [1643] forasmuch as this Court hath by sad experience found that this new change and alteration of electing M*aste*rs or Gouernours hath (contrary to his Ma*jes*ties gracious Intentions) proved very hurtfull and preiudiccall to this Company, there haveing thereby arisen great Controversyes and discontents betweene severall the Members thereof, and other great inconveniences occasioned tending to the disquieting of the ancient peace and amity among the Brethren of this Company, and to the greate Impovering of the same. And whereas among other the Acts or Ordinances of this Company examined and approved . . . according to the statute of the 25th of January In the 19th yeare of King Henry the Seaventh [1503] . . . It is ordained that all such ffreemen of the said Mistery and Com*m*inalty as shall use or exercise any Art Science Trade or Mystery whatsoever (Except the said Art or Science of Surgery) shalbe accepted, reputed, adjudged and taken for Barbers and shall and may be (as by the Custome anciently they have bin) eligible into the office of the said M*aste*rs or Governours Or into any such other place or office as Barbers without any penalty. . . . This court doth therefore for the remedy of the p*re*sent Evills and p*re*vencion of greater Mischiefs that may befall this company, and calling to mind the happy condic*i*on and flourishing estate of the same when the said recited Ordinance was duely observed Doth this day upon the whole matter thinke fitt, and soe order . . . that hereafter in the Election of M*aste*rs or governours ffreemen of this Company of whatsoever Trade, Science, Art or Mysterye (except Surgery) shall be reputed and elected as Barbers." The union, however, was not a happy one, and in spite of this return to the old order the good times could not be recalled.

The Company's dinners.

As we have had occasion to notice, the statutes prescribed that "a dinner should be made" upon certain occasions of festivity, such as the admission of a new assistant, or the translation of a brother from another company. This statute was religiously enforced; thus, "ffredericke Stevenson appeared before the *Masters* of this Company, and prayed to be translated to this Company from the Company of sadlers. Whereupon it was decreed that he shall make his suite at the next Court of Assistaunce, And payeinge for the dynner of such of the Assistaunce as shalbe at such Court, and the ordinary charges due to the howse he shalbe made free of the same." At the next court of assistance accordingly, no less than nineteen members felt it incumbent upon them to be present, whilst at the ten preceding courts the average numbers of assistants in attendance was but eleven. At this court the unfortunate Stevenson was further mulcted, according to custom no doubt, for in addition to "payinge for the dynner of this whole court of Assistants and paying his ordinary charges," he is to give "a spoone of sylver dubble gylt."

In 1645, however, "Mr. Ralph Foster being complayned of by the present Governours, for refuseing to make his Dinner upon his admission into the Assistants, did utter certain Words to the disparagement of the Government of this Court, and tending to the moveing of separac*i*on and affection betweene the Barbers and Chirurgeons. Whereupon this Co*u*rt proceeding to pronounce sentence against him for the same, according to the Ordinance in that behalfe, Mr. ffoster confessed his Errour and humbly submitting himselfe to the Judgem*en*t of this Court, promising conformity hereafter in all things concerning an Assistant and to make his dinner at the appointment of the present Governours. This Court doth forbeare to prosecute the Ordinances against him; and doth Order that none henceforth shall speake anything to the disturbing of the Union now settled betweene the Barbers and Surgeons, upon paine of fforfeiting of vj[s] viij[d] of lawfull money of England."

The Arris bequest.

In 1643 "It is ordered That in respect of the greate troubles and distractions of these times, there shall be noe publique Anatomy this yeare dissected." The troubles, however, were national rather than corporate, for in the following year "this Court doth think fitt and soe Order That a Sermon be made on the next Election day, of thankesgiveing to Almighty God for peace and amity which is now begun to be restored among the members of this Company." And that M[r] Sharpe be desired to performe the same. And in 1645, as if to commemorate this period of peacefulness, "M[r] Edward Arris," who had deserved well of his Company on more than one occasion, and had been at divers times reviled by cholerick persons, "acquainted this Court that a person, a friend of his, (who desired his name to be as yet concealed), through his greate desire of the increase of the knowledge of Chirurgery, did by him freely offer to give unto this Corporac*i*on for ever the sum of 250[li] to the end and upon condic*i*on that a humane Body be once in every yeare hereafter publiquely dissected, and six lectures thereupon read in this Hall if it may be had with Conveniency, and the Charges to be borne by this Company. The said worthy Overture is thankfully accepted by this Court." This benefaction forms the basis of the lectureship on Anatomy, which is still delivered yearly at the Royal College of Surgeons in Lincoln's Inn Fields. The benefactor lived for many subsequent years, and was no less a personage than the worthy Alderman himself, for in 1675 "Mr. Edward Arris a worthy member of this Company having formerly settled by deed £30 a year for a dissection of a body yearly and reading on the Muscles desired that deed might be given up to him in consideration of £500 he is willing to give the company to go on with that work themselves." The Court decided to leave it to himself to give what he thought fitting, and he offered "freely to give to them £10 in addition, and the court thereupon ordered the £510 to be accepted and his deed to be delivered up to him cancelled." The reason for this change appears to have been that the old alderman did not repose any very great trust in the charitable disposition of his son, Dr. Thomas Arris, and as the sequel showed he was correct.

In 1676 Edward Arris died,[1] and was succeeded by Thomas Arris, one of his 23 children, who was a doctor in Physic, fellow of the College in London, Justice of the Peace in the county of Hertford, and a member of the honourable House of Commons. This worthy appears to have quarrelled at once with his father's old Company, for in 1677 the masters and governors made the following reply to his proposals:

"This Court having heard and considered the proposals of the 5th of December in the presence of the

[1] He had been examined and admitted to practise surgery on April 30, 1629.

honourable the Master of the Rolls, in the Cause then depending in Chancery between Dr. Thomas Arris Plaintiff, and this Company, Defendants, do answer and say That Mr. Edward Arris, deceased, the Plaintiff's father, did voluntarily and of his own accord, without any request made to him by the said Company, or any Person in their behalf, offer them £510, to deliver up and cancel a Deed and Settlement of the Lands mentioned in the Proposals for Payment of the rent charge of £30 per annum for a muscular dissection yearly urging this as his reason, Viz. that his only Son and Heir, the now Dr., had and did then receive the profits of the said Lands to his own use, upon Condition, and under Promise, to pay the said £30 per annum for the said dissection, but he found that he did never pay one Penny of it, or ever would do, when he, their Benefactor was dead without Trouble, and suit: with some severe and sharp expressions, which we will by no means mention, although they were the very words of the Father spoken of the Son. And further we say that Mr. Arris, our pious Benefactor, did nor could expect any other security from us for the performance of the said muscular dissection, than a Covenant under our Common Seal, for that he knew, being an antient member, and sometimes Governor of our Society, that we could not secure it by any Lands of our own, nor had we Stock enough to buy Lands sufficient to make any other Security for the five hundred and ten pounds, the settlement we parted with, being really worth a hundred pounds more. And if we should to pleasure the plaintiff, although against his Father's express Will and Intent, who was our Benefactor, return the said Five hundred and ten pounds again, it would be looked upon as a great neglect, perhaps a Breach of Trust, as we humbly conceive to agree to take other than the same Lands, and less than the whole, which was formerly settled by the cancelled Deed, which Premises considered, we do hope that the honourable the Master of the Rolls will give us that Credit, which the Donor of this worthy Gift was earnest with us to do when alive, though his son be not so willing to trust us, and that the *Doctor* may be decreed, without return of the money, to take a Covenant, under the Defendant's Common Seal, for securing the Trust reposed in us, according to the Agreement made with his Father at a Court of Assistants Feby. 29, 1675, and likewise pay the Charges of the Company, which they have or shall be put to by this troublesome and unnecessary suit, which doth so much shew what they must expect from him hereafter, if they should part with the Five hundred and Ten pounds."

Expenses of dissecting an anatomy.

To my friend Mr. Sidney Young I am indebted for the following extract, which shows that the expenses of dissecting a body at the period of Mr. Arris' benefaction were very considerable. The entry occurs in the book of expenses during the year $164\frac{6}{7}$, and is headed :

"THE CHARGES OF THE ANATHOMYE BETWEENE MICHAELMAS AND CHRISTMAS LAST.

Paid for Carryeing the Cophin to Newgate 00 00 06

ffor horsehire to the place of Execuc*io*n .	00	02	00
ffor the ffees at the place of execuc*io*n .	00	05	06
ffor expences at St. Gyles xij^{d} to the Carman xij^{d} and for washing the bodye xij^{d}	00	03	00
ffor Perfumes xij^{d} wax candles ij^{d} and soape j^{d}	00	01	03
ffor lynnen for the Bodye	00	06	08
To the Beadles Assistant in taking the Bodye	00	01	00
Paid the Parsons dutye for the buriall ij^{s} for ye grave xij^{d} for the Clerke and Sexton xxij^{d}	00	04	10
To the Bearers ij^{s} and espended at the buriall ij^{s} vj^{d}	00	04	06
ffor a Cophin to burye the bodye in .	00	03	04
To Doctor Godard for reading six lectures	06	00	00
To Mr Nicholas Brothers and Mr William Watson who desected the bodye xl^{s} appeece	04	00	00
Paid for 3 dynners for the M*aste*rs or Governo*u*rs Assistants Reader and desectors	10	00	00
ffor Candles for 3 mornings	00	11	11
To the twoe Beadles their ffee for three dayes attendance	00	10	00

With the two following extracts we shall close this chapter. The first relates to the barbers:

"Fforasmuch as —— Midleton barber hath com*m*enced a Suite in the Maior's Court against Monsieur de Roth for useing the art of Drawing of Teeth within the ffreedome and hath prosecuted it very farr before he acquainted this Court with his proceedings; This Court doth thinke fitt that he may goe on with his said Suite if he please without the Assistance of this Court. And this court reprooved the said Monsieur de Roth for his exerciseing the said Art and hanging forth his banner of significac*i*on against the Ordinances of this howse and required his reformac*i*on thereof." The second relates to the practice of midwifery, for "Thomas Bowden was this day elected to be one of the Stewards of the Anatomy for 2 yeares . . . . who appeareing in Court humbly prayed that he might be discharged from that place as alsoe from M*aste*r of the Anatomy, for that although he was willing to doe the duty, yet his imploym*en*t in Midwifery being of uncertaine accidente would be a greate hinderance in his performance thereof and submitted to this Court." Finally Mr. Sidney Young has sent to me the following interesting extract copied from the book of the Company's expenses for the year 163$\frac{8}{9}$. It shows that the barber-surgeons possessed the nucleus of a good library, and it would be a matter of great interest to know what has become of the MSS. and printed books here mentioned, for the present College of Surgeons has in its library no volumes belonging to its predecessors, nor even one of the 500 copies of Morus' tables. They were all sold for a few pounds.

by the Barbers' Company after the secession of the surgeons.

1638—1639.

"THE CHARGE AND SETTINGE UPP OUR BOOKES AND AUNTIENT MANUSCRIPTES IN OUR NEW LIBRARY.

Paid for 36 yards of chaine at 4^d the yard and 36 yards at 3^d the yard cometh to	xxij^s	vj^d
Paid to the Coppersmith for castinge 80 brasses to fasten the chaines to the bookes	xiij^s	iiij^d
To porters at sev*er*all tymes to carry these books	ij^s	
Paid to the booke bynders for new byndinge 15 books	xlviij^s	vj^d
Paid for Claspinge 19 large and small bookes and fasteninge all the brasses to the chaines to Three score and foure bookes 8^s, setting on old bosses j^s mending ould Claspes ij^s .	xxxj^s	viij^d
Paid for makeinge Ringes, swiffles, and fittinge all the iron Chaines .	xij^s	
Somme is .	vj^{li}	xviij^s"

CHAPTER XI.

TEACHING OF ANATOMY IN THE SEVENTEENTH AND EIGHTEENTH CENTURIES—RISE OF THE MEDICAL SCHOOLS IN LONDON — SEPARATION OF THE BARBERS AND SURGEONS.

THE records of the Barber-Surgeons' Company during the latter half of the seventeenth century are unfortunately missing. Its history has therefore to be eked out from a variety of sources. One of the greatest troubles of the Company appears to have been the constantly increasing difficulty which was experienced in obtaining the subjects necessary for the annual dissections required by the Charter. Many and bitter were the representations made to the Court of Aldermen about the conduct of the sheriff's officers, and others in charge of executions, as the following extracts will show.

Difficulties in obtaining anatomies.

In 1639, by an order of sessions for "Goale Delivery within the Citty of London . . . upon the humble petition of the Masters or Governors of the Mystery and Co*min*alty of Barbers and Surgeons of London, shewing that notwithstanding the Statute made, 32nd, Henry VIII. (1540) concerning Anotomyes and the Pet*ione*rs Charters and Ordinances, diverse persons inhabiting in London and the Suburbs, and diverse Aliens, fforeyn*e*rs, Mountebanks, Imposters, and Empiricks have of late indirectly obteyned many dead bodyes from the place of Execucion without any lawfull warrant or power for the same, but only by bribing or giving money to some Officers at the place of Execucion, who convey those bodyes into diverse partes of this Citty, and dissect them in private Dwelling-houses to the shame and Scandall of the Governm*en*t of this Citty and to the Damage of the Pet*ition*ers and the Generall Grievance of the Com*m*on-wealth by Smothering men murthered in private places and dissecting them in private houses, giving the word forth that they are Anotomyes which may prove of dangerous consequence.

"It was there ordered by the said Court that noe Officer or other person whatsoever thereafter instructed with or attending the Execution of any Malefactor or attainted person should presume to cause or permitt any Dead body to be brought or conveyed from the place of Execution to any private house or other place to be dissected but only to the Pet*ition*ers Com*m*on Hall being a settled publique place and fittest for the purpose."[1] In 1673 it was found that as this order of sessions had not been effectual "by reason of Sheriffes officers who attend such Malefactors to their Execution refusing to countenance and assist the person employed by the said Company to bring to their Com*m*on Hall the dead bodyes. . . . . It is thought fitt and ordered by this Court [of Aldermen] that from henceforth all such officers . . . . shall at the request of the Beadle of the said Company or other person by them employed to carry off

[1] Repert. lxxix. fol. 67.

the dead bodyes of such Malefact*ou*rs without faile assist him or them for that purpose, and goe along with and attend every such dead body to the Com*m*on Hall of the said Company, as they will answer the same at their perills. For which the said Company are content to allow and pay each time unto such officers vj^{s} viij^{d} more than the ordinary ffees."[1]

Even this order was not effectual, for in the very next year, 1674, "Upon complaint made unto this Court by the Company of Barber Surgeons against Edward Barber and Will*ia*m Jacob, the Sheriffes officers that notwithstanding severall Orders of this Court . . . the said Edward Barber and Will*ia*m Jacob attending the Malefactors at the last Execution refused to execute and comply with the said orders; and that the said Will*ia*m Jacob did publickly trafficke for their Bodyes and disposed of them to other private persons,[2] who would give most money for them." Which matter being now clearly proved . . . "this Court doth therefore

[1] Repert. lxxix. fol. 68.

[2] The following entry shows that the private persons here alluded to were in some cases distinguished surgeons, who occasionally allowed their thirst for knowledge to outstrip their consideration for the Company of which they were members. "At a Court of Assistants of the Company of Barbers and Surgeons held on the 25th March, 1714. Our Master acquainting the Court that Mr. William Cheselden, a member of this Company, did frequently procure the Dead bodies of Malefactors from the place of execution and dissect the same at his own house, as well during the Company's Publick Lectures as at other times without the leave of the Governors and contrary to the Company's By law in that behalf. By which means it became more difficult for the Beadles to bring away the Companies Bodies and likewise drew away the members of this Company and others from the Public Dissections and Lectures at the Hall. The said Mr. Cheselden was thereupon called in. But having submitted himself to the pleasure of the Court with a promise never to dissect at the same

thinke fitt to suspend them from their said places, and the Execution and proffitts thereof."[1]

These difficulties in the way of obtaining bodies for dissection continued for many subsequent years, and as time went on the disorderly scenes enacted at executions became more and more scandalous. Thus in 1706, "Upon a Complaint now made unto this Court by severall Members of the Barbers' and Surgeons' Company That their Beadle was at the place of execucion on Wednesday last, and demanded one of the Bodyes of the Malefactors then executed for a Public Dissection having a warrant from the Sheriffes as usuall grounded on Act of Parliament for his doing thereof. And required the Assistance of the Sheriffes officers, who were prevented therein by about one Hundred and ffifty soldiers as the said officers now Declared there gathered together and Armed with Swords and Staves (and as it appeared to this Court) corrupted to do the same, who did in a Riotous and tumultuous manner cutt down all the executed Bodyes and carryed them away in coaches. It is ordered that . . . . the officers that were present at the said execucion do use their utmost endeavour to find out and discover the names of the said Soldiers or as many as they can, together with what Regiments they belong to, that Care may be taken to bring them to Condign

times as the Company had their Lecture at the Hall, nor without leave of the Governors for the time being, the said Mr. Cheselden was excused for what had passed with a reproof for the same pronounced by the Master at the desire of the Court."

[1] Repert. lxxix. fol. 193.

Punishment and prevent the like Disorders in time to come."[1] On October 12th, 1708, "Charles Bernard, Esq., Sergeant Surgeon to Her Majestie and other Members of the Company of Barbers and Surgeons," complained to the Court of Aldermen that . . . "the officers suffered other persons to take the body away." This appears, however, to have arisen from a misunderstanding, for "the Court was informed that the Company was desirous to have the Body, but had given leave to ye Relations of the Malefactor to take it away, which was the reason the officers suffered them so to do." It was, therefore, ordered to prevent "the like Disappointment and misunderstanding hereafter that whenever the Company is minded to have a Dissection they do timely acquaint Mr Sheriffes therewith, that so the Body may be delivered to their proper Officer and not otherwise disposed of by them; and Mr Sheriffes now present were required to take care that this Order be punctually observed for the future."

In 1728 the same trouble was experienced, and a precept was obtained from the Mayor, Sir Robert Baylis, and the Court of Aldermen, wherein it is set forth that "Great numbers of loose and disorderly Persons do often assemble themselves at the Place and Times of execution of the Condemned Malefactors, and that the proper Officers appointed to attend such executions are frequently insulted and molested in the Performance of their Duty. . . . . This Court doth therefore order that the Under Sheriff of Middlesex and the Officers of the two Compters who are to attend to see the Execution

[1] Repert. cxj. fol. 83*b*.

performed do take special care to prevent any such Disturbances and Insults. . . . And that if any person or persons do in any manner oppose or Hinder the carrying away any such Body that the said officers do seize and apprehend such person or persons and carry him or them before a Magistrate to be dealt with according to Law. And if any Persons to the Number of twelve or more shall obstruct or hinder the said officers in their duty, that then the Under Sheriff of Middlesex shall read the Proclamation in the Act of the First of King George the First to disperse such disorderly persons. And if they continue together in such disorderly manner for the space of an hour after such proclamation, that then the said officers do apprehend such offenders. . . . And it is further ordered that this order be printed and published in some of the Publick Newspapers and affixed at Newgate, the place of Execution, and other Publick places within this City and Liberties thereof." Such a method of dealing with disorderly persons should have been effectual, but that it was not so is shown by the fact that the following remonstrance was shortly afterwards presented to the Court of Aldermen.

"Some Few Facts humbly represented To your Lordship and Worships whereby It is apprehended The Companys Beadles and Officers are obstructed in the Execution of their Duty at the Place of Execution.

"First: The Masters or Governors of the said Company do take the Liberty to represent That the night before the Execution several Sand Carts Bricklayers Carts and other Carts and Waggons are planted round the Gallows On pretence of Receiving and Giving the spectators a Place on the Day of Execution And for which they pay Twopence Threepence and Sixpence apeice. From this Practice the Gallows is so blocked up That your Petitioners Beadles cannot bring away the Body without danger of their Lives nor the Sheriffs Officers assist them in the Doing of it.

"That there seldom are, as your Petitioners are informed by their Beadles, above Six or Seven Officers of each Compter who constantly assist your Petitioners in Taking the said Bodys, the Rest dispersing themselves into divers places Tho' the whole Number would be hardly sufficient for that Purpose.

"That the High Constable of the Division is so infirm a Man as not to be able to come out of his Coach to comand his inferior Officers to assist your Petitioners Beadles in the Execution of their Duty.

"That it is become a late Practice to sell the Bodys of such Malefactors to private Surgeons and others, which always stirs up a Contest about the Body to be delivered to your Petitioners Beadles And is a means whereby the knowledge of Anatomy will be confined to a very few hands.

"That the Common Executioner himself is likewise as your Petitioners have been informed frequently engaged in such unwarrantable Sales Whereas He or such other Persons have no Property in the Bodys of such Malefactors.

"That the Middlesex Officers are of little or no use to your Petitioners Beadles. And one in particular whose name is Brock At the last Execution very much abused your Petitioners Beadles and Officers in Doing their Duty And swore they should not have the Body they had then pitched upon And encouraged and protected those Persons who took the same from them with open Violence And even Mr Watson the Sumoning Bayliff in Middlesex do's (as your Petitioners are informed) permit the said Sand and other Carts to be planted about the Gallows Tho' your Petitioners give him a Fee on pretence of his Assisting their Beadles in procuring a Body for their public Dissections."

Even this remonstrance failed, however, for in 1739 a fresh petition was presented to Sir John Salter, Mayor, and his brethren the Aldermen, wherein, after citing the various charters granted to the Company, it is shown that "numbers of riotous and disorderly Persons have frequently Assembled themselves at the place of Execution, and with open Violence forced the Dead Bodies from your Petitioners Beadles, altho' Assisted by the Sheriffe of the County. . . . That your Petitioners have prosecuted sundry of the said Rioters at Law from time to time, but it is so very difficult to find out the Names and places of Abode of the Persons who thus interrupt your Petitioners Beadles in the Execution of this Right . . . and such prosecutions are attended with so great an Expence . . . that they cannot hope to suppress this growing evil by any Method within their own power." The Company therefore prayed that the laws might be better enforced for the future, basing their request on the ground that from a want of anatomical knowledge on the part of surgeons, the common weal would suffer. This petition probably had no better issue than the former, and these disorderly scenes continued, until the supply of subjects became so scarce that recourse was had to resurrectionists. Fortunately in our own days, the Anatomy Act properly administered has done away with the necessity of obtaining subjects by stealth, and anatomy is the most thoroughly taught of all the sciences cognate to medicine.

The Gale Lectureship.

The readership of anatomy[1] founded, as we have seen, in the early part of the reign of Elizabeth, was for a long series of years an office conferred exclusively upon physicians. As late as 1711, Dr Mead held the post of reader of the muscular lecture, as the Arris lecture was called, with Dr. Freind as his coadjutor for the visceral, and Dr. Wadsworth for the osteological or Gale lecture. The Gale lectureship was founded by Dr. Gale, who left an annuity of sixteen pounds a year for this purpose. On June 30, 1698, it was "ordered that there be an Anatomy lecture called Gale's Anatomy. Dr. Havers and Dr. Hand being put in nomination for reading of the same, Dr. Havers was chosen for three years, and to read on the second

[1] For an incomplete list of the readers of anatomy the Editor is indebted to the kindness of Mr. Sidney Young, who obtained it for him from the records of the Company. As it is too long to insert here it will be found as Appendix M, page 365.

Tuesday, Wednesday, and Thursday in July next by 3 o'clock in the afternoon, and to have 30s. for his pains." Dr. Havers was specially qualified to fill the post of osteological lecturer, since his work on the structure of bone ("Osteologia Nova"), in which he described those canals which have since been associated with his name, was published in 1691.

Glimpses of the methods of procedure at these lectures can be obtained from the Company's records; thus, "such Anatomy readers as shall be present at any Public dissections, shall sit one on each side of y[e] table, and if one reader only be present he to be placed on y[e] Barber's side on y[e] left hand of y[e] *Master*." In addition to the dearth[1] of subjects, the Company had to contend with

[1] The expenses attendant upon a dissection were, as has been shown, not inconsiderable; this is corroborated by the following account, for which the Editor is also indebted to Mr. Sidney Young:

"ANATOMY.

"1725. The charges to be paid by the Masters and Stewards of Anatomy for procuring a body:—

	£	s.	d.
Horse hire	0	2	6
For a coach	0	6	0
For expences in fetching the Body ...	0	2	6
To the sheriffes' officers	0	13	4
To the Beadle's assistant	0	1	0
For washing the Body	0	1	0
For a Coffin	0	5	0
To Parson, Ground, Clark and Sexton ...	0	5	10
To the Bearers	0	2	0
Funeral expenses	0	2	6
For a certificate	0	0	6
The Clark's fees	0	10	0
The 2 Beadles' ffees	0	10	0
For a Link	0	0	3
To the Chairwoman	0	5	0
	£3	7	5"

a variety of minor troubles before they could render their lectures satisfactory. Great difficulty was experienced in procuring fit persons to fill the subordinate posts of masters and stewards of anatomy, and heavy penalties had to be inflicted to enforce the acceptance of these offices upon persons who had been duly selected in their turn to fill them. After officers had been obtained, their petty thefts were a constant source of annoyance to the Company, so that it was enacted that "No Mast*er* or Stew*ar*d shall carry away any part, member, fflesh, or skin of any body dissected [in the common Hall], without the consent of 2 af*oresai*d Masters, nor shall publickly or privately dissect any An*ato*my out of the common Hall upon ye penalty of 10^{li}." In 1739 it was further ordered, that "the demonstrator shall not for the future continue his lecture longer than three days after the public lecture is over; and on no account longer than six o'clock at night on those days." The following document, dated 17th September, 1734, will be read with interest, as it shows the method of conducting a course of anatomical and surgical teaching at this period. It is endorsed

"ORDER FOR THE NEW REGULATION OF THE DEMONSTRATIONS.

"The Court taking into Cons*ideratio*n how necessary it is as well for the benefit of Mankind as for the Honour and Dignity of the Company and the advantage of each particular Member That all that are Educated in Surgery

Anatomical and surgical teaching in 1734.

should be thoroughly versed in the Knowledge and Exercise of Anatomy and sufficiently acquainted with the manner of performing all the Chirurgical Operations in practise for the better Encouragement and Advancement of their Members in Knowledge of those parts of their Profession do order

"That one or more examined Surgeons be chosen yearly by the Court of Assistants on the Election day out of the Livery to be Demonstrators or Teachers of Anatomy and Surgery for the ensuing year.

"That he or they in their Turns do assist ye Masters and Stewards of Anatomy for the time being in preparing the bodies for the publick and private Anatomys if they so desire it or do not make the preparations themselves.

"That every day after the bodies have been made use of for the publick and private Anatomys he demonstrate the parts more particularly and such chirurgical Operations as shall not deface the Body and make it unfit for the publick Lectures for 2 hours ea*ch* day at least.

"That the sev*eral* ensuing days so long as the Body can be preserved he shall dissect in the presence of his Auditors all the parts that are to be demonstrated and work and read upon them during 2 hours ea*ch* day at least Showing not only their sev*era*l uses but also as much as may be the Diseases they are lyable to and the Operations consequent upon them.

"That in Order to Illustrate his Demonstrations the System of the Arteries and Veins or as many of them as shall be found necessary shall be injected with colour'd wax and that he prepare also the Apparatus necessary for every resp*ecti*ve Chirurgical Operation to be demonstrated at the same time with the Operations.

"That as oft as may be the Dissection of the parts shall be showed upon one side of each Body and the Chirurgical Operations on the other so that the parts rote[1] upon may be compared, their Uses and Functions showed, and the Distempers and Operations they are liable to.

"That the Subject Matter of the Anatomical and Chirurgical Lectures be changed every Lecture till the whole Doctrine of Anatomy and Surgery is discussed according to a Syllabus and that when $\frac{e}{y}$ Flesh is removed from the Bones he then shew upon each $\frac{e}{y}$ Ligaments of the Bones their respective Articulations and their Glandulæ. Mucosæ, and save such of the Bones as may be necessary to show the Receptacles, Vessells, and Diseases of the Marrow and Cavernous parts of the Bones; and that he also treat of such distempers of the Bones with the Method of Cure as the circumstances will admit of.

"That in the Summer Season he, and if more than one each, shall shew the human Skeleton and all and every distinct Bone, the Fractures, Dislocations and other Distempers of the Bones with the Apparatus for each and their respective methods of Cure: one Demonstrator giving as many Lectures as he shall think necessary for such Introduction before the Public Lectures of Osteology and the other in the same manner after them.

"That each of the 2 Demonstrators shall have the

[1] wrought.

Dissecting and preparing one private and one publick Body, and shall not therein interfere with each other otherwise than to assist each other for the publick service and that ye first in nomination shall have ye 2 first Bodies.

"That all the Demonstrations shall be made in the publick Theatre, to which all the Members of the Company of the Surgeons side may be admitted, as well as all [such] Apprentices as have served 3 compleat years and have been bound at the Hall, they bringing a note signed by their respec*ti*ve Masters and date[d] with the day of the month and year that he gives them leave to come: the notes are to be filed and left open for the Inspection of any Members of the Company that desire it, and that no other person be admitted Except the Apprentices and Pupills of the Demonstrators, so that the aforementioned Apprentices be not hindered thereby. Whereof Notice shall also be given in the Daily Post and in the St James^s Evening Post as oft as those Demonstrations are made when also the publick Lectures may be Advertized.

"That according to his Discretion at convenient times he prepares such parts of the Bodies as shall be most useful at the public and private Demonstrations; and that such Demonstrations, and such others as shall be given to this Company, be kept in ye Livery Gallery to be inspected by any Member of the Co*mpany* at such times of Demonstrations, but not at any other without leave of the Masters or Governors, and that the name of the Donor shall be written in every such preparation and preserved by the Company.

"That all the Members of the *Company* of the Surgeons side as well Foreign Brothers as ffreemen, and any other Surgeon or Friend as any of the Court of Assistants shall bring with them shall be at liberty to be present at such Demonstration, but the Gentlemen of the Barbers side of the Court of Assistants are not to bring ye same person more than once to the Demonstration.

"That Notice of the Publick Lectures and of ye Demonstrations shall be Inserted in the Daily Advertizer and London Evening Post from time to time for the year ensuing, and afterwards in such newspapers as the Master for the time being shall think fitt.

"That the Beadles of the Company, the Instrument Maker, and other Serv*an*ts attend and assist as at the private Anatomies.

"That for the encouragem*en*t of such Demonstrator and Demonstrators as shall duly and Zealously discharge the Trust reposed in them and distinguish themselves in the service of the Company, a Medall shall be given yearly to such Demonstrator or Demonstrators with a ffine Stamp in Relievo of the most excellent Picture of this Company by Hans Holbein of King Henry the 8th giving the Charter to the Company on one side, On the Reverse the fine Anatomical Theatre of the Company built by Inigo Jones with a Body dissected on the Table and a proper Inscription Expressing the Intention and Motive for Establishing the same.

"These Orders to be put in fframes in the Hall and a copy of them to be sent to every Demonstrator."

In 1742 a notice occurs that "the advertising of dissections as formerly used shall be discontinued;" the anatomies were thenceforward employed more for teaching purposes than for shows, as had previously been the case.

The surgery lectures were at times read by physicians and at times by members of the Company, though there appears to have been a bias towards electing the latter, since whenever a vacancy occurred a surgeon was always proposed to fill the post. In 1646, however, the matter was settled by an order that "the eldest assistant" should read the lecture on the first Tuesday in each month.

The authority of the United Company appears to have been on the wane for many years previous to its final dissolution. In 1684 the Company, in common with many companion guilds, was obliged to surrender its Charter "praying his Majesty to pardon wherein they have offended, and wholly submitting themselves to his Majesties Royal will and pleasure as to their charter." In reply "His Majesty is graciously pleased to refer the consideration of this petition to Mr. Attorney-General to report upon to his Majesty." About the same period the Company formed an alliance with the Periwig makers, by which the latter were to be admitted under certain conditions into the Barber Surgeons' Company.

Rise of independent medical schools.

In October, 1695, complaints were made by many brethren "against breeding soe many Illiterate and unskillful pretenders to Chyrurgery att St. Thomas's hospitall or wherever else ye like ill

Practises are used. Where contrary to ye oath of our Company and of ye Citty of London ye Chyrurgeons pretend to qualifie any person how unfitt soever in half a year or a year for the expert practise of our art. Which injustice in them being attended with a considerable profitt is maintained by a pretence of service to the publick, Whereas in reallity itt is directly contrary, and an absolute overthrow to our Company, subverting the very fundamentalls of ye Legall Education by way of apprenticeship." The Company no doubt suffered considerable pecuniary loss from the establishment of medical schools at St. Bartholomew's and St. Thomas' hospitals. "For," as the Memorial just quoted proceeds, "is it reasonable to suppose y^t any person will give soe much money as we generally require and a seaven years apprenti*ce*ship, when for a fourth or less and in the Space of one year at furthest they become such proficients as to practise for themselves either about the town, in the army or navy or elsewhere with ye reputation of being bred in an Hospital. Nay soe far has this mischiefe extended ittself *th*at . . . one if not two of ye very master Chyrurgions of *th*at Hospital had noe Education but what was Spurious and of this sort." The matter being thus brought formally before the Company, some notice had to be taken of the practice. Accordingly, before the next meeting of the Court, the matter had been fully investigated, and the results of the inquiry are formulated in the following minute.

"At a Court held at Barbers and Chirurgeons Hall London the 5th of November 1695.

"The Court being informed that the Chirurgeons of St Thomas's Hospitall in Southwarke did teach the Art or Mistery of a Chirurgeon contrary to the Lawes of the Company, and they appearing and alledging that they nor any of them ever did take an apprentice for less terme than seaven yeares, and that the cause of complainte ag*ains*t them as they supposed was that they admitted young men to dresse under them being not bound to them or any of them, as an apprentice though bound to another Chirurgeon or such as had served a considerable time to a Chirurgeon in the country and for bettering their Judgements in the Art came to London to see the practice of the Hospitall; and whereas it appeares that young men bred in the Country have not that thorough knowledge in Chirurgery as they ought to have whereby to preserve the lives and limbes of his Ma*jes*ties subjects, and the said Chirurgeons engaging for the future that they will not take any person to dress under them or any of *the*m other *tha*n as above speccified, and will before they admitt them to their practice perduce good certificates under the hands of two or three credible persons of the service of the said party or parties, w*hi*ch said certificate to bee affiled w*i*th our Clarke, as also that good security shall be given to the Govern*ours* of the Company for the time being that such person or persons shall not at any time hereafter (unless first admitted into the Company) practice or use the art of Chirurgery within the Citty of London or seaven miles thereof. Upon due consideration had, its Ordered that they shall have liberty to make use of such young men as above, they performing the severall particulars before mentioned, as also that they shall before they admit an apprentice to a freeman of the Company in the Hospitall to see their practice perduce in Court his Indenture that an Entry thereof may be made by our Clarke for the time being for which purpose its ordered a booke be kept, etc."

Indorsed

"Case of Chirurgeons of St Tho*mas*' Hospitall in regard to taking Pupills."

In 1702 the Governors of St. Thomas's Hospital took the matter of the teaching into their own hands, and whilst recognising the right of the surgeons to take pupils, ordained that "none shall have more than three cubbs at one time, nor take any for less than a year."

The Bishop's licence.

In 1713 the Company became embroiled with the Bishop of London, who pretended "to a concurrent power with the Governors of this Company to licence[1] persons to practice surgery." But the inconvenience of such a practice was represented to his lordship, who promised not to license any persons for the future except such as had been examined in the ordinary manner. In order to prevent the Bishop "from being misinformed touching the Qualifications of any Persons who shall

[1] A copy of such a licence is in Appendix N, page 374.

apply for his licence," it was determined that five guineas should be annually paid to the Registrar as caveat money. Thus was an old grievance renewed, although, as we have seen, efforts had been made by the Company to supersede the power of the Bishop in this respect. In 1689, when a proposition was made to obtain a new charter, a special clause was to have been introduced to the effect "that in case the Bishop of London or Dean of St. Paul's shall oppose the Company in having full power and Authority in licencing chirurgeons, that then application be made to put in a clause to this effect: viz. That no person whatever shall be capable of being allowed by them untill they shall produce their letter of admission under the seal of the Company subscribed by two or more of the Masters or Governors." In course of time the payment of caveat money appears to have been stopped, for in 1715 a fresh attempt was made by the Bishop to bring the surgeons to his visitation. The attempt, however, was frustrated by Mr. Thos. Bernard, who addressed the following letter to his lordship:

"S^R,—All the Members of our Company as well those who already have the Bishopp of London's Licence as those who have only our Diploma are sumond to attend the Bishop of London's Visitation the 13th instant. Now S^r I thought (as to those who have our Diploma) and who I am pretty sure you cannot oblige to take your Licence It had been agreed between you & I that I would send you all I could by persuasion & that you would not endeavour to compell 'em The Company is so alarum[d] at this extraordinary proceeding that if you persist in it we must of necessity engage in a Suite at Law whereby to settle this point

"I am Sir

"Y[r] most obedient Servt

"THO BERNARD

"Barbers and Surgeons Hall

"Oct: 8th 1715."

Frequent collisions occurred throughout the whole of this period between the Company and the College of Physicians. In 1663 "The Court hearing that the College of Physicians are about renewing their charter and confirming it by Act of Parliament do order that the Master and Wardens for the time being . . . be a committee to consider and advise with counsel how to prevent any mischief which may come to the surgeons of the Company by it." In 1689 the clerk of the Company was directed to attend Mr. Trevor, and "have it under hand to answer whether the surgeons of London have power by Law to give internal medicines to their chyrurgical patients. And also whether the Bill now passing in Parliament by the College of Physicians be a confirmation of their Acts of Parliament and the charters they now enjoy? or whether it be prejudicial to the chyrurgeons of London?" Again, when counsel was employed to draught a new charter for the Company in the same year, it was proposed that every person who

Quarrels with the College of Physicians.

had been duly examined and admitted, should be authorised "to give all sorts of internal medicines to his chyrurgical patients, as he shall think most conducive to the health and quicker recovery of the said patients, . . . and that the giving internal medicines in all such cases shall be taken and deemed practising chyrurgery." This bill was never proceeded with, but as time went on it became daily more evident that some restriction of this nature was needed to check the growing authority of the physicians. In the Annals of the College such notices as the following are constantly recurring: "M[r] Atkinson appeared and confest that he had given physic to M[rs] Brent and M[rs] Corner. . . . Then the president told him the penalty was five pounds a month which he ought to pay or be prosecuted for it.

"Ordered that Atkinson be prosecuted at law."

The Company, however, never obtained free licence to prescribe for their patients, and there are persons still living who remember the time when the name of the physician was placed conjointly with that of the surgeon upon the board of a hospital patient who required more than a black draught. It was not until the opening years of this century that the surgeon was emancipated from such thraldom, and it is to the energetic protests of Abernethy upon this subject that he is now permitted to have entire control over his own case.

Another grievance of the Company was that the physicians declined to consult with surgeons. The

Company, therefore, desired that "no one of the College of Physicians of London shall at any time hereafter refuse to consult with any chyrurgeon examined and approved as aforesaid in any case of Physic whatsoever." These quarrels attained such a height that in 1688 the Lord Chancellor Jeffries visited the College, and instituted a searching inquiry into the alleged grievances of the surgeons. The result does not appear, but that it was not wholly favourable to the physicians may be surmised from the fact that on the next occasion of a surgeon being complained of for giving pills, boluses, diets, etc., for which he charged 16^{li}, as no ill practice was laid to his charge, the censors did not hold themselves judges of his demands.

Separation of the barbers from the surgeons.

As early as 1684 the surgeons were feeling their union with the barbers to be disadvantageous, and they presented a petition to the king praying for a dissolution of the United Company. This petition and the king's answer run as follows:

"Upon the Petition of the Surgeons of the Citty of London praying in regard it is found by experience that the Union of the Surgeons with persons altogether ignorant of the Science or Faculty of Surgery (as the Barbers are) who were heretofore a different Company from the Surgeons doth hinder and not promote the ends for which they were united, that his Majesty would make them and all Surgeons within 7 miles distance thereof a Body Politick under such Regulations as his Majesty shall think fit.

"AT A COURT AT WHITEHALL MAY 15, 1684.

"His Majesty is graciously pleased to referre this Petition to the R^{t} Honble the Lord Keeper of the Great Seale of England and the Lord Chief Justice of the Kings Bench to consider thereof and report what may be fit for his Majesty to do in it Whereupon etc

"SUNDERLAND."

The movement, however, does not appear to have gone further than this petition, for, as has been shown, in 1689 the United Company proposed to obtain further powers, the additional clauses being, however, entirely in regard to points affecting the surgeons, who were by this time an influential body, whilst the barbers cared little or nothing for the common good. Things, however, pursued their old course, and time progressed until the surgeons found themselves sufficiently strong to make a fresh attempt, which was in the end successful. The final effort was made in 1744, when, on the 20th of December, the gentlemen on the surgeons' side of the Court of Assistants of the United Company of Barbers and Surgeons gave notice of their desire to separate, and produced a case which they proposed to lay before the House of Commons. A committee from both sides of the court was appointed to report upon these proposals, and their report was read in the succeeding January court. The barbers objected to the separation of the two crafts, and drew up the following protest:

"THE CASE OF THE BARBERS OF LONDON.

"The *Barbers*[1] of *London* were a *Fraternity* before the Time of *Edward the Second*, and by Letters-Patent of *Edward the Fourth* were made a *Body Corporate*, and invested with several Powers and Privileges. In the 32d Year of *Henry the Eighth*, the *publick* Policy thought proper to *unite* them with another Company (not then *incorporated*) called the *Surgeons of London*, in order, (as is most probable) to *transfer* those Powers and Privileges to the *latter*, without directly appearing to *wrest* them from the *former* to whom they had been *originally* granted.

"This *Coalition* of the two Companies having now subsisted above *Two Hundred Years*, the *Barbers* are surpris'd to find an *Attempt* made by the *Surgeons* to *dissolve* it, by Authority of *Parliament*, without their Participation or Consent.

"The *principal Reasons* assigned by the *Surgeons*, in their printed Case, to induce the Legislature to this *extraordinary* Act of Power, are,

"*First*, That the *Barbers*, in the Time of *Henry the Eighth*, were *all Surgeons*, and that the Parliament, by *uniting* them with others of *superior* Abilities, intended *their* Improvement in *that* Profession; but that they having, long since, *ceased* to intermeddle with any Branch of *Surgery*, this *Intent* of the Act is frustrated, and the laudable *Purpose* of the *Union* at an end.

[1] Throughout this document the words here printed in italics occur in the original in italics.

"*Secondly*, That by this *Alteration* of the Circumstances of things, the Junction of the two Companies (how advantageous soever in *former* Times) is *now* become highly *inconvenient*.

"*Thirdly*, That the *Surgeons*, if *distinctly* incorporated, would be encouraged to *meet* and *communicate* to one another their Experiments and Successes.

"And *Fourthly*, That the like Separation has taken place at *Paris*, *Edinburgh*, and *Glascow*.

"BUT the *first* of these *Reasons* is grounded on a *Mistake* in point of *Fact;* for tho' it be *true* that the *Barbers* were *all originally Surgeons*, and incorporated *as such*, yet long *before* the Union in question, *most* of them had *quitted* the *actual Exercise* of that Profession, and the *Right* itself of exercising it *in virtue* of their Charter, had been[1] *taken away* by Parliament: And tho' in the *Preamble* of the uniting Act, *both* Companies are stiled *Surgeons*, yet from the[2] *Enacting* Part (which expresly *restrains* the *Barbers* from occupying any part of *Surgery*, except *Tooth-drawing*) it is evident the *Legislature* did not consider them as *real* Surgeons, nor could intend their *Improvement* in a Science they were *forbid* to *practise*, so that the Circumstances of Things are *not* altered from what they

[1] 3 *H.* 8, *ch.* 11. No Person within the City of *London*, nor within seven Miles, shall occupy as a *Surgeon*, except he be first examined, approved, and admitted by the Bishop of *London* or Dean of St. *Paul's*.

[2] 32 *H.* 8, *ch.* 42, *sect.* 3. No Person within the City of *London*, Suburbs of the same, and one Mile Compass of the said City, using any Barbery or Shaving, shall occupy any Surgery, Letting of Blood, or any other Thing belonging to Surgery, Drawing of Teeth only except.

then were, or from what they manifestly were *designed* to be; and therefore the *Barbers* having no Relation to the *Surgeons*, or their Art (as it was *then* deemed no *Objection* to their *Union*) cannot *now*, with any Propriety, be insisted on as a *Reason* for their *Separation*.

"With regard to the *Inconveniences* complained of, as the Charge is *general*, this *general* Answer only can be given, That the *Barbers* have always, with the greatest Deference, *submitted* to the *Surgeons* in all Matters *peculiar* to them, and chearfully contributed, out of their *common* Stock, towards every Expence which *they* have declared necessary for the Honour or Advancement of *their* Profession. And since *none* of these Inconveniences have been of Consequence enough to deserve being *particularly* pointed out, we may venture to pronounce them *inconsiderable*, and unworthy the Attention and Redress of *Parliament;* and the rather, as *all* of them put together, have not prevented the *Surgeons* of *London* from carrying the Improvement of their Art, both in Speculation and Practice, to a greater Height than has been done in any other Place or Nation.

"That the frequent Meetings of ingenious Men, and their free Communications on the Subject of their Profession, may tend *to the Benefit of Mankind in general, and to the Honour of their Country in particular*, is not denied: But surely the Constitution of the *united* Company is no obstacle to these laudable Purposes. The *Barbers* have for many Years, at their Monthly Courts, submitted to *withdraw* at a stated Hour, and *resigned* the Parlour to the Surgeons: And if this *Condescension* is not supposed to afford them *sufficient* Time for Conversation on these *particular* Days, nothing *hinders* them from holding *separate* Assemblies at the Hall almost *every* other Day in the Year.

"As to what is said to have been done at *Paris*, *Edinburgh*, and *Glascow*, no *particular* answer can be given, unless it appeared by what *Means*, for what *Reasons*, and upon what *Terms* the *Separations* in those Places were brought about. In *London* there are but two Instances of Separations of Companies, *viz.* that of the *Feltmakers* from the *Haberdashers* in 1604, and that of the *Apothecaries* from the *Grocers* in 1617; but both these were effected by *mutual* Consent, without the Intervention of *Parliament;* and it may be proper to observe, that the Feltmakers miscarried in a former Application for an exclusive Charter in 1576 for *want* of the Haberdashers Consent.

"Upon the whole, therefore, the *Barbers* humbly hope the foregoing Reasons will be deemed *insufficient* to induce the *Legislature* to destroy an *Union* they *themselves* thought proper to *form*, an Union which two hundred Years Existence has rendered *venerable*, and which, by the *Improvements* above-mentioned, appears to have answered all the *Purposes* for which it was established.

"But if for *other* Reasons (which the *Surgeons* say *may* be given) the Parliament shall be inclined to favour them in *this part* of their Request; the *Barbers*, from the scrupulous Regard and Tenderness which that

August Assembly has always shewn for *private Property*, cannot but hope they shall be *continued* in the Enjoyment of *all* their present Possessions, without any *Division* whatsoever, and that, for the following Reasons:

"*First*, Because most of the *united* Company's Lands and Tenements, (particularly the Site of their *Hall*, *Parlour, etc.*) originally belonged to *them*, and by the uniting Act[1] seem with great Justice to have been intended to remain to their *sole* and *separate* Use, tho' in fact the *Surgeons* have hitherto been *indulged* in the *equal* Enjoyment of them with the *Barbers*.

"*Secondly*, Because the *Surgeons Share* of what may have been acquired *since* the Union, will scarce be an *adequate* Recompence to the *Barbers* for the above *Indulgence*, much less for the unmerited *Loss* of a Brotherhood now so *honourable* and *advantageous* to them.

"*Thirdly*, Because the Expences of the *Barbers* when *distinctly* incorporated, can fall but very little, if any thing, short of those of the *united* Company, so that a Reduction of *Income* must subject them to very great Difficulties, which (considering that *they* neither desire, nor have given the *Surgeons* just Cause to desire a Separation) would be extremely hard and unreasonable.

"*Lastly*, Because the present *flourishing* Condition of

[1] 32 *H.* 8, *ch.* 42. The united Company shall have, possess, and enjoy, to them and their Successors for ever, all such Lands and Tenements, and other Hereditaments whatsoever, which the said Company or Commonalty of *Barbers* have and enjoy, *to the Use* of the said Mystery and Commonalty of *Barbers* of London.

the *Surgeons*, (the only *real* Alteration in the Circumstances of Things) will sufficiently enable them to support the Dignity of their *new* and *favourite* Institution with becoming *Splendor*, without distressing their *less happy* Brethren the *Barbers*."

Separation of the surgeons from the barbers.

The surgeons, however, were steadfast in their purpose, and on the 31st of January, 1744, a petition was presented to Parliament by the surgeons of London praying that they might be separated from the barbers. This petition was referred to a Committee under the charge of Dr. Cotes. On February 27th Dr. Cotes reported to the Commons that the Committee had examined the matters contained in the petition, and had heard the surgeons as well as the barbers through their counsel. He then read the report of the Committee, and leave was given by the House to bring in a Bill, Dr. Cotes, Mr. Carew, Mr. Knight, and Mr. Bacon being ordered to prepare it. On March 8th "Dr. Cotes presented to the House, according to order, a Bill for making the surgeons of London and the barbers of London two separate and distinct corporations, and the same was received and read the first time." On the 13th of March it was read and discussed a second time. On the 20th inst. it was ordered that "it be an instruction to the Committee . . . . to receive a clause exempting the surgeons from parish offices, etc." Subsequently, provision was made that "the examiners of the said Company are to examine all persons who shall be candidates to be surgeons or surgeons mates in his Majesty's army." On the 26th of March the Bill was read a second time, after which it was adjourned till the following day, and it was "ordered that the Bill with its amendments be engrossed." On the 28th the engrossed Bill was read a third time, "ordered that Mr. Carew do carry the Bill to the Lords and desire their concurrence." The Bill was read on the same day in the House of Lords, and was committed on the following day. Lord Sandys reported from the Committee on the 2nd April, and the Bill subsequently receiving the Royal Assent became law on May 2nd, 1745.

The Surgeons' Company.

It does not appear distinctly who were the prime movers in this change, which, although seemingly important, really consisted in little more than a simple separation of the surgeons from the barbers. The provisions of the old Act of Incorporation were retained in nearly every point by the new Act, which established the Company of Surgeons. Mr. Cheselden, now surgeon to Queen Caroline and to Chelsea Hospital, and Mr. Ranby, sergeant surgeon to King George II., with whom he had been present at Dettingen battle, were probably the mainsprings of the movement. The exertions of Mr. Ranby in promoting the separation were rewarded by his nomination as first Master of the newly founded Surgeons' Company, a special favour, as he had not been a member of the old corporation. Mr. Sandford and Mr. Cheselden took office under him as wardens.

From the evidence offered to the Committee of the

House of Commons during the passage of this Act, the following facts have been gathered as to the state of the old Company at the time immediately preceding its dissolution. The Master was elected yearly, and was alternately a barber and a surgeon. He with the wardens present at the examination of a candidate signed the diplomas, the examination itself being conducted by one or more of the ten examiners. Two barbers were always present, but they never interfered in any way with the examination. The Master if a surgeon interposed his judgment in the examination of surgeons, but if a barber he only put the formal question, "Whether the person examined shall be allowed?" This passive action of the barbers in such matters was borne out by an independent witness, Mr. Burril, a surgeon, who stated "that two barbers were present at his examination and signed his diploma. That one of the said barbers (Mr. Fradin) expressed some resentment against him (the witness) for not previously waiting upon him, but that he did not interfere in the examination, the whole being managed by the surgeons." At the monthly meetings, too, the barbers, as they alleged in their petition, always withdrew at a certain hour, that the surgeons might discuss professional topics with the greater freedom.

Inconvenience attending the union of the barbers and surgeons.

Certain ludicrous inconveniences sometimes attended the holding of the United Company's licence to practice. Mr. Neil Stewart, however, had better tell his own story as he related it before the Committee of the House of

Commons. "Being examined, he said that he was appointed surgeon of his Majesty's ship, the *Looe*, and being in his return home taken prisoner in a merchant ship by the French and put into the common prison at Brest, he petitioned the superintendent of the place to be removed to an open hospital at Dinan along with the surgeon and his mate of the *Northumberland* man-of-war. That he enclosed his warrant from the Navy office in his petitionary letter. And some days afterwards inquiring of the linguist about the success of his petition, he was told that the superintendent did not know by his warrant whether he (the witness) was a barber or a surgeon. That upon desiring the witness to read the warrant over, by which it would appear he was a surgeon, the linguist replied it might be so, but that if the witness had been taken on board one of the King of Great Britain's ships, it would have been out of doubt. Whereupon the witness, concluding his warrant would be of no service to him, made application by another channel, and was ordered to go in a draught appointed some time after."

The barbers were usually present at the four public lectures of Anatomy, two of which were delivered at the expense of the Company, and the other two at the expense of the surgeons. The annual expense incurred by the Company on account of these lectures amounted to about one hundred guineas, towards the defraying of which they received the interest of £510, given by Mr. Alderman Arris for a muscular lecture, and an annuity of £16 bequeathed by Dr. Gale for an osteological lecture.

For the visceral lecture the master and stewards of anatomy, who were chosen for two years, were bound to find four bodies during their term of office; in many instances, however, they were only able to procure one or two, and in such cases the Company and stewards alternately provided the others.

Barbers were obliged to become free of the Company before they could commence their business in London; with each apprentice they commonly received about £10, whilst each apprentice to a surgeon paid £250. The great expense of the feasts and the charges attendant upon the taking of the freedom of the Company deterred many surgeons from joining the Livery, and compelled them to remain as foreign brethren.[1] It was, therefore, thought (and as the event showed with good reason) that if the surgeons could be constituted a company distinct from the barbers, many of the foreign brothers would join the new company, thereby rendering its freedom more reputable as well as less expensive.

Licensing naval surgeons.

The licensing of naval surgeons had always been an important branch of the United Company's work even from the time of its incorporation in Henry VIII.'s reign: The Charter of Charles I. expressly empowers "the *Masters* to appoint in any part of England proper surgeons for the King's ships to be sent out to sea, to take medicines, etc., from such as are not sufficient to serve personally

[1] For the various payments required of barbers, surgeons, and foreign brethren, see Appendix O, page 376.

on board, paying a proper price, and to examine surgeons chests for sea service. And no person who practices within London or seven miles shall go or send out any apprentice from the port of London as a surgeon in the King's ships or in the ships of any merchant unless he is approved of by the Masters or Governors, and his medicines, plaisters, etc., have been examined." Ample provision was thus made for the securing of competent surgeons for both the navy and the merchant service, and the powers entrusted to the Company appear to have been, on the whole, well employed.

The following is one amongst many entries of a similar nature occurring in the minute books of the United Company: "On the tenth day of March, 1606. This daie Thomas Garrett intending a voyage at sea was by Mr Warden Mapes examyned in surgery, and his chest and furniture liked and allowed of." The service which the Company thus rendered to the country was sometimes utilised as an argument for obtaining a better supply of subjects for anatomical purposes. In a petition to Sir J. Salter, Lord Mayor in 1739, upon this matter, it is stated that "your petitioners do without any Reward or Benefit to themselves, employ a great Portion of their time in the Service of his Majesty, as their Predecessors have done for some hundred years past in the services of his Majesty's Royal Predecessors Kings and Queens of this realm, by Examining at their Public hall all the Surgeons and Mates who serve on board the Royal Navy; and by viewing all their Chests of Medicines, Instruments, and Journals, as also by viewing all such Officers as are maimed in Fight at Sea and for Superannuation." These examinations, although wholly acts of charity, as the Company would have us believe, brought in a considerable sum of money annually to the corporate coffers. Mr. Hayward, a former Master of the Company, "being examined" by a committee of the House of Commons, "as to the money usually given to the poor's box by sea surgeons at the time of their receiving a qualification, said that it is always expected, but that if they are not in a capacity (to pay) it is not insisted upon; and that he never heard of any persons being denied a qualification for refusing to pay it. That the qualification is delivered to the party sealed up, to be carried to the Navy office, and that the Master commonly signifies to him what fees are expected. That the said box is examined every month[1] and distributed by the master and three wardens in about nine parts in ten of the money to the poor of Barbers. Being asked what proportion the barbers pay to the poor's box, he said that the greatest part of the

[1] In it was found as derived "from the sea surgeons alone" during the year:

	£	s.	d.
1739	34	11	0
1740	110	2	6
1741	95	2	6
1742	85	2	6
1743	46	8	0
1744	68	9	0

For these six years the Company obtained an average revenue of £73 per annum from this source.

income applied to that use arises by examinations of sea surgeons, but that the apprentices of the barbers (who are as twenty to one) always pay when bound at the hall, and at their admission to their freedom, as well as the surgeons."

In a few cases the results of the examination were unsatisfactory, if we may judge from the following letter sent to the Company:

"Navy Office 6 June 1740

"Gentlemen

"Captain Vincent of his Majestys Ship the St Albans, having given Admiral Cavendish an Account of the Insufficiency of Samuel Cowling his Second Mate, We send you enclosed an Extract of his Letter, and, if what it contains is fact, you must have been to Blame We are

"Your affectionate friends

"R^d^ Haddock Fred Acworth Tho Sharpe
J Hawler G Crowe Jno Phillipson

"Qualifyed the 15th of Aprill 1740 for 2nd Mate of a 4th Rate.

" (Extract of a Letter from Capt Vincent of the St Albans dated 3^d^ June 1750)

"The Surgeons Second Mate (who is now on Shore as bad as any of the rest) being a raw unexperienced Lad not capable of even letting blood and the Men concealing their Complaints rather than come under his hands, so that it appears to me, that his Relations or Friends obtaining a Warrant for him, intended only by it to receive the benefit of his pay and the further one of putting him in a good way at no Expence of Learning his Business from the Practice and Experience of the Surgeon.

"Navy Office. *Directed*

"To the Master and Wardens of Surgeons Hall."

CHAPTER XII.

THE SURGEONS' COMPANY.

The Surgeons' Company.

By the Act of Parliament passed in 1745 (Stat. 18, Geo. II., c. 15), the surgeons of London and the barbers were made two separate bodies. The surgeons were incorporated by the name of the Masters, Governors, and Commonalty of the Art and Science of Surgery, with a power of holding lands in mortmain to the amount of 200[ll] per annum. The corporation was to consist of twenty-one assistants, of whom one was master, two were wardens, and ten were examiners. The master and wardens were selected either from the examiners or the assistants, the examiners from the assistants, and the assistants from the freemen. There was an annual election for the choosing of master and wardens. The examiners and assistants were appointed for life, unless removed for some valid reason. The power of making bye-laws and elections was vested in the master, wardens, examiners, and assistants.

The actual gain to the profession of surgery by the establishment of a separate company was that the fees were fixed upon a much lower scale than they had been amongst the united barbers and surgeons, whilst the more expensive offices in the older Company were abolished, thereby enabling the corporation to reduce the fines to the younger members. Those who joined the new Company were thus entitled to all the privileges and advantages enjoyed in the older corporation, with certain additional ones, for less than a quarter the sum of money they would have been obliged to pay as members of the Barbers' and Surgeons' Guild. Under the old Company the expenses of each member seldom fell below one hundred pounds before he had served or been fined for all the offices, or in other words, before he had become exempt from further service in the Company. The privileges enjoyed by the members of the united craft having been granted in different reigns, and being often supported by insufficient documentary evidence, were during the latter years of the Company found to rest upon a very insecure basis. The courts of law several times decided against the Company, and compelled the members to serve expensive city offices, from which they had formerly considered themselves exempt. Under the new Act these privileges were especially claimed, and the members of the new Company were exempted from the several offices of constable, scavenger, overseer of the poor, and all parish, ward, and leet offices, as well as from being put into or serving upon any jury or inquest. The expenses of the sea surgeons were also very considerably reduced, and instead of the numerous examinations which they were formerly obliged to submit to, however well qualified each man might at first appear, the Court of Examiners was now to grant to every man to the full extent of his merit. The examiners were also called upon to select surgeons for the army, as they had hitherto done for the navy.

Pending the erection of new buildings, the Stationers'

Company generously placed their hall at the disposal of the surgeons. In Stationers' Hall, accordingly, on July 1st, 1745, the first meeting of the Court of Assistants, of the Master, Governors, and Commonalty of the Art and Science of Surgeons of London was held, Mr. Ranby, the master, being in the chair, with Mr. Sandford and Mr. Cheselden as his wardens. At this court Mr. Ranby was presented with the freedom and livery of the Company, taking the usual oath of a freeman as well as that of principal master. In return for the favour thus shown to him, he presented a handsome silver cup to the new corporation, for which he received due thanks. For many subsequent years Mr. Ranby served the Company well and truly; he was re-elected Master in 1751, when the Company entered into occupation of their new theatre in the Old Bailey, and for a third time in 1752.

John Ranby.

Mr. Ranby was appointed sergeant-surgeon and principal surgeon to the king in 1740; he died in 1773. In 1744 he published "'The Method of Treating Gunshot Wounds,' a little piece" which, he says, "was penned in a camp," and which he subsequently speaks of as "a bare recital of his own practice." It is an eminently practical work, and such as we should expect to come from the pen of a surgeon who had actually seen service in the field. In after years Mr. Ranby gave great offence to the physicians by publishing a narrative of the last illness of Sir Robert Walpole, in which he utterly condemned the use of such edged tools as the Lithontryptic Lixivium; an opinion in

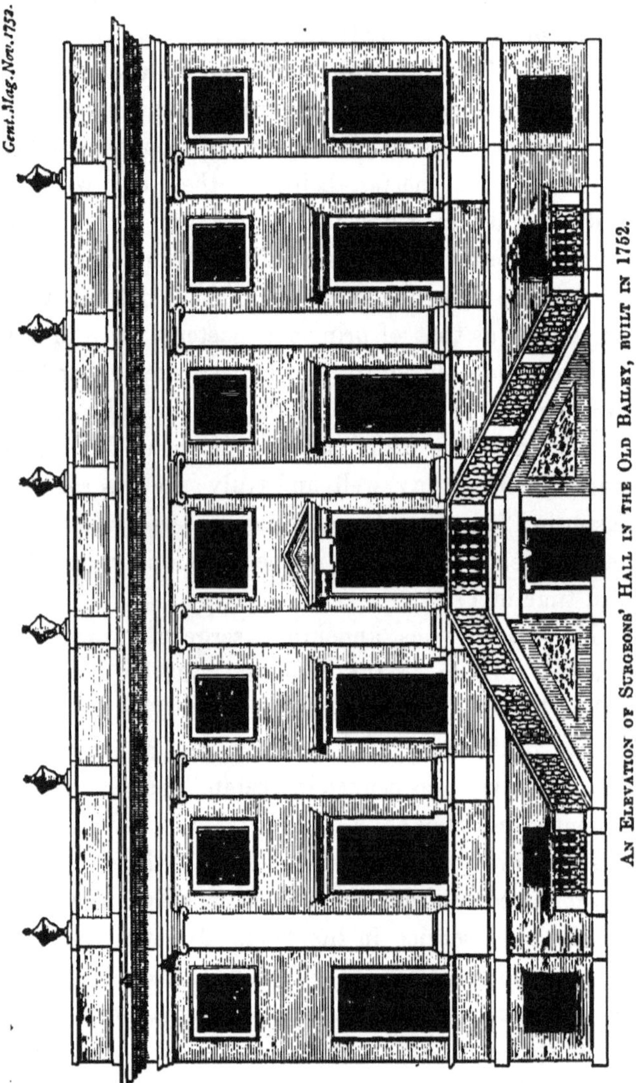

Gent. Mag. Nov. 1752.

An Elevation of Surgeons' Hall in the Old Bailey, built in 1752.

which one is inclined to agree, on hearing that Dr. Jurin assured his lordship that it was four times stronger than the strongest soap lye.

The surgeons' theatre.

The surgeons, when they separated from the barbers, seem to have carried absolutely nothing away with them except the Arris and the Gale bequests, to which they were clearly entitled. Hall, library, plate, everything remained with the barbers, and the new Company had to make an entirely fresh start for themselves in the world. As soon as possible, therefore, after their foundation, negotiations were entered into to secure a proper meeting place for the members. A site in the Old Bailey was leased from the City authorities, at the comparatively low rental of £53 6s. 8d. per annum, and the Court of Assistants ordered "unanimously that a theatre be the first part of the new intended building, and that the same be erected with all possible dispatch." The theatre was built ("the ground being first piled") from designs originally furnished, as it appears, by Mr. Kent, though it was erected under the superintendence of a Mr. Jones, "who was chosen our Surveyor, and who has full power to light the Theatre after his own manner, and give Orders for such proportions in all parts of that Building . . . without being controuled by any power but that of a Court of Assistants." He was "to receive fifty Guineas for his former and future Care of this Building, by making drawings and measuring." He was not, however, so successful as his illustrious namesake, Inigo Jones, who had built the theatre for the United Company in 1636 in such a way as to make it one of the sights of London. Workmen of the last century appear to have borne a close resemblance to those of our own time, for "it was resolved that when the Scaffolding is erected for the Plaisterer's work at the Theatre, Mr. Steere be desired to re-measure the Bricklayers' and Carpenters' Work already done . . . . and make his report thereof in writing to the Court . . . . and if it shall appear that any person employed has wilfully defrauded the Company, they be Immediately discharged all further service." This suspicion of roguery, however, was not confirmed, as the suspected persons continued in the employment of the Company. Every effort was made to urge on the completion of the theatre, and all other building was postponed till it should be ready for use. The cost of the buildings erected amounted to £4,000, which the Company raised by the issue of bonds to those members who chose to take them, bearing interest at the rate of 4 per cent. per annum.

Provisions for teaching anatomy.

The first Court of Assistants was held in the newly built theatre in August, 1751, though it was not until 1753 that the first masters of anatomy were selected. The result of the first election was a brilliant one, Mr. Pott and Mr. Hunter being the masters, Mr. Crane and Mr. Paul the wardens, and Mr. Hewitt and Mr. Minors the stewards of anatomy. Of these officers, Mr. Paul desired to be excused on account of ill health, and was exempted on paying his fine of £21; whilst Mr. Crane desired to be excused as he had been fined for not serving the same office in the old Company; his excuse was held to be valid.

As soon as these elections had been made, the Court proceeded "to consider of the Disposal of the Bodies of three Persons who, it was expected, would in a few days be Executed for Murther, and sent to the theatre pursuant to the late Act of Parliament (1752), when It was Resolved that one of them should be delivered to Mr Hewitt to be by him dissected. And he be desired to make such Preparations of the bones as to render them useful in any future Osteological or other Lectures, and to return them after they had been so prepared to ye theatre. And that the Masters, Wardens, and Stewards of Anatomy be desired to dissect the other two, and to make such preparations thereout, either Muscular or Vascular, as may hereafter be of service at any Public Lectures." At the next Court "The duties of the Masters, Wardens, and Stewards of Anatomy lately elected were determined, and the Court came to the several following Resolutions, Vizt. That it be the particular Duty and Business of the Masters of Anatomy (unless otherwise ordered) to read such Lectures in Anatomy as shall be appointed by the Court of Assistants. That it be the particular Duty and Business of the wardens of anatomy to be Demonstrators to the Lecturer, and to take Care that everything be Conducted during the time of the lecture with Decency and Order. That it be the particular Duty and Business of the Stewards of Anatomy to dissect and prepare the Bodies for the Lecturer."

The utmost difficulty appears to have been experienced in getting persons to fill these offices, for even after the first election the Court had "to insist that Mr Hewitt the Junior Master of Anatomy do read the lectures," and the clerk was "ordered to endeavour to procure one of his colleagues to read them in case of Mr Hewitt's refusal or non-attendance;" and finally, it was "ordered that the master and wardens have power to dispose of the Body in case no Person can be procured to read the lectures." The excuses offered to the Court to evade these duties were as varied as they were ingenious. Thus "Mr Dowdall, who was elected one of the Stewards of Anatomy [in 1762] . . . requested that he might be excused either from serving or fining for such Office, on account of his having been called upon to serve several offices in the Barbers' Company; and likewise of his having quitted the business of surgery, and residing entirely in the Country. Whereupon, being desired to withdraw, the Court took such his Application into Consideration, and Resolved that the reasons urged by him in support of his request were not sufficient to induce the Court to comply therewith." The other steward of anatomy on this occasion requested "That his serving of the said office might be dispensed with for the remainder of the ensuing year, on account of some particular Law Business he was engaged on. . . . The Court, having duly considered his request, were unanimously of Opinion that it could not be complyed with, Of which resolution he, being called in, was acquainted by the master." In July, 1776, Mr. David

Irish, who had been elected one of the stewards of anatomy, desired that "on account of his being afflicted with the Gout, and the Distance of his residence, he might be excused from serving the said office for one year. Resolved that the clerk do acquaint the said Mr Irish that as there will not be any business for him to do till Michaelmas next, they hope his disorder will be removed. And that his place of residence is not at a greater distance than that of Several other Members who have been chosen into the same office, nor will be less another year, and . . . they expect he should abide by the Election, and either Serve the said Office or ffine for the same." At another Court "the clerk read a letter from Mr Francis Tompkyns . . . desiring to be excused on account of his being in the army on half-pay; and the same being read it was ordered that the clerk do acquaint the said Mr Tomkyns that this Court can't admit of his plea, and expects he should either serve or ffine for the said office." So great was the difficulty experienced in this matter, that at length it became the custom to elect a number of persons to be successively stewards in the place of these defaulters, and in this way alone could the succession be ensured.

In 1776 it was decided to elect a committee "to consider of an effectual Plan for improving the Lectures, and rendering them more creditable to the Company and more easy to the Members." As a result of the labours of this committee the two masters of anatomy were abolished, their place being taken by "a member of the Company of known abilities, who should be annually appointed, under the character of professor of anatomy." His office was to teach anatomy three days in each week throughout the year to all such of any profession who should enter with the consent of the master and wardens. The salary attached to the post was £120 per annum, with a share in the profits derived from the fees of the pupils.[1] He was further required to read six lectures on the muscles and six lectures on the viscera between Michaelmas and Lady Day, the lectures, as it is expressly stipulated, "to be read in a gown." Mr. Henry Watson was elected the first professor of anatomy in 1766, and he was succeeded by Mr. Joseph Else. As might have been expected, this change did not meet with universal approval. "A memorial was presented to the Court of Assistants by Mr Arnaud, one of the present Masters of Anatomy, setting forth his apprehensions that if, in consequence of the late resolution of the Court to elect a Professor of Anatomy, he should be excluded from reading Lectures, his reputation in his profession might be injured. And that as he had received notice to prepare Muscular Lectures, and had prepared them accordingly, his not reading them might be prejudicial to him." Mr. Arnaud, therefore, had leave given him to read his lectures between Michaelmas and Christmas, provided any subject could be obtained within that period.

[1] The details of this scheme, which probably originated with Cheselden, will be found in Appendix P, page 378.

Amongst the bodies brought to the theatre for dissection was that of Lord Ferrers, who was executed for the murder of his steward in 1760. In reference to this event the following minute occurs in the books of the Company: "The Master having laid before the Court a Letter he had received from Lady Huntingdon in respect to the treatment of Lord Ferrers body,[1] And the same and part of the late Act of Parliament relating to Murderers having been read . . . It was resolved that it be left to the Master and Wardens to dispose of the Body of Lord Ferrers. And to give such directions in respect to applying for a Guard and other particulars as they shall think proper." The guard here mentioned was probably for the purpose of keeping order in the hall during the exposure of the body after the execution.[2] It does not appear that any dissection was performed upon him, and he was buried in Old St. Pancras Church.

The army and navy surgeons.

During the earlier years of the Company's separate existence they were constantly disputing with surgeons who had retired from the army and navy, and who sought to make practices for themselves without taking the diploma of the Company. The old army surgeons

[1] The sheriff, Vaillant, had promised before the execution that the Earl's body should not be stripped.

[2] Of this ceremony there still exists a print, lettered "Lord Ferrers, as he lay in his coffin at Surgeons' Hall." It appears that the public were admitted to the hall to view the bodies previous to their dissection, in much the same manner as is common in the Morgue at Paris. Special provisions were made for preventing the undue intrusion of the mob, as may be read in Appendix Q, page 381.

sheltered themselves behind an Act passed in the second year of the reign of George the Second, entitled "An Act to enable such officers, mariners, and soldiers as have been in his Majesty's service since his accession to the throne to exercise Trades." In 1782 the matter was set at rest by the Company stating a case for the consideration of the law officers of the Crown, who decided against them.

The examination of surgeons for the army and navy long continued an integral part of the functions of the Company, and in its capacity of examiners for the services two persons of note came before it. The one, in real life, was unsuccessful; the other, in fiction, was happier. "Oliver Goldsmith presented himself in a new suit (not paid for) to be examined as to his qualifications for being a surgeon's mate, on the 21st December, 1758, and in the minutes of the Court of Examiners held on that date the entry occurs: James Bernard, mate to an hospital; Oliver Goldsmith, found not qualified for ditto."[1] The second and no less famous person, although non-existent, is Roderick Random, who has left us a most interesting and graphic description of what passed at the examination; an account which is the more interesting when we reflect that his creator, Smollett, had received his qualification as surgeon's mate from the Company, probably after undergoing a somewhat similar ordeal.[2]

[1] "Life and Times of Oliver Goldsmith," by J. Foster, vol. i. p. 166. Ed. ii.

[2] "Works of Tobias Smollett," vol. i. chap. xvii. pp. 97 *et seqq.* Ed. v. in six vols.; Edin. 1817.

Finances of the Company.

The Company does not appear to have done more than pay its way during the earlier years of its existence, for in 1780 it became almost insolvent, owing to the bankruptcy of its clerk, to whom the Company had incautiously advanced £300; a large sum, considering that its available funds at the time only amounted to £518. The clerk resigned, and Okey Belfour was appointed in his place. At the first audit after this change the Company's balance was £75 2s. 6d. Under the able management of Mr. Belfour the former position was rapidly regained, and was soon far surpassed, since in 1795, within fifteen years of this date, the Company had £15,000, and the receipts during the quarter had amounted to £1,750. This large sum was in great part acquired by the energetic action of the clerk, who compelled all the members of the Company punctually to discharge their quarterly dues; whilst the examination fees constituted a steadily increasing revenue. The affairs of the Company appear to have been grossly mismanaged before the appointment of Mr. Belfour, for in 1781 it was "ordered that the Clerk do search in the Books of the Company, and make such other enquiries as may be necessary respecting the donation of £510 given to this Company by the late Alderman Airis (Arris), to see how the same has been disposed of and what is become thereof, And report the result of such inquiries to the next Court of Assistants." The very name of the worthy alderman was in danger of perishing, and his benefaction, which had been made over to the surgeons at the separation of the united Company, had been misappropriated; but not beyond recall, for it still exists, and forms a part of the stipend of the Arris and Gale lectures, which are delivered yearly. The clerk, as a result of his investigations, reported "that in the Company's Accounts for the year 1746 he found an Entry that the sum of £510, with a further sum of £15 17s. 1d. for Interest thereon, was received from the Barbers' Company, in pursuance of the directions of the Act of Parliament."

In 1786 a general meeting of all the members of the Company residing within seven miles of London was convened to receive the pleasing intelligence that as the debts of the Company incurred for building the theatre and other works had been paid, the quarterly dues would in future be reduced to half-a-crown. To ensure punctuality at the meetings of the Court of Assistants, it was decided as early as 1746 "that every person who attends punctually at the time of his Summons at any Court of Assistants before St. Paul's clock strike the hour mentioned in such Summons shall have half-a-crown apiece. And the same rule shall be observed at any Court of Examiners when it shall happen there shall be no examination for the Grand Diploma." In 1787, as the finances of the Company became more flourishing, this sum was raised to half-a-guinea. In 1782 the same means were taken to ensure punctuality at lectures, for it was decreed that "every Member of the Court of Assistants who should attend at any of the Lectures hereafter to be read at the Theatre should be entitled to and receive the Sum of five shillings out of the cash of the Company."

The rules of the Company in regard to translation were strictly observed. In 1784, "William Osborn, a Member of this Company," obtained a licence from the College of Physicians to practise midwifery without being disfranchised from the Company or previously obtaining the leave and consent of the Court. By this action he incurred a penalty of £20, which he declined to pay, endeavouring to justify himself. Counsel's opinion was taken upon the subject after much debate, and the matter then dropped, probably because the opinion was against the Company.

John Hunter.

Occasional entries occur in the minute books relating to John Hunter. In October, 1786, "the Clerk reported that he had received from Mr John Hunter 3 Books as presents from him to the Company, two of them being the Natural History of the Human Teeth, and the other a Treatise on the Venereal Disease, with a letter from Mr. Hunter, which was read. It was resolved that such present be accepted, and that the thanks of this Court be given to Mr. Hunter for the same." In November, 1793, "The Master reported that since the last Court of Assistants, Mr. John Hunter, a member of the Court, had departed this life, and declared a vacancy in the Court of Assistants in pursuance of such death." The vacancy thus caused was filled by the election of Mr John Heaviside. Originally both brothers were members of the corporation, but in 1756, "Dr. William

Hunter, a Member of this Court [of Assistants], was desirous of being disfranchised on such terms as the Court should agree on." It was thereupon "ordered that the Clerk do deliver him the Instrument of Disfranchisement under the Seal of this Company, on his paying down 40 Guineas for the same." It does not appear, however, that this payment was ever made, as in 1758 Dr. Hunter paid a fine of £20, "being the penalty he had incurred by becoming a member [*i.e.* licentiate] of the College of Physicians without the previous consent of the Court." It appeared, however, that the doctor was ignorant of the bye-law which enforced this point, and his fine was therefore remitted to him.

Mr. Gunning.

At the election in 1789 Mr. Gunning was declared master of the Company, in place of Mr. Watson; he had for his wardens Mr. Grindall and Mr. Lucas. Mr. Gunning had been elected steward of anatomy in 1773, but had paid his fine rather than serve; in 1789 he had been elected an examiner, to fill the vacancy caused by the death of Percival Pott. He was, no doubt, an active and sturdy opponent of abuses, but until the year 1789 no signs of his activity remain. His accession to the mastership, however, was signalised by a firm effort to re-organise the Company. He first attacked the system of dining in the following memorial to the Court of Assistants: "The expenses of your annual dinner, as well as that of the other Dinners held on the Meeting of your two Courts, the Court of Assistants and the Court of Examiners, having been of late years so increased as to make an Enquiry into Them necessary, to retrench for the present and to prevent an accumulation. . . . To begin with the dinners provided for the entertainments of these two Courts. . . It appears that they have been summoned to such Dinners from no printed or written Authority, but by the bare discretion only of the Masters and Wardens. That they have been unlimited in number. That whenever they have met on ye business of the Company a Dinner has been ordered of course. . . . That the highest number of Meetings within the year has been 25, and the lowest 13, each of which has been attended by an Expensive Dinner at the sole charge of the Company. That from this mode of proceeding as many Members have been summoned and as many Dinners directed for those members, Strangers, and others, as the Master and Wardens have thought proper. That for many years, from the Establishment of the Company in 1745, the Members belonging to the two Courts were content to be entertained at the moderate Expence of 4s. and 5s. per head, and less. That in the years 1774 and 1775 it exceeded 10s. per head; that it was contracted again in 1779 and 1780 to 8s. per head, and that in the subsequent year it fell back to 10s.; and that from the years 1781 and 1782 to the years 1787 and 1788 it has been increasing rapidly from 10s. to 19s. . . . That during the last 8 years and one half the Expence of these Dinners for the Court of Assistants and the Court of Examiners only have exceeded the sum of £1,300. . . . Respecting the meeting of the Master and Wardens or

Audit, it is an Appointment for the regulation of the Annual Dinner only, and entertains the four Gentlemen concerned, the Master, Wardens, and Clerk, at the Expence of between 20s. and 30s. per head." The result of this memorial was that "the same having been read and debated, it was moved and seconded that for the future there shall be no more than Twelve Dinners in the course of the year at the Expence of the Company, that is to say, four for the Courts of Assistants and Examiners, and eight for the Court of Examiners only; and that for the future the Company shall not in any case pay more than £12 12s. for the Courts of Assistants and Examiners together, nor more than £6 6s. for the Dinner of the Court of Examiners only. And that if the expence upon any occasion shall be greater than these sums, the surplus shall be paid by the Gentlemen themselves."

At the end of Mr. Gunning's year of office he delivered a philippic, which is so interesting, and at the same time so important, as showing to how low a condition the management of the Company had fallen, that I have thought it worth while to transcribe the whole, lengthy as it is.[1] In this scathing address the following lines occur: "You have a theatre for your lectures, a room for a Library, a committee room for your Court, a large room for the reception of your communities, together with the necessary accommodations for your Clerk. . . . Your Theatre is without Lectures, your Library room without books is converted

[1] Appendix R, page 382.

into an office for your clerk, and your committee room is become his parlour, and is not always used even in your common business, and when it is thus made use of it is seldom in a fit and proper state." The reproof was taken in good part by the Company, and a committee was appointed to inquire into the truth of the allegations, and a series of resolutions were ultimately embodied reforming the more flagrant abuses.

Professorship of Surgery.

In 1790 it was deemed expedient to appoint a professor of surgery, and Mr. John Gunning was selected as the first occupant of the new chair, with Mr. Abernethy as his colleague in the chair of anatomy. After holding the professorship of surgery for a short time, Mr. Gunning resigned it, on the plea that it occupied too much of his time, and there is no evidence to show that any new appointment was ever made.

In 1793 Pennell Hawkins, a former master of the Company, died, and Mr. David Dundas, of Richmond, was appointed to be one of the king's sergeant-surgeons in his stead. Mr. Dundas being only "a surgeon and apothecary," or, as we should now say, "a general practitioner," his appointment to so important a post gave great offence to the Company. By the traditions of the Company the person appointed sergeant-surgeon was elected into the Courts of Assistants and Examiners at the first vacancy, and was afterwards made principal master of the Company at the election next ensuing upon his appointment at Court, unless he had already passed the chair. By a bye-law of the corporation, however, it was enacted that "no person practising as an Apothecary, or following any other trade or occupation besides the profession or business of a surgeon, shall be capable of being chosen into the Court of Assistants, or if he be one of the Court of Assistants, be eligible to the still higher office of Master." The Company was therefore in a dilemma. A special Court of Assistants was called, and, as might have been expected, the exclusive party carried the day, and it was ruled that no apothecary could hold office in the Company, and that Mr. Dundas was ineligible for the posts of assistant, examiner, or master. Mr. Dundas protested against the decision, and the Company took steps to defend themselves in case the matter should be tried at law; but it was never followed up, and we may therefore suppose that Mr. Dundas was content to hold his appointment without intruding upon the Company.

The following notice, occurring in the books of the Company during the year 1793, carries our thoughts to France during the period of the Revolution. "The Clerk produced a letter from the parish officers of S[t] Martin's, Ludgate Hill, requesting a contribution from the Court towards the relief of the French refugees. . . . Resolved, that as the Members of the Court have all of them contributed to that fund at their respective residences, they do not think fit to comply with the request contained in such Letter."

On the 19th of May, 1796, "The Master informed

the Court that in consequence of a Survey and Examination made sometime since by M[r] Neill, a surveyor, called in for that purpose, it appeared that the Hall and Theatre were very much out of repair, and that the first Estimate for these repairs exceeded £1,600. That the Tenure by which they are held is only about 55 years, subject to a ground-rent and taxes amounting to £240 a year. It had frequently been a subject of Consideration among the Members of the Court of Examiners whether it would not be for the benefit of the Company to dispose of the Hall and Theatre, and to erect new premises upon freehold ground." This report of the surveyor appears to have decided them upon the subject, and directions were accordingly given to the clerk to sell them by public auction, if a profitable bid could be obtained. A committee of six members, endowed with plenary powers, was selected to carry out the business. The property was offered for sale, but in July, 1796, Mr. Gunning reported from the committee that as no one had bid within £200 of the price fixed upon, the premises had been bought in upon the Company's account. At the same court at which Mr. Gunning made this announcement Mr. Cline was elected a member of the Court of Assistants, in place of William Walker, who had died whilst holding the office of warden. This court, held on July 7th, 1796, is remarkable in that it terminated the legal existence of the Corporation of Surgeons.

The Corporation of Surgeons annulled.

By the Act of Incorporation, 18th Geo. II., the

Court of Assistants of the Company was to consist of a master, or chief governor, and two governors, or wardens, with other members, of whom it was enacted that the master and one governor, together with one or two members, should form a court for the despatch of business. It happened that William Walker, one of the governors, died in May, 1796, whilst the other governor, John Wyatt, was lying blind and paralysed in Warwickshire, and though his son was sent, at great expense, to bring him to London, he was too ill to be moved. At the meeting of the court, therefore, there were present the master, William Cooper, and seventeen members of the Court of Assistants, but not one of the governors, as ordained by the Act. The meeting was therefore not legally a court. The persons present, however, determined on proceeding to business, and, as just mentioned, they elected Mr. Cline into the court, besides transacting the usual business of an election day. This was not, however, the first occasion that such an informal court had been held, for a similar occurrence took place at two successive courts in 1762, when the master was absent, though both wardens were present. In 1784 the corporation had become more careful, for "A Quarterly meeting was summoned to meet at the Theatre as usual on the 1st day of April, 1784. But neither of the Wardens being present, and it being Impossible to hold a Court without the presence of one of them, the Master adjourned the Court to the Shakespeare Tavern, in Covent Garden, to be there assembled Immediately, and the Beadle was sent to desire the attendance of the Wardens, or one of them. Whereupon both of the wardens immediately came to the Shakespeare tavern, and in consequence thereof the usual quarterly meeting was held." In 1796, on the last occasion of holding an informal meeting, the Company soon found that they had got into a very serious scrape, and on laying a case before counsel, there was no doubt that their corporation was destroyed by the illegal construction of the Court of Assistants.

Endeavour to reconstruct the Surgeons' Company.

A bill was therefore brought into parliament to legalise those acts of the corporation which, though not morally, were legally wrong, and to give the corporation greater power over the profession. This latter attempt excited great wrath in those who practised without the diploma of the corporation, and a violent opposition to it was set up. It is probable, however, that the opposition would have been overcome, for the bill had passed through the Commons and got into committee in the Lords, where it was lost by the influence of Lord Thurlow, owing, it is believed, to the hatred he bore the Surgeon-General Gunning, who, in reply to that noble brute's observation, "There's no more science in surgery than in butchery," had promptly and spiritedly answered: "Then, my lord, I heartily pray that your lordship may break your leg, and have only a butcher to set it, and then you'll find the difference between butchery and surgery." Well, the bill went into committee, the Bishop of Bangor (Dr. J. Warren) was in the chair, and

Thurlow, soon finding that he was not strong enough at that time to oppose, urged on the honest churchman the propriety of further consideration of the bill at a future day, hoping then to carry his point. The college secretary (Belfour) being present watching the bill, and guessing the object of Thurlow, urged the bishop to bring the matter at once to a decision, especially as he had the Book of Numbers in its favour. "Why," said the bishop, "you don't suppose my Lord Thurlow will play me a trick, do you?" "Tricks have been played in this House, my lord," was the secretary's quick reply. The bishop, however, yielded to Thurlow's suggestion; the consideration of the bill was deferred, and when it was next taken up in committee, on July 17th, 1797, Thurlow had taken care to be better supported, and threw it out by carrying his motion, that the third reading of the bill be put off to this day three months.

The ready passage of the bill through the Lower House was, no doubt, in great measure due to the able advocacy of Mr. Erskine, "who, as being of Counsel for the Court, had declined bringing the bill into the House of Commons, but had signified his readiness to assist the Court and the Company, and by all the means in his power to support and promote the success of the Bill." For these services he would take no fees. On the refusal of Mr. Erskine to introduce the bill, Mr. Mainwaring undertook to do so, but being unexpectedly prevented from attending at the proper time, Mr. Rose actually laid the bill before the House. At the same time, Mr. Earle obtained from the king the privilege that the college should be called The Royal College of Surgeons.

New premises.

After the failure of the sale by auction, the Company cast about for some means of getting rid of their property in the Old Bailey, and it was sold to the City authorities for the sum of £2,100. In the meantime they had purchased for £5,500 a freehold house in Lincoln's Inn Fields, belonging to a Mr. Baldwin. Part of the opposition in the House of Lords to the new bill was based on the fact that the Company occupying these premises being so far from the usual place of execution, great inconveniences, it was apprehended, would arise from the conveying of bodies through the streets, and dissecting them in the neighbourhood of Lincoln's Inn Fields. To obviate this objection, a clause was added to the bill giving power to the Company "to provide a convenient house or building as near as can be procured to the place of execution, to which house the bodies of all persons who shall be executed for murder, and shall be sentenced to be dissected and anatomised . . . according to the Act entitled, An Act for the better preventing the horrid crime of Murder, shall be conveyed." The clause, however, did not save the bill. In the meantime, as the Company were not provided with a place for dissection, "Mr. Chandler," in July, 1797, "a member of this Court, in the most polite and ready manner offered his Stable for the Reception of the Bodies of the two murderers who were executed last month: which offer the Court thankfully accepted."

Consequences of the rejection of the new bill.

The loss of the bill entitled "An Act for erecting the Corporation of Surgeons of London into a College, and for granting and confirming to such College certain rights and privileges," exposed the Surgeons' Company to great mortification, and not a few insults. Many of the members of the Company declined to pay their quarterly dues, and the routine of business came to a standstill, since no assistants or examiners could be elected. The examination of the navy surgeons was undertaken by the Sick and Hurt Office, and the Company found itself in a most deplorable and degrading situation. As an instance of this, "Mr. Lucas informed the Court on Oct. 5th, 1797, that a gentleman who had been examined at a Court of Examiners on the 7th day of September last, and had been passed for a 2nd mate of a third rate, had called upon him, and informed him that the day following on his going to the sick and hurt office for his qualification he had been required to undergo, and had undergone, another long examination in Surgery there, and that the Commissioners of that office had certified him to be qualified for a first mate of any rate, and that he had actually gone to sea in that capacity."

Foundation of a College of Surgeons.

At a Court of Assistants held at the Company's house in Lincoln's Inn Fields on November 22nd, 1797, the master reported that a committee from the Company had met a deputation from the committee appointed by the members who had opposed the bill, and had received from them the conditions upon which they were inclined to withdraw their opposition, and to co-operate with the Company in obtaining a new Act of Parliament. The names of the opponents do not appear. The chief clauses in the proposed new Act were that the Company be converted into a college, with a council of thirty members, a president, and four vice-presidents. That the council were to be divided into three examining committees: one for surgery, a second for midwifery, and the third to examine the army and navy surgeons in medicine. That the jurisdiction of the college should be unlimited in point of area, and that all practitioners in England and Wales should be subject to the examination of the Court of Examiners or to a deputation from that body. That the lectures on anatomy and surgery should be on a more extended scale, and that there should be no disqualifying bye-law so far as regards the practice of midwifery and pharmacy. That a library and museum should be formed, and that surgical transactions should be published periodically. The members who petitioned against the former bill thereupon consented to contribute towards the expense of supporting Mr. Hunter's museum.

After numerous committee meetings and much deliberation, it was proposed to present a new bill during the ensuing session of Parliament, when it occurred to a member of the Court whose name is not recorded that "a charter from the Crown will be preferable to a Bill in Parliament." A petition for this purpose was therefore presented to the king through the Duke of Portland, and a draft of the intended charter was submitted to the consideration of the Attorney and Solicitor-General. Some opposition, however, was made to this proposal, for "a caveat was lodged by the committee of members who had opposed the late Bill," but it was unsuccessful.

On March 22nd, 1800, the Royal College of Surgeons in London was established by a Charter of George the III., which reinstated the Company in its former position on condition that it resigned its municipal privileges. The titles of master and governors, however, which had belonged to the older corporations, were retained, and only gave place in 1821 to the more high-sounding titles of president and vice-presidents when the college received a supplemental charter from George IV. It was not until the year 1843, by a charter from the Queen, that the name of the College was changed to that which it still retains of The Royal College of Surgeons of England, with its present constitution of President, Vice-Presidents, Council, Fellows, and Members.

APPENDIX A.

ORDINANCE OF THE BARBERS THAT NO UNLICENSED PERSONS SHOULD ACT AS BARBERS WITHIN THE CITY OF LONDON. TWO OVERSEERS OF THE BARBERS TO BE APPOINTED.

On the sixth day of October, in the 49th year of the reign of King Edward the Third, after the Conquest [A.D. 1375] John Warde being Mayor. To the honorable and wise Lords the Mayor and Aldermen of the City of London; the good folk Barbers of the said city show that from day to day there come from Uppelande,[1] Men, Barbers, little skilled in their craft, into the said city, and take houses and intermeddle with barbery, surgery, and with the cure of other sicknesses, whereas they know not how to do such things nor ever were qualified in that craft to the great damage and cheating of the people and to the great scandal of all the honest barbers of the said city: wherefore the said good folk pray that it would please your honorable lordships for God's sake and in the work of charity to ordain and establish that henceforth no such stranger coming into the said city from Uppelande or from any other part, whatever be his condition should occupy house or shop of Barbery in the city itself before he has been found hable and skilled in the said art and craft of barbery and that by trial and examination of the good folk barbers of the city itself. And that you would please to ordain and establish that from henceforth there should be for all time two honest persons of the said trade chosen by common assent to be guardians of the said craft. That these two should be presented to the Mayor, Recorder and Alderman of the said city and sworn before them well and loyally to rule their mystery to the best of their power and skill. And that the masters should oversee the tools of all the

Ordinance of the Barbers.

[1] The general name for country places (Riley).

said art that they be good and fitting for the use of the people to avoid the peril which might happen. And that on the complaint of the two masters all rebels from the said craft shall be made to come before you and whoever shall be found in defiance of this ordinance shall pay to the chamber xl^{s}. And that henceforth no men of this craft shall be received into the franchise of the city if he have not witness for honesty and ability by good examination before you, and that no stranger shall hold house or shop of this craft within the said city nor within the suburbs thereof. And that this ordinance be enrolled in the chamber of the Gyhalle [Guildhall] of London to endure for ever.

Which was granted to them.

And therefore Laurence de Westone
John de Grantone

were chosen Masters of the Barbers and were sworn well and truly to govern their mystery etc and duly present defaults.[1]

[1] Letter Book H, fol. 27*b*.

APPENDIX B.

REGULATIONS FOR THE GOVERNMENT OF A CONJOINT COLLEGE OF PHYSICIANS AND SURGEONS IN THE CITY OF LONDON.

The Ordenaunce and Articles of Phisicions withinne *the* Cite of London and Surgeons of *the* same Cite.

THE xv Day of may ye yere of kynge Henry ye sixte after ye conquest ye first (1423), Maister Gilbert Kymer maistre of art Doctour of medicyns and Rectour of medicynes in ye cite of London maistre John Sumbeshete Comensour in medicyns and maister Thomas Southwell Bacheler in medicyns petic*ioners* Surueiours of ye faculte of Physik in ye same Cite: And Thomas Morstede and John Harowe ye two maistres of ye crafte of cirurgy with alle ye cirurgeans wirkyng in ye crafte of cirurgy Withinne London comen here and putten up to ye maire and Aldermen a bille or a petic*i*on in Englissh concenyng ye honeste of ye ffaculte of Phisyk and ye honeste of ye crafte of cirurge and ye co*mmon* proffit of ye cite in these Wordes

Noble Lordes forasmoche *th*at ye glorious konnyng of Phisyk and the crafte of Cirurgy er[1] fro Day to Day gretlich Disclaundred and so*rro*wfully skorned and grete parte of ye peple spillide[2] be Wreeches and *pre*sumptuous practisours in phisyk nought knowyng ye treuthe or ground of *tha*t ffaculte of phisyk And be Unkonnynge Wirkers in cirurgy nought knowyng ye welbe*ing* [of the] crafte of cirurgy, like u*n*to your lordships for ye Disclaundre of so high a ffaculte of Phisyk and so worthy a crafte of cirurgy to be putte awey And that mankynd be nought begiled from hennes forward bi ye Disceites of Unkonnynge practisours in phisyk and unkonnynge Wirkers in ye crafte of cirurgy withinne ye boundes of your ffraunches[3] to stable yordinance underwriten eu*er* more to be obse*r*ued

In ye first please it you Ordeyne *th*at fro hennes forward [for]

[1] are. [2] failing. [3] liberties.

alle Phisicians and cirurgeans withinne ye libertees of London practisyng in Phisyk and Wirkyng in cirurgy as oon Comminalte be oon Rectour of Medicyns and two Suruciours of ye ffaculte of Phisyk and two Maistres of ye crafte of cirurgye mowe[1] be gouerned in maner and fourme suynge[2] That is to say that oon place be hadde withinne ye cite of London contenyng atte lest[3] thre howses seuerall. Whereof oon be rehaued and desked for redyng and Disputacons in Philosophye and in medicyn And that other for congregacions eleccions and counseils of alle phisicians practisyng in Phisyk for all manner of articles to be decided oonly pertenynge to ye ffaculte of Phisyk And ye third for congregacions eleccions and counseils of alle cirurgeans wirkyng in ye crafte of cirurgy for all manner of articles to be dealed oonly pertenynge to ye crafte of cirurgye So that the Rectour of medicynes be at bothe if he be present in towne as president and Rewler And if he be not present than ye ffaculte of Phisyk and ye crafte of Cirurgy procede as he ware present eche by hymselfe oonly in alle manner of poynts that longeth to ther konnynge.

Also please you to ordeyne that ye said comminaltie of all Phisicians and cirurgeans of London euereche[4] yere of hemselfe mowe chose and presente to ye mair of the cite for ye tyme beynge oon Rectour for ye ffaculte of medicyn by hym to be rewled And ye saide phisicians euereche yere of hemself mowe chose and presente before ye mair of the cite of London for ye tyme beyng two Surueiours for ye ffaculte of Phisyk to be gouerned And ye cirurgeans of London euereche yere of hemself mowe chose and presente to ye mair of the cite of London for ye tyme beyng two maistres for ye Crafte of Cirurgy to be gouerned by Alwey excepte that none be chose Rectour of medicyns bot he be Doctour of Medicyns maistre of arte and Philosophie or a bachiler in medicyns of long tyme in Vertu and konnynge approued if eny suche may be found. And if non suche may be found ne be present than ye ffaculte of Phisyk be gouerned oonly by ye Surueiours of ye same ffaculte. And ye Crafte of cirurgy by ye two maistres of ye same crafte eche by

[1] may. [2] ensuing. [3] least. [4] every.

hymself unto tyme *th*at suche a Doctour may be found or a Bachiler. So *th*at ye Bachiler ne occupie y office of ye Rectour langer *th*en suche a Doctour of ye condic*io*ns afore rehersed may be founde Also excepte *th*at non be chose*n* Rectour ne Surveiours of Phisike ne maistres of cirurgye bot *th*et be born withinne ye Roiaume of England wisest ablest and most discrete of long tyme also in vertu and konnynge yproued

Also please it you to ordeyne that no *per*sone withinne ye liberte of ye Cite of London *pre*sume in eny wise to *pra*ctise in Phisyk *on*lesse *th*an he be examined and found able *th*ereto bi ye Rectour and two Surueiours of Phisyk and ye holer partie of *th*at ffaculte And *th*an admitted bi ye mair and Aldermen on peyne of C^s^ to paie to ye Chambre of Gyldhalle for ye co*mmon*e proffit of ye cite [and] *th*at no *per*sone withinne ye saide liberte of London *pre*sume to wyrke in ye Crafte of Cirurgie o*n*lesse *th*an he be examined and found able if to be ye seide Rectour in medicynes and two maistres of ye Crafte of Cirurgie And ye more and holer *par*tie of ye same Crafte And than admitted be[1] ye Mair and Aldermen on ye peyne before rehersed paiable in ye same maner

Also please it you to ordeyne *th*at no Phisician withinne ye ffraunchise of London resceive no cure upon him Desp*er*ate or Dedly bot he showe it with alle ye circu*m*stance to the Rectour of Medicyns or to oon of ye Surueiours of Phisyk withinne two or three dayes *th*at it may for the remedy *th*erof if eny be possible be comuned with alle ye Co*mmin*alt*i*e of Phisicians ne do no*th*ing be way of medicyne to no paciente by ye whiche it is lyke to hym or doubte *th*at ye paciente myght stande in *per*ille Also *th*at no Cirurgian withinne ye boundes of London resceive no cure into his hande of ye whiche may folowe Dethe or mayme without *th*at he shewe it to ye Rectour of Medicyns if he be *pre*sent in towne and to oon of ye two maistres of cirurgy withinne thre or foure Days *th*at hit may be comuned with ye Discrete parte of cirurgians nor he shal not make eny kutyng or cauterizac*io*n Where of may folowe Deth or mayme without

[1] by.

*th*at he shewe it afore to ye Rectour of Medicyns if he be *pre*sent in towne and to ye two maistres of cirurgy *th*at it may be comuned With ye discrete *par*te of Cirurgeans for saluac*i*on of ye paciente and worship of ye Crafte of Cirurgy With that the Rectour Surueiours and Maistres aforeseid be alwey redy when that they be required to ouerse and decide ye matters aforesaid without eny thing takyng for her[1] labour on peyne of xx^s Alwey obserued *th*at ye Rectour of medicyns gif no dome[2] in eny case of cirurgie W*ith*out consent of ye two maistres of cirurgie or of oon of hem with ye discrete p*arti*e of ye Co*mmin*altie of ye Crafte of Cirurgyans nor he shal not make non ordeyna*u*nce nor no constituc*i*ons that *per*teneth to ye Crafte of cirurgy withoute ye consent of ye two maistres of cirurgy or of oon of hem and ye discreteur *par*tie of cirurgeans And nether of ye two maistres of ye Crafte of cirurgy Do ye same without hym and he be *pre*sent And ye same article be obserued in ye ffaculte of Phisyk

Also please it you to ordeyn and stable *th*at if eny phisician before ye Rectour of medicynes and ye two Surueiours of Phisyk trewly and lawfully be conuicte of false practicke in Phisyk or of any other open Defawte Disclau*n*dres and worthy accusac*i*on by two or thre trewe men this Doo*e*n anone relac*i*on *there*of made to ye mair of the cite of London he be punysshed by ye saide mair without delay with peyne pecunier or prison or puttynge out fro*m* alle practyk in Phisyk for a tyme or for euermore after ye quantite and qualite of his trespas as to ye mair and Aldermen by examinac*i*on of ye treuthe and informacion of ye Rectour and Surueiours of Phisyk may be found. Also if any Cirurgian before ye Rectour of medicyns and ye maistres of cirurgy trewly and lawfully be conuicte of false wirkyng [etc.].

Also please it you to ordeyne *th*at e*u*ereche seke man nedyng ye practyk of Phisyk or ye Wirkyng of cirurgy fallen in such pouerte *th*at he sufficeth not to make good for ye labours of his Phisician or of his cirurgean if ye cause perteygne to physik if he will pleyne hym to ye Rectour of medicyns or to oon of ye Surueiours of Phisik a gode practisour shal be

[1] their. [2] opinion.

assigned by oon of hem besily to take hede to hym without eny expense resceivyng for his labour upon resonable peyne to be sette by ye counsell of Phisyk And if it be a cause of cirurgy if he will pleyne him to ye Rectour of medicyns or to oon of ye two maistres of cirurgy a good worker of ye same crafte shal be assigned by oon of them besily to take hede to hym without eny expense resceivyng for his labour upon resonable peyne to be sette by the Rectour and ye two maistres of cirurgy and ye holer *par*tie of ye same Alway outake[1] *th*at none of ye Phisicians ne cirurgeans take o*u*er moche more or unresonabely of eny seke bot after ye power of ye seke man and mesurabely after ye deseruyng of his labour.

Also please it you to ordeyne *th*at if eny false medicyns or sophisticate or made untrewly be found to selle be ye Rectour of Medicyns and ye two Surueiours of Phisyk and ye two maisters of cirurgy and two Apotecharyes assigned *there*to in ye Shoppe of eny Apotechary or elleswhere withinne ye boundes of London in e*uer*eche o*uer*seyinge of her medicyns *th*at *th*ei be demed[2] alwey to be caste awey by ye Rectour and ye six *per*sones afore rehersed And ye Apotechary or other seller holdyng with hym suche medicyns be punisshed by ye mair as to ye mair and Aldermen be examinac*i*on of ye trouthe and informac*i*on of ye Rectour and six *per*sons afore rehersed may be found.

Also please it you to ordeyne *th*at non be admitted at London for a graduate man in ye ffaculte of Medicyns into ye Comunaltie of Phisicians bot he bryng before ye Rectour and two Surueiours of Phisyke lettres of Recorde of sufficiente auctoritie or other sufficiente Witnesse of his graduac*i*on And *th*anne nedeful *th*inges done asked before y admission he be admitted to practise And after ye Worthynes of his Degre and tyme of Admission holde a place as other men don*e* in ye conseil of Phisicians so *th*at ye names of e*uer*eche admitted for a graduate man be sent be ye saide Rectour to ye mair.

Also please it you to ordeyne *th*at ye Rectour of Medicyns

[1] except. [2] judged or condemned.

of London and bothe ye Surueiours of the ffaculte of Phisyk and bothe Maistres of ye Crafte of cirurgy in ye time of *th*air admission be bound to swere *th*at inasmoche as in *th*em is alle and *euer*eche constitucions to *th*air office belongynge shal obserue or Do to be obserued alle loue hate fauour and negligence lefte as God and ye holy gospels *th*em shall helpe And ye Rectour be sworne to be indifferente[1] to bothe ye konnynges And also *th*at *euer*eche Phisician in his admission to ye practyk of Phisyk in London before ye mair of ye cite in Guyhall swere *th*at he shall practise in phisyk well and trewly not in gevyng wityngly nocious medicyns to eny man nor consentyng to ye geuer ne he shal entermete of eny sekenes after his trewe estimacion unknowyng to hym in eny maner nor in medicyns doyng or makyng he shal not use eny vigilyng eny evyle sophisticacion or untrouthe so god hym helpe and ye holy Eu*au*ngeiles And if he knowe eny man*ne* use eny untrouthe of ye forseide malices or not admitted to ye practise of Phisyk withinne ye boundes of London hym that he knoweth so to practise in Phisyk he shalle shewe without caryinge to the Rectour of medicynes and to ye two Surueiours of Phisyk and to ther counsel And he shal appere without eny gensayinge at ye callyng of ye Rectour and ye two Surueiours of Phisyk and her[2] counseill in alle maner of causes lawfull and honest to her konnynge *per*tenynge And *euer*eche cirurgean in his admission to ye Crafte of cirurgye without fraude welle and trewly nat in gevyng layng or usyng eny noious medicyns to ye crafte of cirurgye *per*tenynge nor he shall nat entermete of sikenes sore or hurte *per*tenynge to ye Crafte of Cirurgy after his trewe estimacion unknowynge to hym in eny maner nor he shall nat use eny vigilyng eny evile sophisticacion or untrowthe so god hym helpe and ye holy gospels And if he knowe eny *per*sone use eny untrouthe of ye forseid malices or nought admitted to ye crafte of cirurgy withinne ye boundes of London hym *th*at he knoweth so to wirke in cirurge he shall shewe without carynge to ye Rectour of Medicyns and to the two maistres of ye crafte

[1] impartial. [2] their.

of cirurgye and to her counseil And he shalle appere without eny gensaynge atte callynge of ye Rectour of medicyns and of ye two maistres of ye crafte of cirurgy and her counseill in alle maner of causes lawfull and honest to her crafte *per*tenynge Sauynge alwey priuileges statutes and custumes of London comendably vsed.

Also please it you to ordeyne *th*at ye halvendele[1] of ye money comynge of ye peynes that ben recei*v*ed be a Sergeant[2] *par*te ordeyned for forfetes made in ye ffaculte of Phisyk t'one halvendele to ye chambre of ye cite of London And that other halvendele to ye ffaculte of Phisyk as it best semeth to ye Rectour and to ye two Surueio*r*s of Phisyk and to her co*mmin*altie to be don And halvendele of ye mone[3] comynge of ye peynes that ben received by ye Sergeant aforseid for forfetes made in ye crafte of cirurgye tone[4] halvendele to ye chambre of ye cite of London and *th*at other halvendele to ye crafte of cirurgy as it best semeth to ye Rectour and to ye two maistres of cirurgy and her co*mmin*altie to be done.

The whiche bille after *th*at hit was redde herde and diligently understande by ye mair and Aldermen*ne* forasmoche as all and e*uery*e the articles contened *there*inne semed good and honest and acordyng to open Reason Therefor it is graunted by ye forsaid mair and Aldermen *th*at ye forseid articles from hennesforward be holde[n] stedfastly and be kept without varia*u*nce and putte to execuc*i*on Outake alway *tha*t if eny tyme to come hit seme here to *th*is Courte eny article aforseide to be unproffitable or harmefull So *tha*t in alle or in *par*celle it nedeth to be corrected or amended or hoolly to be adnulled be ye discrec*i*on of ye mair and Aldermen for ye tyme beyng be hit lefull whenn *tha*t same article by way of correcc*i*on to adde or take away *tha*t fro or all to putte away as hit to hem most nedefull and spedefull semeth.

This Ordinance speedily came into action, as appears by the following entry :

[1] half. [2] *i.e.* belonging to the City sergeant, an officer of the Mayor.
[3] money. [4] the one.

MAGISTER GILBERT KYMER Rector ffacultatis phisicorum

THOMAS MORSTEDE } Superuisores Cirurgie
JOHES: HARWE }

Jurat: XXVIII die Maii, A° H. VI p° jur.

The next entry in the 3 Henr. VI. presents the swearing in of the rector and supervisors of the physicians:

MAGR: GILBERTUS KYMER Doctor Medicinarum et Rector Medicorum presentatur: per phisicos et cirurgicos et jur: XXVII die septembr. Anno R.R. henrici sexti post conquestum tercio.

MAGR: JOHANNES SUMBRESHEDE } Superuisores Jurati eodem die
MAGR: THOMAS SUTHUELL } et presentantur per phisicos.[1]

[1] Letter Book K, fol. 62*b*, *et seqq*.

APPENDIX C.

REGULATIONS MADE IN THE YEAR 1435 FOR THE GOVERNMENT OF THE GUILD OF SURGEONS.

i. In p*ri*mis *th*at ȝeerli[1] *th*e craft come togidere.
ij. It*e*m of quarterage.
iij. It*e*m of *th*e quarter daies.
iiij. It*e*m noon schal take ano*th*eris cure.
v. It*e*m of gouernaunce of foreyns.
vj. It*e*m of schewynge of curis.
vij. It*e*m of euene porciou*n*s of *th*e maistris auau*n*tag*es*.
viij. It*e*m of foreyns resseyued.
ix. It*e*m of prentisis maid free.
x. It*e*m how noon shal enplete ano*ther*.
xj. It*e*m of visitynge of housholders.
xij. It*e*m how peines schul*en* be moderat & bi whom.
xiij. It*e*m of amending & addicions of *th*e composicion.
xiiij. It*e*m of payment to *th*e dyner.
xv. It*e*m of condiciounis of *th*e boondis.
xvj. It*e*m of *th*e charge and ooth.
xvij. It*e*m of peyne of mysgouernau*n*ce.
xviij. It*e*m of *th*e ooth of *th*e maistres.
xix. It*e*m *th*at noman p*re*sume breke *th*is ordinau*n*ce.
xx. It*e*m of tymes & houris sett of comy*n*ge togideris.
xxj. It*e*m for the*m* that laketh[2] on St. luke is day at masse.
xxij. It*e*m for the*m* *th*at laketh on *th*e day of cosme & damyan.
xxiij. It*e*m for the*m* that [lake] on *th*e day of *pre*sentac*i*on hauy*n*g know[ledge].

[1] ȝ is a corruption of the Anglo-Saxon *g*. It is found in English MSS. written after the twelfth century, and sometimes answers to our *g*, sometimes to *y*, and sometimes to *gh*, and also to a mute consonant at the commencement of a word. In the middle of a word it occasionally stands for *i*.

[2] lacketh, *i.e.* is absent.

In the name of God amen In the tenthe dai of may the yeer of oure lord a thousand cccc & xxxv in the ȝeer of kyng herry the vj the xiii Bi the good advys of the worschipful men of the craft or science of cirurgie in the citee of london & al the com*m*ounalte of the same craft a composicioun or an ordinau*n*ce in this mater is maid & assentid stabilli to stonde euere here aftur

¶ Ffirst that ȝeerli the seid craft come togidere on the dai of Seint Cosme & Damian afore noon & chese hem[1] iiij maistris for the ȝeer as oold custum was to rule and gouerne wel & truli the seid Craft And that tho[2] maistris haue the tresour & comoun godis of the same craft or feloschip i*n* gouernaunce the seid ȝeer and thei to be bounden therefore aft*er* the valu of the seid godis to iiij men of the seid craft chosen bi the assent of al the felowschip And that the iiij men deliuere the seid comoun godis of the craft to the forseid Maistris for the ȝeer whanne thei be*n* bou*nden* therfore And at the eende of the ȝeer aforeseid or in the dai of hem asigned that is to seie euere withinne the dai of seint luke next folowinge the seid maistris to come & ȝeld accountis of the godis aȝen to hem & trewe rekenyng therof & of alle other mercementis & dutees longing to the seid craft of the which it bilongith hem bi this composicioun to gadere & to ȝelde accountis of as it is ordeined & seid of withinne And that tho iiij men aforseid chosen for the felowschip be*n* bounden also to the forseid iiij maistris & ech of hem bi hi*m*silf i*n* vj^{s} & viij^{d} to come at the dai asigned to resseiue & to heere the acounte of the seid maistris Alle maner of costis or expensis doon resonable for the seid craft or felowschip & bi her[3] assent to be alowid hem & the ouerplus to be delyuerid to her successouris if newe be chosen for the ȝeer following aftir And so *th*at euerie ȝeer aftir othir contynueli al the seid craft come togidere at the dai aforseid asigned of Seint Cosme & Damian And if it like thanne the seid felowschip to chaunge her maistris or sum*me* of hem that thanne the seid maistris chese[4] two newe

[1] them. [2] the. [3] their. [4] choose.

of the seid felowschip & after that if it like the felowschip chese thei ij men of the olde And othir wise also if the felowschip like the olde maistris or sum*me* of he*m* for her good gouernaunce to stonde a ȝeer lenger thei to chese tho*se* of hem newe And also if the seid felowschip biseine spedeful to chaunge alle the olde maistris at the ȝeeris eende, or at the two ȝeeris ende to be leueful to hem to chese alle newe i*n* the maner forseid And in al maner wise as it is ordeined or is spedeful in chesinge of the iiij maistris for the ȝeer aforseid in the same maner of wise be the chesing of the iiij men for the felowschip on the forseid dai or withinne viij days aftir And whoso euere hath been maistir two ȝeer togidere that he be not compellid aȝens his will to be stille the iij ȝeer aftir But whatsoeuere he be that is newe chosen & w*a*s not the ȝeer tofore & he that is chosen of the olde i*n* the maner aforseid thei to be presentid & to take her charge withinne X daies aftir the chesinge And whoso eu*er* it be of the seid maistris thus chosen as it is biforseid & wole not obeie hi*m* to h*is* charge of maistirschip withinne the daies forseid that he paie to the box of the craft xiij^{s} & iiij^{d} withouten prouable cause founden of the contrarie And thanne anothir to be chosen in his stede for whiche cause of eithir it be bi the iiij men chosen for the felowschip for the seid ȝeer And also that tho iiij men for the felowschip forseid so chosen & the olde maistris wit*h* hem presente alwei the newe maistris to her charge withinne the seid X daies & ellis ech of hem fauti bi himsilf to paie to the box of the craft iij^{s} & iiij^{d} And i*n* what maner of wise that thei ben chose*n* as it is rehersid aboue that eueri ȝeer thei be newe presentid & chargid as ordinau*n*ce of this Citee is

Of Quarterage

Also it is ordeined & assentid in this composicioun that eueri Cirurgian of the felowschipe i*n* the craft of Cirurgie j
to paie ȝeerli ij$_{d}$ a quart*er* to the box that is viij^{d}
aȝeer to the profit & worschip of the craft in helping and releuying the nede of pore men of the same felowschip And the iiij

Maistris to gadere it or do gadere it of al the forseid felowschip with oon of the iiij men with hem & eithir partie to haue a book to counterolle ech othir of hem And so to the same entent gadere thei alle othir mercementis peynes of fynes maad[1] or doon for ony defautis doon or ony persoone withinne the seid felowschip

Of the Quarter daies

And also it is ordeined in this same composicioun that al the craft come togidere oonys a quarter of dutee that is
iij to seie iiij tymes a ȝere wit*h* outen the dai of chesinge aforsed to heers lerne the good ordinauncis rulis and gouernaunns of the seid craft And as ofte as it is nedeful othir tymes the maistris to calle the seid craft to come togid*e*r thei to come And he that is of the craft forseid & cometh not at the quarter day asigned to paie to the seid box vj^{d} And for ech other tyme that he is warned to come & cometh not that he paie iiij^{d} except resonable cause prouable & due warnyng had afore And if the maist*ri*s warne not neither do warne the seid felowschip to come togidere oonys a quarter of dutee as it is aforseid Or if thei warne hem to come & thei come n*o*t there hemsilf for ech of these ij defautis ech of hem fauti bi hem silf to paie to the seid box ij^{s} & iiij^{d} And for ech othir tyme that thei warne the felowschipe to come togidere & the seid maistris come n*o*t there hem silf that thanne the seid maistris or tho fauti of hem to paie at ech tyme xij^{d} But euermore at alle tymes resonable cause except with due warnyng had afore the tyme And if ony sich cause falle to ony of the maistris or her deputees or to ony of the othire iiij men chosen for the felowschip whether it be to oon or ij of hem of whethir partie that it be the othere of hem to procede for all the othere present for the tyme beinge with deputees present of the seid maistris absent And if ony of hem viij or ony othir of the felowschip be proued untrewe or feyned in his excusacioun or in his absentynge or comynge togidere rehersid afore & aftir bi ij or thre witnessis that he paie double of his peyne sett afore

[1] made.

How noon schal take anotheris cure

Also in this composicioun it is ordeined that noon of the iiij maistris neithir ony othir persoone of the seid
felowschip of the craft of Cirurgie putte ony iv
mán of the felowschip out of his cure Otherwise than honeste of the craft wole but that ech of hem be redi if nede be or by ony of the parties called th*er*eto thanne honestli to helpe ech other wi*th* counseil or dede that worschip profit & honeste of the craft & helpinge of the sike be had & doon on alle sidis & that as wel to be seen to of the nedeful helpers therto of the seid felowschip or of the maistris whether that it be of hem as resoun wole hem to be rewarded And if ony of the seid craft do the contrarie that ech sich doere contente the owner of the cure to the valu of al the cure & ouer that to paie to the box vj^{s} viij^{d} & for his trespas ¶ Or if ony persoone of the seid felowschip of the craft disclaundre or dispraue ony of the felowschip vnriȝwysli[1] or unhonestli if it mai be proued on hi*m* bi two or thre witnessis that he thanne paie to the box iij^{s} & iiij^{d} & ouer that make amendis to the seid persoone that he hath disclaundrid aftir the iugement of the honest maistris & her felowschipe nott fauti herinne

Of gouernaunce of foreyns

Also moreouer it is ordeined and assentid in this composicio*u*n that no freemen of the craft of Cirurgie resseuie no
Cirurgian that is a foreyn ouer a monethe to occupie v
hi*m* in the craft of Cirurgie bi no maner of colour but that he bringe him first withinne the dai forseid tofore the maistris of the seid craft And that thei with her felowschip examyne hi*m* dueli of his craft And if thei fynde him able[2] i*n* kunnynge of the seid craft than thei to suffre him thre ȝeer to abide with the seid Cirurgian bi couenaunt maad aftir the advys of the maistris & aftir that he to be rulid & gouerned bi the craft And if ony persoone of the seid

[1] unrighteously. [2] skilful.

felowschip do the contrarie with ony sich forein that he paie to the box xx^{s} And also forthwith to leeue the forseid seruaunt or Cirurgian

Of schewynge of curis

And more ouer if ony persoone of the seid felowschip haue ony cure disperat of the which is lykli to falle into
vj deeth or mayme or to him unknowen that he schewe it to the maistris or to su*m*me of hem withinne foure or fyue daies upon peine to paie to the box xiij^{s} & iiij^{d} And if ony of the forseid maistris be required bi ony of the felowschip to se sich a cure disperat & he wole not come there to se it either for comfort of the sike and honeste of the craft that he paie to the box at ech tyme & as ofte as he is herinne fauti vj^{s} & viij^{d} And if ony man of the maistris forseid for the ȝeer falle thus i*n* ony peine aboue ordeined if he wole not paie it withinne the dai of his offise of maistirschip that thanne hise successouris that is to seie[1] the maistris aftir hi*m* gadere it of him as thei gadere it of othir personys of the same felowschip or craft.

Of euene porciouns of the maistris avauntages

And if ony profit or avauntage of gifte come to ony of the foure maistris whethir it be thoruth[2] callinge &
vij visiting of ony persoone as it is aforseid or thoruth ony othir cause aparteynynge to the office of the seid maistris that euermore it be departid in euene porciouns amongis he*m* iiij And if ony sich caas thoruth nede falle ony tyme to ony of the iiij deputees for the seid maistris absent that thanne the seid depute haue his part thereof as oon of the maistris *pre*sent ¶ And eueri examinacioun or iugeme*n*t that is to be had or to be doon at ony tyme of ony mater aperteynynge to the craft of Cirurgie that it be doon & parfoormed euermore honestli bi the iiij maist*ri*s or ther deputees & fremen of the same felowschip

[1] say. [2] though.

How foreyns ben ressyued

And if ony man shall be resseyued into the craft & maad free bi redempcioun that it be doon bi the asse*n*t
of alle the iiij maistris for the tyme beynge & at the leeste ij of the iiij men chosen for the felowschip with he*m* & that the seid persoone ȝeue[1] to the iiij maist*ris* her fees & a certein to the box & a dyner to the craft. And also that no persoone be made maistir of the craft withinne vij ȝeere after his entrynge neither ony tyme but that he bi proued good & honest of gouernaunce & secreet at the seid teerme of vij ȝeer viij

Of prentisis maad free

And eueri prentys whanne he is maad free that thanne the maistris of the craft for the tyme bei*n*ge schal calle
the seid persoonne to he*m* & ther to ȝeue hi*m* ix
his charge to be rulid & gou*er*ned bi the seid craft And thanne as he goodli mai to ȝeue iij^s & iiij^d to the box And also that no sich persoone be ma*de* mastir of the craft withinne the teerme of six ȝeer aftir his entringe & in al the seid terme he be proued wel gouerned & honest wys & secreet & ellis he not to be resseiued to bere the office of maistirschip withi*n*ne other vj ȝeer And if he be not founden bi tho xij yeeris wel rulid in man*n*er forseid he neuere to be chosen maistir.

How noon of hem schal enplete[2] *ano*thir

And also it is ordeined that no persoone of the same craft of Cirurgie schal enplete another of the same felowschip for no cause longynge to the seid craft on the x,
peyne of xx^s to the box at ech tyme that he so doith withouten that it be maad knowen first to ye maist*ris* with her cause in bothe parties And than the maist*ris* to take it into her hand & thei dueli & truli to examine it & redresse it riȝtwysli[3] & consciensli for bothe *par*ties if thei withinne fourti daies at the

[1] give. [2] implead. [3] righteously.

moste or ferthest And if ony sich caas falle ony time bitwixe ony of the maistris & ony othir of the seid felowschip that thanne the seid maister chese for *him* in this cause to the othir maistris oon of the foure chosen for the felowschip and thei to make an eende therof *in* the maner as it is aforseid And moreouer if ony sich persoone or persoones so discordid of the felowschip as it is bifore rehersid haue the maistris or ony of hem suspect of untrouthe that thanne in ony sich caas the maistris & the foure men chosen for the felowschip to gouerne it in the man*ner* forseid And if ony of the foure men chosen for the felowschip be partie in ony sich cause of discord aforseid that thanne he chese for him & for his cause if he wole oon of the felowschip to the othir persoones that schulen trete for his cause And if the maistris in the maner forseid mai not make an eende for the *par*ties neithir with the other me*n* chosen for the felowschip withinne the seid fourti daies that thanne if ony of the parties have not unresonabli absentid hem for to varie the seid arbiterment thanne thei to lete the mater stonde at large alwei except the maters & causis aparteynynge to the secretis of the same seid *cra*ft

Of the visitynge of housholders

And also it is ordeined that the maistris eueri ȝeer & as ofte as it is nedeful visite the householderis of this seid
xj
craft or tho[1] of hem that hau seruauntis how thei haue hem or holde hem whethir it be prentys or couenaunt man and that the holderis of siche schewe the maistris the couenauntis & dentouris[2] of the seid seruauntis so that thei moun wite and knowe that all siche seruauntis and prentisis & ther maistris be rulid & gouerned after the fraunchise of this citee and her ooth & if thei fynde ony sich persoone or persoones of the craft that wole not obei this ordinaunce or ony othir leueful or laweful ordinaunce afor writen thei to make it knowen to the mair or chaumbirleyn as custum & maner of the Citee wole

[1] those. [2] indentures.

How paines shal be moderated & bi whom

And if ony ordinaunce that is m*ade* or hereaftir is to be maad of ony peyne sette aboue or hereaft*ir* is to xij
be sett is ony tyme spedeful to be moderatid that thanne the foure maistris with the foure men chosen for the felowschip thei to moderate it aftir her best advys to the fortheryng of the present profite helthe and welfare of goddis peple & the kingis

¶ And if it so be that thei mai not in this maner of wise acorde withinne the seid felowschip the maistris thanne to haue recours to the mair or Chaumbirlein as fredom & ordinaunce of this citee wole & so i*n* this maner wise for to correcte he*m* that ben misgouerned aȝens the good ordinaunce of the craft aforseid & also untrewe worchers[1] i*n* the craft of the same felowschip and so proued

Of amendynge and addicioun of the composicioun

And if ony tyme to come hereaft*er* it biseme to the craft ony thing in this forseid ordinaunce & composicioun xiij
to be to miche or to litil that thanne the seid craft bi common assent & aftir her good advys & discreciouns it to be comowned discreetli that is to seie that the mater be dueli examined bi good advys i*n* comunicacioun of the felowschip bi foure or v daies & that bi a copi had out of the original of the mater & it to be answerid bi profitable resoun & writinge & otherwise not to be resseiued at ony tyme to come hereaftir And if ony ordinaunce i*n* this man*ner* wise is thus aproued afterward be it n*o*t empungned

Of payment to the dyner

And also it is ordeined that euery free man of the craft of Cirurgie paie ȝeerli to the dyner of the craft that xiv
is to seie oonys aȝeer on the dai of Seint luke ech man lich mich[2] whether he be p*re*sent or absent except noun[3]

[1] workers. [2] equally. [3] no.

power[1] & the ouerplus thereof not spendid if ony sich is be it kept & spendid on the nexte [quarter?] dai And also that eueri free Cirurgian mainteine & supporte in alle tymes aftir his kunninge & his power the honeste of the seid craft: & neithir for occacioun of displesaunce or ony othir cause neuere the seid craft or the honest felowschip thereof to leeue ne to forsake in ony wise withoute resonable cause & openli knowen to the seid craft

Condicioun of the boondis

And also in this forseid ordinaunce & composicioun is specified
xv withoute ony fraude the maner of the boondis & condiciouns how & in what maner & for what cause the maistris schulden be bounden for the comoun goodis of the craft forseid & also of the foure men chosen for the felowschip to be also bounden to the maistris that is to sei the maistris to be bounden in a plain obligacioun of the summe or sumwhat more of the value of the seid goodis for noon other entent but thei schulden truli kepe it to the uss of al the craft al the seid ȝeer & at the eende of the ȝeer forseid or in the dai asigned the seid maistris to ȝelde her acounte of tho goodis to the forseid foure men for the felowschip & thei contente aftir the truthe & maner of this forseid ordinaunce that thanne the forseid boondis of the maistris to be broken or to stonde for nouthe & ellis it to stonde in strenkthe & vertu And in the same manner wise be bounden the foure men chosen for the felowschip to the seid maistris in an obligacioun of ij mark for noon other entent but to heere & resseue aȝen[2] the seid goodis of the craft & heer her acounte & to alowe hem that that is nȝt[3] and so fulfille the composicioun with the seyd maistris for the whiche thei ben chosen for the ȝeer forseid And this ordinaunce content that thanne the seid boond stonde for nouȝt & ellis it to stonde in strenkthe & vertu And thus continueli euermore her aftir ȝeerli this rule & ordinaunce to be kept & fulfillid at eueri chaunge of the maistris or of the

[1] poor. [2] again. [3] nought.

chaunginge of the foure men for the felowschip *wit*houten ony variaunce in al maner wise as it is rehersid & bifore write*n* i*n* this composicioun

The charge and ooth

This is the foorme & maner of the charge that the maistris schulen ʒeue to the persoones in the bigynnyge
that ben newe resseiued into the craft ¶ ʒe shal xvj
swere that ʒe schal wel & truly bihaue you in worchinge of the craft of Cirurgie in sauinge of goddis peple & the kyngis And alle the good ordinauncis & rulis & secretis of the seid craft ʒe schule*n* wel & truli kepe withinne the seid craft And to alle the leueful & laweful biddingis of the maistris of this seid craft that now ben & here aftir schulen be ʒe to be therto euere continueli obedient whanne ʒe be*n* callid & neuere it to forsake but to fulfille so god ʒou helpe and alle seintis.

Of peynes of mysgouernaunce

Also it is ordeined & assentid in this composicioun that whanne the maistris at ony tyme sitte in iugement
or in examinacioun or in cominicacioun of the seid xvij
craft with the hool felowschip or parcel thereof as place cause & tyme axith that thanne eueri persoone of the seid craft *that* tyme present kepe scielence at the firste biddinge or comaundement of the seid maistris & not withoute license of hem had aʒen to speke And if ony of he*m* wole not at the firste biddinge ceesse (for the secund tyme boden to ceesse) to paie for the seid faute xij^{d} And if he wole not ʒut ceese thanne for the iij warnynge ij^{s} And for the fourthe tyme a noble And if he wole not thanne ceesse at the fifthe tyme to be take for rebel ¶ And moreouer that no persoone of the seid felowschip or craft reuile[1] ne lieue ne schewe occasioun of malice ne stiringe to malice ony to othir [or do whatsoever] miʒte be cause of disturblaunce of the good pees among the felowschip of the seid craft upon peyne to paie to the box xij^{d} at ech sich

[1] revile.

defaute ¶ Also furthermore it is ordeined that if ony of the seid felowschip drawe ony wepene i*n* violence or unlawfulli manace ony persoone of the seid craft to paie therefor a noble ¶ And if ony of hem smite anothir of the seid craft to paie to the box xx^{s} and ouer that the parties to be iustified aftir lawe or bi ordinaunce of the seid craft as it is bifore ordeined ¶ And if the mastr*is* or ony of hem trespas in ony sich unresonable cause & unlawful aforseid thi*n*[1] ech of hem fauti to paie the double of the peine sett afore And if ony of the seid felowschip reuile or manace either false ony of the maistris or upon he*m* unskilfulli complayne he to paie ij nobles at eche tyme & as ofte as ony hem so doith ¶ And so it is to knowe that the entent of this ordinaunce aboue writen is thus for to undirstonde that al the seid craft & eueri persoone therof be wel rulid & gouerned withinne hemsilf that is to seie bothe the maistris & her felowschip & alle thingis that schal amongis hem be doon or seid that is to wite the maistris pacientli her maters to heere & wysli & truli *th*in therto seie & the seid felowschip i*n* tyme resonable resonable to axe & in her complayntis & seyngis[2] honestli to be mesurable and to scielence mekeli to obeie aftir the discrecioun of the maistris as it is aforeseid bi vertu & ordinaunce of this citee ordeined to maistris & wardeynes to craftis ¶ And to alle these rulis & ordinaunces bifore writen euery persoone of the seid craft bi hi*m*self & alle thei togidere hau consentid that it schal be holde & kept alwei contynueli fro tyme to tyme & fro ȝeer to ȝeer & as longe as the maistris for the tyme beinge & the felowschip of the seid crafte bisemeth it spedeful & profitable to be kept.

Of the *charge & the ooth* tha*t* the *olde mast*er*s ȝeue to* the *newe*

This is the ooth that the olde maistris eueri ȝeer shal ȝeue to
the newe whanne thei ben chosen ¶ ȝe schal swere
xviij that ȝe schal wel & truli gouerne the craft of
Cirurgie & the felowschip of the same craft aftir ȝoure kunnynge & ȝoure power as longe as ȝe be*n* maist*er*s for this ȝeer And

[1] then. [2] sayings.

also ȝe to kepe & se to be kept alle the good rulis & ordinau*n*cis of this seid craft now maad & that noon of ȝou ony other ordinaunce to make neithir to ȝoure knowleche suffre to be maad withouten the assent of al the felowschip & that also with condicioun & maner as it is bifore ordeined & writen

Th*at no man presume to breke* the *ordinance.*

The conclusioun of this composicioun aboue writen is this
*th*at no persoone of the seid craft presume in ony xix
wise to breke this seid ordinau*n*ce neither ony
other to breke it on peyne of an hundrid schilingis i*n* to the time that a better ordinaunce be founden or maad & so of the craft ressyued

¶ The names of the forseid ordinau*n*cis grauntynge ben these william Bradwardin John hatfelde John Corbi John fforde Robert wiltone William Wellis John Cosyn John Barton Thom*as* hertford John Polley Thomas Warde herri Stratforde Gefferei Costantyn Robert Braunche Richard Saxton herry Arschbourne And Thomas Morstede

Of tymes & houris sett of comynge togidere

And as for tymes & houris sett of cominge togidere it is to
be noted & also the maner of speche thereof as xx
if it be seid to come at oon of the clocke or
at ij or bi oon or bi ij etc it is than alwei to understonde to be there at the same hour or bi half an hour after & not to passe & if it be seid bitwene oon of the clocke & ij etc it is to be kept alwei bi the laste hour named or anoon[1] upon And so of dute half an hour is to be abiden if it be nede And that is more is of curtesie of ponyschable

In the name of god ame*n* In *th*e 28 day of septe*m*b*er* *th*e
ȝere of owre lord 1503 and in *th*e ȝere of kyng xxj
harry *th*e 7th *th*e 19 by *th*e aduice of *th*e worship-
full men of *th*e science of Surgery i*n* all *th*e comynallte ys made & to sto*n*de for eu*er*

[1] immediately.

First it is ordeyned that from hensforth is alowed owte of *the* comon box *in the* worship of god and seynt Luke for *the* syngers

Also it is ordeyned that at masse ij^{s} what *per*sone of *the*
xxij sayde felishipp beyng *in* towne *that* lakes[1] of the
day of saynt luke at x of the cloke at the masse & so contynue to masse be done and to brynge *the* wardens wher *they* shal dyne & they shal pay for this defaute ij^{s}

Also what person of *the* seyde feliship beyng *in* towne lakes
xxiij at Corp*us* Xρι[2] chapell at *the* day of cosme &
damya*n* therfor to chese thee warde*ns* at x of the cloke shal pay for this defawte viij^{d}

Also what persone beyng *in* towne *that* lakes at the pres-
xxiv sentasion of ther warde*ns* knowyng wher or whe*n*
shal pay for this defaute xij^{d}

Also at the presentac*i*on eu*ery* man that hath a wyfe shal pay for his dyner ij^{s} And he that hath non xx^{d}

[1] lacks. [2] Christi.

APPENDIX D.

PETITION OF THE GUILD OF BARBERS FOR THE REGULATION OF THEIR FRATERNITY.[1]

MEMORANDUM qd xxvto die ffebruarii Anno regni Regis Henrici sexti post conquest*um* vicesimo nono [1450] venerunt hic in Curiam Regis in Cam*er*a Guyhald Ciuitatis london: coram Nich*ola*s Wyfolde Maiore & Aldermannis eiusdem Ciuitatis magist*ri* & Gardian*i*: necnon et probi homines mister*iæ*: barbitonsor*um* Ciuitatis *pre*dictæ & porrexerunt d*i*cto Maiore & Aldermannis quand*am* billam siue supplicacion*em*. . . . Unto the ryght Worshipfull and Worshipfull lord and Soueraignes Mair and Aldremen of the Citie of London

Besechen most mekely alle the *per*sones enf*ra*unchised in the craft and mistier of Barbours wythin the said Citee That it please unto your lordshipp*es* and Right wise discresions for to consider howe that forasmoche as certein ordina*u*nces been establisshed made and ent*er*ed of Record in the Cha*u*mber of the ȝeldhall of the said Citee all *per*sonnes of the said Craft haue fully in opinion to obeye observe and kepe them and noon other in eny wise So been there many and diu*er*se defaultes often tymes amonges your said besecchers not duely corrected for default of such other ordinances to be made and auctorised sufficiently of record in the said Cha*u*mber like it therefore vnto your said lordshippe and grete wisdom of your blessed disposicon for the *per*petuell Wele and good Rule of the said Craft for to establish and make these ordinaunces here following *per*petually to endure and soe to be observed and putte in due excecucion in the same Craft And to be auctorised of record in the said Cha*u*mber for eu*er* And your said besechers shall pray God for you

[1] Letter Book K, fol. 250.

ffirst that eu*er*y barbour enf*rau*ncheised householder and other occupier of the same craft holdyng eny shoppe of barberye wythin the Citie of London shall be redy att all man*ner* [of] som*m*ons of the Maisters and Wardeins of the same Craft for the tyme being that is to say the Kyng the Mair or for eny worshippe of the said Citee And ȝef eny man occupying the said craft in manner and fourme aforesaid absent him from eny such sommons wythoute cause reasonable And thereof duely convict than he to pay at the chaumber of the ȝeldhall xiij[s] iiij[d] That is to say vj[s] viij[d] to the same cha*u*mber and other vj[s] viij[d] to the almes of the said craft Also eu*er*y man enfrauncheised under the fourme above said that disobeyeth and kepeth his house of his som*m*ons at eny time w*i*thoute cause reasonable and ȝef duely convict shall paye to the almesse of the said craft at eu*er*y tyme *tha*t he maketh such defaute ij[d] And ȝef eny of them whatsoe*u*er he be of the same craft that disobeye this ordina*u*nce he shall paye to the Cha*u*mber of the ȝeldhall iij[s] iiij[d] at euery tyme that he maketh such default

Also that ȝef eny mat*ter* of debate or difference be betweene eny *per*soones of the said craft which God defend that none of them shall make eny pursuyt at the com*m*on lawe unto the tyme *tha*t he that findeth him aggreved in that *par*tie hath made his compleint unto the Maister and Wardeins of the same Craft for the tyme being and they to ffynyssh the mat*ter* and the cause of the said compleint Wythin vj dayes after such compleint made and ȝef the[y] conclude not and ffynyssh the same mat*ter* Whithin the said vj dayes that then it be lefull to either *par*tie to take the benefice of the Common Lawe Wythin this Citee so alwayes that the partye ageinst whom the compleint is made be not fugityf And what *per*sone of the said craft doth contrarie [to] this Ordina*u*nce shall paye at eu*er*y tyme at the Cha*u*mber of the ȝeldhall xiij[s] iiij[d] that is to say vi[s] viij[d] to *th*e said Cha*u*mber and other vj[s] viij[d] to *th*e almesse of *th*e said craft

Also that noon able *per*sone of the said craft enf*ra*uncheised shall refuse eny man office or clothing *per*tinent to the said craft whan and what tyme that he be by his bretheren beabled and elect *there*to upon pein to paye at the cha*u*mber of the ȝeldhall xl^{s} that is to wete xx^{s} to the same chaumber and other xx^{s} to the said almesse. Also what man of the said craft that absente him fro the said eleccion Wythoute cause reasonable or absent him fro the dyner to be made the same day and will not paye therto his *par*te thanne he shall paye at the said cha*u*mber iij^{s} iiij^{d} to the Almes of the said craft

Also that the Maisters and Wardeins of the same craft that nowe be or in tyme to come shall be shall not take admitte or resseiue eny *per*sone into the bretherhede or clothing of the same craft Wythoute the com*m*on assent of the bretheren of the said craft or the more *par*t of them upon pein of fyne for such maisters or wardeins that doth contr*ar*ie to this ordina*u*nce xx^{s} that is to say x^{s} to the Cha*u*mber and x^{s} to the Almesse of the said craft

ffurthermore it is ordeigned that from hens forward ȝef eny man occupying the said craft be imfouled & of euell Will and malice so be unavised to revile or reproof eny man of the same craft that is to seye for to lye him on wyth other dishonest Wordes misgou*er*ne him in presence of the said Maisters and Wardeins or in eny *othe*r places and ȝef by the report of the said Maisters and Wardeins [he] be duely convict Whatsoeu*er* he be of the same craft that is so misgou*er*ned at eny tyme shall paye at the said Ch*a*umber for euery such default v^{s} viij^{d} that is to say to the same Ch*a*umber iij^{s} iiij^{d} and to the Almes of the said craft iij^{s} iiij^{d}

Also it is ordeigned that e*uery* man*ner* [of] man enfra*u*ncheised of the said craft under fourme aforesaid shall assemble with his ffelaship*pe* of the same craft by thassignement of the said Maisters and Wardeins being for the ȝeer in a certein place limited by theim at iiij tymes of the ȝeer And at e*uer*y such quarter day in the ȝeer eu*ery* brother enfr*au*ncheised and being of the clothing thereof shall paye to the almes abovesaid iiij^{d} And euery man that is so enf*ra*uncheised of the same craft and is not of the clothing of the same shall paye to the same almes j^{d} Which iiij dayes be these that is to saye the tewesday next after all hallows day the tewesday next after candelmas day the tewesday next after Trinite sonday and the tewesday next after lammas day to thentent that the said Maisters and Wardeins shall enquere amongs the said compaignye so assembled that ȝef eny default ranker or discord be hadd or amongs theim that thanne the said Maisters and Wardeins shall sett theim at rest accord and in vnite to that they canne or may after the fourme and custoume as have been before vsed And what parsoune of the said craft be absent eny of the said dayes wythoute cause reasonable he shall paye for e*uer*y such day iiij^d to the expenses of the said Maisters.

Also it is ordeigned and establisshed that no barbour nor other able *per*sonne vsing barbourye shall enfourme eny foreyn nor him teche in no wise in eny man*ner* [of] point that belongeth to the craft of barbourye or surg*er*ye Whereby the same foreyn shall *per*ceyve and take by his own capacite and ex*er*cise unto the tyme that the same foreyn be bounden app*re*ntice to a *per*sone barbour or other *per*soune able enf*ra*unchcised using the same craft wythin the Citee of London upon peyn to paye at the ch*a*umber of the ȝeldhall [for] *eue*ry such defaute iiij marc that is to say to the same ch*a*umber xxvj^s viij^d and to the almes of the said craft other xxvj^s viij^d

And also that no barbour nor other able *per*soone occupying the same craft shall take eny Alien nor st*ra*unger into his Ser*u*ice unto the tyme that the same alien or st*ra*unger be examined by the maister and Wardens of the same craft of his abilite and Conuyng and therupon the Maist*er* and Wardeins with other vj or viij of the moost able and Kunnyng *per*sonnes of the craft shuld taxe him after his abilite after that hit semeth that he be worthy to take yeerly for his salarie And also that no barbour shall take eny alien or st*ra*unger that hath been or *ser*uice wyth an other barbour enf*ra*uncheised before that he knowe well that the same *serva*unt hath complete his couen*a*ntes wyth his former maister upon pein to paye to e*uer*y such defaute at e*uer*y tyme that he be founde defective aȝenst eny of these ordin*au*nces at the said ch*a*umber xiij^s iiij^d and to the said almesse vj^{s} viij^{d} and also make restitucion of the damage unto the *par*tie that findeth him greved

And also that no man occupying the said craft shall *pro*cure eny other mannes *seruа*unt oute of *ser*uice upon the peyn aforesaid and damage unto the *par*tie pleintif And also it is ordeigned that from hensforward that no barbour enfr*a*unchised nor eny other able *per*soons occupying the said craft shall not take into his *ser*uice eny str*a*unger or forein for lasse t*i*me than a ȝeere And what *per*sone enfr*a*unchised or occupying the said craft disobeys this ordin*au*nce shall receiue in the pein of xiij^{s} iiij^{d} that to be devided in man*ner* and fourme aboue said

And also that no *per*sons of the said craft of barbours nor other able *per*sone occupying the same craft huyre no ffr*a*unchised man of the same craft oute of his shopp ne dwellyng place upon pein of xl^{s} that is to wete to the said ch*a*umber xx^{s} And to the Almesse of the said Craft other xx^{s}

APPENDIX E.

THE COPPYE OF THE CORPORACION GRAUNTED BY KINGE EDWARDE THE FOWERTHE BY HIS LETTERS PATENTES AS FFOLLOWETHE.

Letter patent F 4 Barbitonsor: London:

EDWARDE by the grace of god kinge of Englande and ffraunce Lorde of Irelande Greateth all those to whome these presencs shall come knowe ye, that we consyderinge howe the welbeloued unto us honest and ffreemen of the mysterye or Crafte of Barbours of our Cytie of London using the Mysterye Crafte or ffacultye of Surgerye as well about woundes Cutts Sores hurtes and other infyrmyties of our leige people there to be cured and healed, as in lettinge of bloode and drawinge of Teeth of this our leige people, haue of longe tyme susteyned and supported and cease not daylye to sustayne and supporte greate and manyfolde attendaunce and Labours Howe also throughe the ygnoraunce neglycence and follye of manye of theise Barbours as well ffreemen of our sayd Cytie as of other Surgeons forrynours and no freemen of the sayed Cytie resourtinge daylye to the same Cytie and in the mysterye or crafte of Surgerye not instructed, throughe suche Barbours and Surgeons very manye (and as you woulde say) infynite evell before this tyme by their defaulte hath happened to dyuers of our liege people in healing and curinge their woundes Cuttes Sores hurtes and other their infyrmyties of the whiche some of our Liege people for that cause have dyed. And some other for the same cause hathe ben lefte and forsakan of all as insauable and incurable. And yt ys greately to be feared that in this case the semblable or lyke evell or rather worser may in tyme to come happen onles convenyent remedye for this by us be quyckely p*ro*vyded for We intending and effectually consyderinge that such evells are wonte to happen to our leage people for the defaulte of dewe ouersighte searche corrections and pounyshement of suche Barbours and Surgions not suffycyentlye learned and instructed in the same Mysteryes crafts or facultyes aforesayed, at the humble petyc*io*n and supplycation of our sayed welbeloued unto us honest and ffremen of the sayed my[s]terye Crafte or facultye of Barbours in our Cytie aforesayed, haue graunted unto them *tha*t the Mysterye and all men of the same mysterye Crafte or facultye of the sayed Cytie maye be bothe in effect and in name one boddye and one p*er*petuall ffelloweship or Communytie And that two of the cheffeste and pryncipallest of the same felloweship or Communytie together w*i*th thassent of xij or viij at the leaste of the same feloweship or Communyte maye elect and chose eue*r*ye yere men verye expert in the mystery Crafte or facultye of Surgerye and take out of the same fellowship or Com*m*unytie two Maisters or Governours verye experte in the mystery or facultye of Surgerye to oversee rule and Governe the mystery Crafte and Communytie or felloweship aforesayed and all the men of the same mystery or Crafte and their affayers forever. And that the same Masters or Governors and Co*m*munytie and ffelloweship maye haue p*er*petuall successyon and a com*m*on Seale for the affayers of the same Comunytie or felloweship to serue foreuer And that they and their Successors maye at all tymes be p*er*sons able and haue Capacytie in the Lawe to purches and haue in possessyon in ffee and perpetuytie Landes Teneme*n*ts Rents and other possessyons whatsouer they be to the value of ffyve marks by the yeare over the reprys*t* And that they by the names of Maisters or Governors and of the Comunytye or ffelloweship of the mysterye Crafte or facultye of the barbours of London maye pleade and impleade before all manner of iudges in all man*ner* of Courts and actyons And that the saide Maisters or Governors and Co*m*munytie and ffelloweship and their Successors

1, Barbors corporacon. 2, Elccon of Mr. & govnors by 2 of *the* cheifest or xij others or viij. 3, 2 Mrs or governors.

Succesion perpetuall comon Seale

pleade be impleaded

Lawfullye with*out* damage occation or impedyment of us our heyers or Successors Justyces Eschetours, Shreaves Crouners or other our Baylyffs or mynisters or of our heyers or Successors as ofte and whensoeuer neede shalbe maye make honest and lawfull assembles or congregac*i*ons of themselues And also ordeyne and make Lawes and orden*aunces* for the wholsome gouernment oversyghte and correction (as nede shall requyer) of the forsayed mysterye or felloweship so that those Statutes Lawes and orden*aunces* be in no wyse againste the Lawes and customes of our Realme of Englande Moreover we wyll and graunte for us our heyres and Successors as moche as in us ys that the Masters or Governors of the said felloweship for the tyme beinge and their Successours maye & alwaye have ouersighte searche correctyon and Government of all and singular ffreemen of the sayed Cytie Surgeons using the mysterie Crafte or facultye of barbours in the same Cytie And other Surgeons fforryners whatsoever they be frequentinge or vsinge by any waye the mystery Crafte or facultye of Surgerye with*in* the sayd Cytie or Suburbes of the same And the pounyshement of the same as well of ffreemen as of fforryners accordinge to their facultyes in not executinge doyinge and vsinge the mysterye Crafte or facultye and also the oversighte and searche of all maner of Instruments playsters and other Medycynes and receipts that by the sayed Barbours or Surgeons are geven layed and vsed amonge theis our leige people for to heale and cure their wounds sores Cuttes and suche other infyrmyties as ofte and whensoeuer neede shall requyer for the com*m*odytie and vtylitie of theis our leige people So that the pounyshement of suche Barbours vsinge the sayed mysterye of Crafte or facultye of Surgerye and of suche Surgeons forryners offendinge in the premysses ys to be executed by ffynes mercementes and impreysonments of their boddyes and by other reasonable

Assemblies
Make Lawes

The laws not impugninge y[e] Lawes of y[e] realme

oversight
Search
correccion
Surgeons
vsinge Barbery
Free
Forrens

Search of medicines

punishm*ent* by fine
amerciam*ent*
Imprisonm*ent*

and convenyent wayes and that no Barbour vsing the mystery Crafte or facultye of Surgerye with*in* the sayed Cytie or Suburbes of the same or any other Surgeon fforryner whatsoeuer he be be admytted hereafter to execute dooe and frequent or exercyse by any waye or means the same mysterye Crafte or facultye of Surgery in the same Cytie or in the Suburbes of the same onles he be fyrste approved learned able and suffycyent in that Mysterye Crafte or facultye by the sayed Maisters or Governours or their Successours. And for his full profe in that behalf let him be presented by these Maisters or Govenou*rs* appointed thereunto to the Mayor of the sayed Cytie for the tyme beinge.

Surgeon to be approved
Surgeon approved presented to the Maior

Also we wyll and graunte for us our heyers and Succesours as moche as in us ys that neyther the sayed Maisters or Governors and the Com*m*vnytie or ffellowship of the sayed Mysterye or facultye of Barbours nor their Successors nor any of them shall at any time hereafter for any maner of thinge in our sayed Cytie or Suburbes of the same be sum*m*oned or be put in any assyses Jurats Inquests attents or recognysaunces hereafter in the sayed Cytie or suburbes of the same before the Mayor Shreves or Crowners of our sayed Cytie for the tyme being, or by any offycer or Mynister of his or their mynisters or offycers be attached or sum*m*oned thoughe the same Jurates enquests or recognysaunce hathe ben somoned by Wrytte or Wryttes of us or of our heyers de Recto that the sayed Maisters or Governours and Com*m*unytie and ffellowship of the Crafte or mynisterye aforesayed and their Successours and eueryone of them bequyte and utterlye discharged againste us our heyers and Succesours and againste the Mayor and Sheriff*es* of our sayed Cytie for the tyme beinge and againste all other their offycers and mynisters for euer by this presence And furthermore we by considerac*i*on of the premysses of our specyall grace have graunted for us our heyres and Successours the forsayed Maisters or Governours and the Communytie and ffelloweship

not to be som*m*oned upon none Inquests Assize Jurats Inquests Attaints Recognizaunce of the sayed Mysterye or Crafte of barbours & their Successors this lybartye to wyte that they at all tymes may admytt and receave into the same Crafte or mysterye able persons and suffyeyently learned and enfourmed in the sayd Mysterye of Surgerye and by the same Masters or Governors of the sayd Mystery or Crafte for the tyme beinge approved in forme aforesayed and to the Mayor of the sayed Cytie for the tyme beinge beinge as ys aforesayed presented to haue and enioye the lybertyes of the sayed Cytie And no other persons whatsoeuer they be not otherwise by any comaundement or request of us our heyers or Successors by letters wrytten or otherwise howesoeuer yt be made or to be made to the contrarye notwithstandinge And thoughe the same Mr or Governors and Communytie and ffelloweship and their Successors shall vse this Lybertye contynually hereafter agaynst any comaundment or request of vs our heyers or Successours or any other whatsoeuer he be in the forsayed forme to be had or made yet they nor noon of them shall by no meanes by that occasyon incurre any ffyne Contempte dommage or into any other evell in their goodes or bodyes against other whatsoeuer they be and that without ffyne or ffee for the premysses or sealinge this presence to vs to be made payed or otherwise to be Delyuered any statute ordynaunce or other to the contrarye heretofore sett out made ordeyned or provyded notwithstandinge In Wytnes herof we haue caused these our Lettres to be made patently Wytnes ourself at Westm[inster] the xxx Daye of ffebruarye in the ffyrste yere of our reigne

quer. whether wee may make of ourselves forrein surgeons free of our Company without a Court of Aldermen Forren brothers

Exd xxvjto Aprylys Anno Dni xvo lvjto cum Originali *per* me Thomam Knot being Mr of the Company

APPENDIX F.

THE WRYTYNG ENDENTED OF COMPOSICYONS MADE BETWIXT THE FFELISHIPPIS OF SURGEONS, AND THE FFELISHIPPIS OF BARBOURS SURGEONS AND SURGEONS BARBOURS.[1]

THIS present wrytyng endentyd of composicyons made the xii day of Juli in the ȝere of owyr lord God mcccclxxxxiij and the viij^th ȝere of the reyne of kyng harry the vii^th William Martyn then beyng mayre of this cyte of london, betwixt the ffelishippis of surgeons, enfraunchessid within the cyte of london on that on partye And the felishippis of barbours surgeons and surgeons barbours enfraunchessid in the seyd cyte on the other parte witnesyth that the sayde felyshippys of ther comon assent and more mocyons ben condescendyd and agreyd together the day and the ȝere abouesayd, in maner and fourme folowyng That is to sey that euery person and persons of the faculte or scyens of surgeons admyttyd and sworne to eythir of the sayde felyshippis from hensforward shall stond and abyde with ther felyshippis as they now do and dyd before thys present composicyon.

Also that from hensforth non of the sayde felyshippis shall admyt nor reseyue into ther felishipyys any alyent straunger or foreyn usyng the sayde faculte or scyens of surgery withowten knowlege or consent of the wardens of bothe the seyd felishippys. Allso for the welth and suerte of the kyngs lege people, and the honour of the seyde felyshippys, It is agreyd betwixt the same two felyshippys that non alyent straunger nor foreyn shall use nor ocopy the seyde faculte or scyens of surgy withyn this cyte or subbers[2] of the same unto suche tyme as he shew hymselfe to the mayer for the tyme beyng, and

[1] This heading is not in the original. [2] suburbs.

by the iiij wardeyns of bothe the saide felishippis that is to say, of eyther of the sayde felishippis, tweyn, and other suche as by theyr wysdomys they will call unto them, be dewly examyned and approuyd to be sufficyent of conyng and habilyte in the sayde faculte And yf any suche person or persoues bi the sayde iiij wardens as is aforesayde be taken reputed and alowed to be sufficient of konyng and habylite in the seyd faculte or scyens of surgery that then the person or personys so knowen and admyttyd shalbe sworne to all the good rewlys and ordenans of the seyd faculte or scyens of surgery and to be under the correccyon of the iiij wardens for the tyme beyng to the entent that at all tymes he may be under due correcyon for the sauegard of the kyngis lege people And if any suche alyent straunger or foreyn of presumpcion refuse to be examyned of the seyde wardens in maner and fourme as is aforeseyd; Or if any such straunger or foreyn so examyned be the seyde wardens be Juged onsufficient of conyng and neutheles takyth upon hym to occupy or use the seyde faculte of surgy withyn the seyde cyte or subburs of the same, Then take the name of hym or them so doyng and present hym by the sayd iiij wardens to the mayer for the tyme beyng, to the entent that by his wysdom and advice of hys honerabyll brethern may set suche direccyon as shall be thought resonabil formacion thereof, restrayne hym from the ocupacion of the same scyens within the sayde cyte. Also it is agreyd and compoundyd betwyxte the sayde felyshippys that from hensforth euery of the seyde felyshippis and seu*er*ally by them selfe, Chese of them selfe two discrete persons usyng the fete of surgery to be seuallẏ wardens of the sayd seuall felishippis and that these iiij wardens for the tyme beyng when and as oftyn as nede shall requere Shall haue the syght and good gounance of the seide faculte of surgery And euery person or personys of euery of the sayde felyshippis that happenys or shall fortune to haue any Jeopde[1] or dowtefull cure dredyng deth or may he or they hauyng euery tyme to come shall shew and present the cure or curys in as short tyme as

[1] jeopardy.

nede shall be required And at the leste at the thyrde dressyd to the said iiij wardens for the tyme beyng or any other person or personys that is to seye, to one of yche of the seyde felishippis. and the same wardens ȝif them seme nedefull shall call unto them ij or iij or more if nede require of the wysest and beste experte men of the said felishippis occupyeng the said faculte of surgery as them semeth most expedyent for the cause or causis aforesayde. Also in this composicyon yt is ordeyned and agreyd that none of the iiij wardens for the ȝere beyng neyther any other person of the sayde felishippis ocopyeng the crafte of surgi Put any man of thes sayde felishipis oute of ther cure otherwyse then the honeste of the crafte wolle but that yche of them be redy ȝef nede be to helpe eche other with counsell or deed that worship, profite and the honeste of the crafte and helpyng of the seke be had and done on all sydes. And if any of thes iiij wardens for the tyme beyng or eny other person or personys of the seyde felyshippis do the contrary that eche suche doer content the valuye of the cure after discressyon and Jugement of the same iiij wardens for the ȝere beyng. And also for his trespas to paye after discression and Jugement of the iiij wardens. Also if any person of the sayde felishippis dislaunder or depute any of the said felishippis onrightfully ounonestly ȝef it so may be prouyd upon hym by two or thre witnesse that he paey for that trespas iij^{s} [& iiij^{d}] and ouer that to make amendis to the seide persons the whiche he hath so dislaundered aftur Jugement of onest men of the seyd felyshippis not founden in non suche defaute. This composicyon was made the daye and ȝere abowe wreton by Roberd taylour, Robt Halyday, Thomas Koppesley, Thomas Thornton, Jhon Herte : John Martun Roberd Beuerley : James Stote : James Ingoldsby : John Taylour Richard Suodenham Nicholas Leueryng John Wilson. Inrowlid in Raffe Osterigis[2] tyme mayer.

Immediately after the Composicyon comes the following rule, which probably had been omitted by accident:

Also that what person or personys of the seide ij craftes or of

[1] dressing. [2] Sir Ralph Astric, Mayor in 1493.

ony of them from hensforward breke or do the contrary to any articules of thordenauns abouesaide The same person or personys so brekyng or doyng the contrary as ys abovesayde shall forfayt and pay at the tyme x^s the on halfe thereof to be applied to use of chambyr of london and the other halfe therof to the use of the seyde craftys and at the second tyme to forfait and pay xx^s to deuyded and applyed to uses afore rehersyd. At the iij^d tyme that any person or persones breke or do the contrary as above saide he or thei so doyng to be punysshyd aftur the discrecion of the mayre and aldyrmen for the tyme beyng.

APPENDIX G.

THE COPPYE OF THE *LETTRES* PATENT*ES* FOR THE CONFYRMACION OF THE CORPORACION GRAUNTED BY KING HENRY THE SEVENTH.[1]

HENRYE by the grace of god kinge of Englande and ffraunce and Lorde of Irelande Greateth all men to whome these present L*ett*res shall come Knowe ye that we haue *per*vsed certaine bull*es* of the moste holye ffather Eugenyus[2] by godd*es* mercye highe busshop sealed under his leaden seale sayinge this: Eugenius Busshop servaunte of the servaunt*es* of god sendethe greatinge and Appostolycall benedyction to the reverent bretherne Archebusshop of Canterburye Busshop of Winchest[er] and our beloved sonne Deane of the churche of London We doo willingly agree to the honest request*es* of our supplyant*es* and convenyently favor the same because of late Marteine of happye memorye of that name the ffyrste our predecessour beinge enfourmed on the behalf of our beloved sonnes Kepers and Wardeyns of the arte or mysterye of Barbours dwelling in the Cytie of London that seing in tymes paste yt was set fourthe in a provyncall Councell kept and celabrated by Thomas of good memorye Archebusshopp of Canterburye and prymate of Englande and Legate of the Sea Appostolicall and his bretherne Archebusshoppes Sufferegans and other prelat*es* of the provynce of Canterburye that the barbours of the Cytyes Townes and plac*es* of the sayed provynce and namelye of the sayed Citie also of the suburbes of the same contrary to the Lawe of god and Canonycall Decrees and publyke honeste had presumed to kepe their howses and shoppes

[1] This is the heading given to the document as it stands at the end of the first volume of the Court Minutes belonging to the United Company, though it is clearly not the document referred to in Appendix H.

[2] Eugenius IV., elected 1431, deposed by the Council of Basel.

for the exercyse of their crafte open publykelye to exercyse the same arte upon Sondayes on the whiche God ordeyned to absteyne from servyle work*es*. The sayd Thomas prymat w*i*th the councell and assent of those Suffrigans prelats and Clarge had determyned and ordeyned that those barbours shoulde be from thensforthe cohersyd from suche presumpc*i*on by publycke inhibycons undere greavous peynes and censures of the Cannon Commyttinge and commaundinge the Ecclesyastycall *per*sones then expressed that they shoulde euery Sondaye and ffestyvall daye publisshe solemplye and cause to be publisshed the same statute and ordyn*au*nce in their Churches Also to inhybyte or cause to be inhibyted the sayd Barbours under the payne of excommunycation that they shoulde not onely on the forsayed Domynicall dayes kepe open or cause to be kept open by anye meanes their howses or shoppes for the sayed exercyse And successyvely our sayed predecessour being humbly requyered on the behalf of the sayd Kepers and Wardeyns that he wolde withe save of the appostolycall bennigenytie supporte wi*th*in the fortyfycac*i*ons of the apostolycall power the sayed statute and ordyn*au*nce Our sayd predecessour through supplycation being inclyned ratyfyed and by the apostolycall authorytie confyrmed the sayed statute and ordyn*au*nce and whatsoeuere shall thereopon followe and w*i*th the deffence of his wrytynge hathe made stronge commaundinge by the processe of his Le*tt*res all & singular prelats and *per*sones of the churches of the sayed provynce that they shoulde solemplye publishe the same authoritie the statute and ordynaunce and the contents and effect of the same Le*tt*res where and when they see it to be expedyent And cause the same statute and ordyn*au*nce yrrefragably to be obserued by ecclesyastycall Censures and other convenyent remedyes of the Lawe Also shoulde publishe shewe and cause to be shewed on the forsayed Sondaye and festyvall Dayes when the greate parte of the people are gathered together to dyvyne service all and singuler of those Barbours being present or shall come whome they certaynely knowe by occasyon of transgressyon of the same statute and ordynaunce have incurred into suche sentence of excom*m*unycation as often as they be lawfully requyered thereunto on their behalf to whome yt belongeth to be excommunycate and to be exempte from all*e* men untyll thys commycates mergyt to obtayne the benefyt of absolutyon for the sayed sentence as yt is conteyned more fullye in the L*ett*res of our sayed predecessour made for that purpose And seying (as the pettyc*i*on made vnto us of late on the behalf of the sayed kepers and wardeyns dyd conteyne) manye of the same arte and mysterye dwelling wi*th*in the plac*es* and lymytt*es* of the Churches and monasteryes of the provynce and Cytie and Dyoces of London and specyallye of the Churche of greate saynt Marteyns of London and of the Monastery of Westm[inster] of the sayed Dyoces Alleging suche exemples that were made as well as by the Appostolycall as by the king*es* authoritie Also vylepending under the pretence and cloke of suche exemption by Statute ordyn*au*nce and mandate doo presume to kepe open their forsayd howses and shoppes on the forsayd dayes for the exercyse of their forsayd Arte We therfor reproving the rasshe presumption of suche Also by the proces of the letters decernyng and declaring the exempte *per*sons and suche exempte places of theyrs to be comprehended under the statute and ordyn*au*nce aforesayd doo geve in charge to your dyscressyons by the Appostolycall wrytynge that ȝoue or ij or one of yowe by your selues or by other or others gooe see the same Statutes and ordyn*au*nces to be kept yrefragablye by suche exempt parsons aforesayed in suche exempte places of theirs or ellswhere according to the mandate forme and tenour of the Letters of our predecessour Dated at Rome.

Exd: the XXVJth Daye of Apryll Anno Dni XV° LVJ^{to} being M^{r} of the Companye And agreeing wth the verye Originall *per* me THOMAM KNOT.

APPENDIX H.

THE COPPYE OF THE CONFYRMACON OF THE CORPORAC*ION* GRAUNTED BY KINGE HENRY THE EIGHTE BY HIS L*ETT*RES PATENT*ES* AS FOLLOWETH

HENRY by the grace of god Kinge of Englande and ffraunce and Lorde of Irelande Greateth all men to whome these present L*ett*res shall come We have *per*vsed the Letters patent*es* of our Souereigne Lorde Henrye the seventhe late kinge of Englande our welbeloued father of confyrmac*i*on made in theis wordes Henrye by the grace of god Kinge of Englande and ffraunce and Lorde of Irelande Greateth all men to whome this present L*ett*res shall come We have *per*vsed the l*ett*res patent*es* of Edwarde the iiijth of moste worthye memorye late kinge of Englande our progenytor made in theis wordes Edwarde by the grace of god kinge of Englande and ffraunce and Lorde of Irelande Greateth all men to whome theis present L*ett*res shall come Knowe ye that we consyderinge howe the welbeloued unto us honest and ffremen of the Mysterye or crafte of Barbours of our cytie of London vsynge the Mysterye crafte or facultye of Surgery w*ith* as before in the other L*ett*res patent*es* unto thende In witnes wherof we haue caused theis our L*ett*res to be made patent*es* Wytnes ourself at westm[inster] the xijth Daye of Maye in the xviijth [1526] yere of our Reigne

KYRKEBYE

Ex^{d} xxvj^{to} aprylys Anno Dni xv^{o} lvj^{to} Concordat: cum Originali *per* me Thomam Knot being M^{r} of the Companye

APPENDIX I.

THE COPPYE OF THE ACTES STATUTES AND ORDYN*AU*NCES CONFYRMED RATYFYED AND ALLOWED BY SIR THOMAS MORE KNIGHTE CHAUNCELOR OF ENGLANDE THOMAS DUKE OF NORFF[OLK]E AND THREASAUROR OF ENGLANDE JOHN FFIT[Z]IAMES AND ROBERT NO[R]WICHE KNYGHTES CHEIFFE JUSTICES OF THE TWO BENCHES

To all trewe Christen people to whome this present wrytyng shall come Syr Thomas More knighte and Chauncelor to our moste drade sou*r*eigne Lorde Kinge Henry the Eighte kinge of Englande and of ffraunce defensor of the ffaythe and Lorde of Irelande Thomas Duke of Norff[olk]e and Threasauror of Englande John Fitziames knighte Chief Justyce of our sayd sou*r*eigne Lorde his benche and Robert Norwiche knighte cheif Justyce of the com*m*on benche Sende gretinge in our Lorde god euerlastinge Where[as] in a certain Acte in the Parlyament holden at west-m*inste*r the xxv daye of Januarye in the xixth yere of the moste noble kinge of famous memorye kinge Henry the vijth [AD 1503] made and ordeyned for the weale and proffyt of his subiects yt was amonge other thing*es* ordayned establysshed and enacted that no Maister wardeins or ffellowship of Craft*es* or Mysteryes or any of them or any Rulers of Gildes or ffraternytics shoulde take upon them to make any act*es* or ordynaunc*es* ne to execute or vse any act*es* or Ordynaunc*es* by them heretofore made in disinherytinge or dymynicion [1] of the king*es* prerogatyve or any other or againste the co*m*monweale and proffyt of the king*es* Subiect*es* and leige people but yf the same Act*es* or Ordyn*au*nc*es* be examyned appointed and admytted by the Chauncelor and Threasauror of

[1] diminution.

Englande and the cheife Justices of eyther benche or thre of them or ells before Justices of Assize in their Circuyte or progresse in that Sheire where suche Act*es* or ordynaunc*es* be made upon payne of forfayture of X *libri* for euery tyme that they dooe there unto contrarye as in the sayd Act*es* yt dothe more playnely appere

The Maister and Wardeyns of the Mysterye or Crafte of Barbor Surgeons of the Cytie of London willinge and desyeringe the sayd Acte in euery behalf to be obserued and kept the xx^th^ Daye of October in the xxij^th^ yere of the reigne of Sou*er*eigne Lorde kinge Henrye the viij^th^ haue exhibyted and presented their petycon therupon made w*ith* a Boke conteyning dyuers Statutes actes and Ordyn*aunces* heretofore devysed ordeyned and made for the ffellowship of Barbours Surgeons and their Successors and for the Com*m*on weale and conservac*i*on of the good estate of the sayd Crafte and Mysterye of Barbors Surgeons aforesayd and for the better Rules and ordyn*aunces* of the same ffellowship establysshed ordeyned and vsed And thereupon instantlye haue desyered us that we all and euerye the sayed Statut*es* ordyn*aunces* and Othes by the same maister and wardeyns *and* their predecessors to the forsayd entent made ordeyned and establyshed woulde ouersee and examyn and the same and e*uery* of them correct reforme and amende after the man*n*er and due forme convenyent and as the forsayd acte in the sayd parlyament made requyereth We wel perceyvyng the sayd Supplycac*i*on to be good and acceptable accordinge to their petycyons and desyers and by auctorytie of the sayed Acte of Parlyament to vs com*m*ytted All and eu*er*eye their Othes Actes and Ordyn*aunces* in the sayed Boke specyfyed haue perfectlye seene the same well and ripelye vnderstande and theym all and eu*er*eye of them examyned corrected and reformed the tenore as hereafter followeth.

The othe of eu*er*eye ffreman ffurste ye shall swere that ye shalbe good and trewe unto our Leige Lorde the king and to his heyers kinges of Englande and obedyent to the Mayor and his bretherne the Aldermen of

fremans othe 1

the Cytie of London And also ye shalbe obedyent to the Maisters or Governors that nowe be or that hereafter shalbe of the Crafte of Barbor Surgeons whereof ye be nowe made free Ye shall also obey kepe and observe all the good orders Rules and ordyn*au*nce*s* of the sayed Crafte heretofore made and not repelled and hereafter to be made, so helpe youe god and all Saynct*es* and by this Booke.

The othe of the Master or Governors Ye shall swere that ye shall observe kepe and mayntayne the worship proffyt and co*m*men weale of the Crafte of Barbors Surgeons in all point*es* lawfull and leafull as good and proffytable Maisters or Governors and Rulers oughte to dooe after your good connynge good dilligence & power Also ye shall kepe and maynteyne and doo to be kept and mayteyned duringe your tyme as farre forthe as youe lawfully maye As well all soche good Vsages Customes Lybertyes and Ordyn*au*nce*s* of the same Crafte and at this Daye vsed approved and contynued and all and singular point*es* conteyned in the premysses duely and treuely when ye shall put in execucion and as often as the case shall requyer Duringe your tyme And also ye shal duelye and treuely make your Serches throughe all the Companye of the same Crafte w*i*thin the Cytie of London and Suburbes of the same And therupon as the case shall requyer all the defaultes and neglygence*s* of comytement*es* and inconvenyence*s* that may happen or ffall to be founde in the Crafte of Barbory or Surgerye in your tyme ye Dilligentlye shall reforme and sett in good Rule and trewelye correct and punyshe accordinge to the power and Rules for the reformac*i*on had & made for the same in the sayed Crafte And for and Duringe your tyme correct and lawfully ponnyshe after the qualytyes and gravyties of and upon the Demearyttes of and default*es* founden in the same after your connynge and power. Also ye shall not admytte any fforrein to be of this mysterye whiche hereafter shall sue to be ffreman of this Cytie by Redempc*i*on and to be enfraunchised in this Mysterye wit*h*out thassent of

Mr & Governors othe 2

Searches

the xxiiij[ti] Assistant*es* of the same Crafte or the more *par*te of them And over that ye shall not chardge the hole bodye of this ffellowship by puttinge the com*m*on Scale of the same mysterye to any maner wryting cause or matter whereby the same Companye in any wyse may be chardged hurted or hindered Also in all the premysses and other thing*es* necessarye concernynge the weale and proffyt of the sayd Crafte *tha*t shall truely lawfullye dilligentlye and indifferentlye behave your self after your connynge and power and nether for Love meede favour affecc*i*on nor for dreade mallice hatred or envye otherwise procede rule or conclude to or w*i*th any *per*son or *per*sons w*i*th whiche ye shall have to dooe by reason of your sayd Offyce then the good Vsages Rules lybertyes and ordn*au*nc*es* for the good order of the same Crafte heretofore made and not repelled and hereafter to be made Also at thende of your Offyce ye shall geve vnto the Maisters or Governors that shall succede youe next in the same occupac*i*on this present othe So that they shall duelye and truelye in all thyng*es* duringe the tyme that they shalbe in lyke offyce performe and fulfyll the same othe so god you helpe and all Saincte*s* and by this Booke.

othe to be geeven by old M[r] to the newe

Also yt ys ordeyned that euery *per*son enfraunchysed in the same Crafte shalbe readye at all maner of Som*m*ons of the Maisters or Governors of the sayed Crafte for the tyme beinge And yf any suche person absent him from any suche Som*m*ons w*i*t*h*out cause reasonable to be tryed by his othe before the Masters or Gouer*n*ors yf they thynke yt necessarye Then he to paye for euery so dooynge iij[s] iiij[d] the one half to the Chamber of London and the other to the Almes of the Crafte Also that euery man enfraunchysed in the sayed Crafte beinge duely warned or sommoned that kepeth not his hower according to the Som*m*ons w*i*thout cause reasonable to be fixed in maner and forme aforesayed for euery tyme so doying shall paye to the Almes of the sayd Crafte vj[d] And he or they that disobeyeth this ordyn*au*nce shall paye for his or their disobedyence

Absens after summons fine 3s. 4d. 3

hower of Summons 4

in that behalf for euery tyme so offendinge iij^{s} iiij^{d} to be devyded in forme aforesayd.

Also yt ys enacted and ordeyned that euery man in the Clothinge of the sayed mysterye shall paye quarterly to the Mayntena*u*nce of the com*m*on Charges of the same vj^{d} And euery man oute of the Clothinge and euery wydowe keping an open Shop iij^{d} And this to be payed quarterlye vpon payne and forfayture at euery tyme offending or doying the contrarye iij^{s} iiij^{d} the one half therof to be applyed to the vse of the Chamber of London and thother half to the Almes of the sayd ffelloweship so alwaies that the sayed quartrage be lawfully demaunded.

Liverie quartridge vjd. not of the liverie iijd. not paying quarteridge iijs. iiijd. to be demaunded 5

Also yt ys ordeyned that no *per*son enfraunchised in the sayd Crafte shall take any Apprentyce unto the tyme that he furste present the same *per*son before the M*aster* and Governors for the tyme beinge that they may see he be clene wi*th*out contynuall Dyseases or grevous infyrmyties whereby the kings leige people mighte take hurte upon payne for euery tyme so doing of xl^{s} to be applyed in maner aforesayed Also it ys ordeyned that euery *per*son of the sayed ffellowe-ship shall paye towards their charge for euery Apprentice that he taketh ij^{s} vj^{d} to be payed at the presenta*ci*on and allowyng of euery Apprentice

everie apprentice to be presented before he be bound payne xls. 6

presentment ijs. vjd. 7

Also yt ys ordeyned that no *per*son of the sayed ffellawship being in the Clothinge or lyverye shall have any mor*e* servauntes apprentices or fforeins vsing the facultye or mystery of Barborye or Surgerye together at ones above the nomber of iiij *per*sons provyded alwaye that wi*th*in half ayere of the goying out or endinge of the terme of one of the sayed iiij *per*sons yt shalbe leafull to euery suche *per*son to take or have another Apprentys or servant the sayd Acte notwi*th*stonding and he that offendeth in Breking of this Acte shall forfayte and paye xl^{s} to be devyded and applyed in forme aforesayed

none of ye liverie to have above iiij servants at one tyme payne xls. 8

9 outo of the liverie above iij

Also yt is ordeyned that no maner of parson of the same ffellowship being out of the Clothinge shall have togithers at ones above the nomber of three Apprentices or S*er*vauntes to occupye the sayed mysterye and facultye *pro*vyded as ys provyded in the latter Artycle and upon lyke payne

10 Ceasinge of servants wages

And also yt is ordeyned that no parson of the sayed ffelloweship shall take to his servyce as servaunte Alowes any Englishe fforein or Aliaunt Straunger to occupye the facultye of Barborye or Surgerye but the said *per*son shall present the same *per*son servaunt wit*hin* iij Dayes next after his comyng to the sayd *per*son to and before the sayd Masters and Governors of the sayed ffelloweship for the tyme beinge to thentent that he before them maye be cessed what wages he shall take And yf he be an Alyaunt straunger borne to paye yerelye of his wages to the Almes of the ffelloweship iij^s^ iiij^d^ And that monye to be taken quarterlye of the Maister of the same Straunger and of his wages and who *that* doeth contrary to this rule shall forfayte at euery tyme so dooyng xl^s^ to be devyded and payed as ys aforesayed

11 Entiseing awaie of servants

Also yt ys ordeyned that no *per*son of the sayd Crafte shall entyse or desyer any servaunt from his maister nor shall take any forrein into his servyce for lesse terme then for one yere and he to be sessed or rated for his wages by the Maisters or governors of the same mysterye and this to be dooen yerelye euery yere upon payne for euery tyme doying the contrarye of xiij^s^ iiij^d^ Thone half to be applyed to the vse of the Chamber of London and the other half to the vse of the Almes of the sayd ffellowship

12 no freemen to open shop before he be lycenced

Also yt ys ordeyned that no parson of the same ffelloweship after that he be admytted and sworne a ffreman of this Cytie afore the Chamberlaine presume to open his shop wyndowes before he hathe presented hymself to and before the masters or governors of the sayd

Mysterye for the tyme being and w*i*th them haue agreed in payinge his Dutye accustomed that ys to saye to the vse of the Company iij^{s} iiij^{d} and to the Clarke xij^{d} to the maynten*a*unce of their com*m*on charge and in taking his othe afore them according to the lawdable custome and order in the same mysterye of olde tyme vsed upon payne to Lose forfayt and paye xl^{s} to be devyded and applyed in forme aforesayd.

Clarkes fee xij^{d}

Also yt ys ordeyned that no *per*son enfraunchysed in the sayed mysterye shall enfourme or charge any fforrein other then his apprentyce any pointe of his crafte belonging to Barborye or Surgerye upon payne for euery tyme so doying xl^{s} to be applyed in fourme aforesayed Also yf any matter of Stryfe or debate hereafter be betwene eny *per*son of the sayd Crafte as god forfend That noen of them shall make any pursuit in the com*m*on Lawe but that he whiche fyndeth him agreved shall fyrste make his complaynt to the Maisters or Governors of the sayd Crafte for the tyme being to thentent that he shall order the sayd matter or cause of complaynte so made yf they can*ne* And yf it fortune that the can not nor doo not order and appease the same matter w*i*thin xiiij dayes then next ensuyng That then it shalbe leafull to the partye agreved to take his advauntage at the Comon Lawe So alwayes that the partye agenist whome the complaynte ys made be not fugytyve and who so dothe the contrary herof shall paye for euery tyme so Doyinge xiij^{s} iiij^{d} to be devyded and applyed in fourme aforesayd

not to teache any his science but his aprentices 13

not to sue one another at the common Lawe before complaint 14

Also that no parson of the sayed ffellowship shall revyle rebuke nor reprove another of the same ffellowship by any vnfytting opprobrius condyc*i*ons or dyshonest wordes in the presens of the Masters or Governors or anye of them nor before any other *per*sons in any other places and he that offendeth in this behalf and due proofe therof had shall paye for euery suche defaulte vj^{s} viij^{d} to be devyded and

not to use any opprobrious wordes 15

applyed in forme aforesayed Also that no parson of the sayed

no *person* refuse to be of ye Liverie Eleccion of *the* Liverie to be by Court of Assistants 16

Crafte shall refuse to be of the Clothinge of the sayed Mysterye or to beare offyce in the same at any tyme when he by the sayd Maisters or Governors and Thassistants of the same mysterye or the more *par*te of them shalbe abled therto upon payne to paye xl^s to be applyed in fourme aforesayed And that the maisters or Governors of the said mysterye for the tyme being shall not take nor admyt any *per*son into the Clothing or lyverye of the same mysterye wit*h*out the com*m*on assent of the xxiiij Assistants of the same or the more *par*te of them upon lyke payne as ys aforesayed for euery tyme so*e* Doying to be devyded and applyed in fourme aforesayed And where by Dyvers highe authoryties for the honor and reverence of the Sondaye it ys

not to barb washe or shave on Sundayes 17

ordeyned and enacted of olde antiquitie that no Barbour dwelling wit*h*in the Cytie or Suburbes of the same nor ells who shall occupye shavinge on the Sondayes neyther wit*h*in their house nor wit*h*oute prevelye nor apertlye It is nowe therfore ordeyned and enacted that no *per*son free of the sayd Companye from hensforthe occupye any Shaving on the Sondayes wit*h*in this Cytie nor lybertyes of

payne xl^s

the same upon payne and forfeyture for euery time so doyinge xl^s the one half therof to the Chamber of London and thother half therof to the Almes of the sayde Crafte

Also yt ys ordeyned that no man*n*er parson being free

to present euery hurt *per*son in *per*ill of his lyfe to the M^r & Governors wit*h*in iiij^{or} dayes 18

of the sayd fellowship shall take any seke or hurte parson or parsones to his Cure whiche is in perrelle of Deathe or Mayme but yf he shewe the same secke or hurte parson by him receaved to the Maisters or Governors of the sayed Mysterye or two of them for the Savegarde of the kinges Liege people and that wit*h*in iiij dayes next after the receyving of the said secke or hurte *per*son upon payne for euery tyme doyng the contrarye of xx^s to be devyded and applyed in maner and forme aforesayed euerie Surgeon to reade his Lecture when his turne cometh or finde a sufficient man 19

Also yt is ordeyned that euerye man enfraunchysed in the sayd felloweship occupying Surgerye shall come to their Hall to the reading of the Lecture concernyng Surgerye every Daye of Assemble therof and euery man after his course shall reede the Lecture himself or ell*es* fynde an able man of the sayd ffelloweship to read [for] hym and not to absent himself at his Daye of the same readinge w*ith*out cause reasonable and w*ith*out he geve lawfull warnynge therof before the daye upon payne to forfayte and lose for euery tyme doyinge the contrarye xx[s] to be devyded and applyed in fourme aforesayed

payne xx[s]

none to deceiue another of his cure 20

Also yt ys ordeyned that no man of the sayed ffellowship shall take any cure from another of the same ffellowship nor supplant one another nor geve or speke any slanderous wordes in disablinge him of his Scyence or conninge but be rather in a readynes to geve good counsell to helpe the king*es* people And euery man offending in this behalf to pay at euery tyme so offendinge xiij[s] iiij[d] thone half thereof to the Chamber of London and thother half to the Almes of the sayd ffellowship provyded that yf the pacyent fynde himself agreved w*ith* his Surgeon that then the same pacyent paying to his fyrste Surgeon reasonablye for his Labour shall and may take and have any other Surgeon at his Lybertye and pleasure

Sunday aft*er* St. Bartholomewe Dinners *the* day of St. Damian and Cosme except it be on saterd*ay*

And whereof olde Custome yerelye upon the Sonday next ensuyinge the feaste of Saint Barthelmewe the Appostell[1] a Dynner ys kept and provyded for them of the lyverye of the sayed Companye in their com*m*on Hall called Barbors Hall And on the daye of Saint Cosme and Damian[2] yf it be not on the Satterdays A Dynner for them of the same Companye oute of the lyverye It is ordeyned and enacted that euery man that hathe been upper Maister or upper Governor of the sayed Companye shall paye at

The upper M[r] xij[d] his wief viij[d] Liverey viij[d] his wief iv[d]

[1] 24 Aug. [2] 27 Sept.

and for the same dynner xij^{d} for himself and viij^{d} for his wyffe yf they come And euery other man beinge of the lyverye of the same Companye shall paye in lykewyse for himself viij^{d} and for his wyffe yf she come iiij^{d} Provyded alwaye that the Maisters or Governors of the sayd Companye for the tyme beeng shall paye nothinge for their wyfes com*m*ynge to the Dynner for that yere forasmoche as their wyffs must of necessitie be there to helpe that euery thyng there be sett in order. And that euery man of the sayd Companye beinge out of the lyverye shall paye at and for his dynner on the said morrowe viij^{d} and for his wyfe yf she come iiij^{d}

Mr & Governors nothinge for their wives

Every other *the* day after viijd. his wief ivd.

Also yt ys ordeyned that at euery assemblie holden in the com*m*en Hall of the sayd mysterye no man being there present shall multiplye language in the assemble tyme that is to saye yf any mans othere speke mor wordes or multeplye more Language in the Assemble then the Maisters or Governors for the tyme being then present thinke to be good and necessarye That then yf they or one of them comaunde him to kepe sylence that then he shall so doo in keping his obedyence

None to multiplie wordes at Assembles 22

And also no man commynge to any of the sayd Assembles shall departe from thens Duringe the Assemble tyme w*it*h*o*ut lycence of the Masters or Governors then present or one of them and the offender in anye of the sayd ij pointes or cases to forfayte and paye at euery tyme so offending xx^{d} to be devyded and applyed in fourme aforesayd

non to depart w*i*thout Licence of *the* Mrs or one of them

Also yt ys ordeyned that no man of the Clothinge of the sayd Companye *pre*sume to gooe one afore another of theym in processyons buryalls or Annyversaryes nor in sytting in their Assembles or in their Hall at Dynner or other repastes their[1] or in any other honest place to be had otherwise than he is in Auncyetie[2] in the same Companye and according to the true enteraunce therof in the

Senyoritie to be observed 23

[1] there. [2] antiquity.

Bedylls Skroll Nor that any of them of any scrupulosytie frowardnes ne pusyllanimytie refuse to take his owne rome or place according to the order aforesayd but that euery man in theis ij cases kepe and occupye his owne roume and place in fourme aforesayde wyll he nyll he in good obedyent maner And he of them that offendeth in breakinge the order in any of the sayd ij Cases shall forfayte and paye at *euery* tyme of offendinge xij[d] to be applyed and devyded in fourme aboue rehersed.

All whiche ordenaunces and othes in maner and fourme afore specyfyed at the request of the sayed Maister Wardeyns and ffellowship by aucthoritie of the same Acte of Parliament we the sayd Chauncelor Threasouror and Chief Justices of eyther benche aforesayd for good La*u*dable and lawfull ordena*unces* estatutes and othes doo accept and admytte by theis *pre*sent*es* as moche as in us is ratyfye lawde and approve Provyded alwayes and forseen that theis ordena*unces* *with*in this booke expressed or any of them in no wyse extende nor be preiudycyall or hurtefull to the kinges prerogatyve nether to the hurte of any graunte or graunt*es* by the kinge or his moste noble *pro*genytors made to the sayd Cytie of London or any other or any lawdable custommes nowe used in the same Cytie and in case any Article or artycles in this boke expressed be preiudycyall or hurtefull as ys abouesayd to be voyed cassate and of noen effect Any thinge by vs the sayd Chauncelor Threasourer and ij Justices in this behalf Dooen or made to the contrary not*wi*thstanding Provyded alwayes that for dyvers consyderac*i*ons as well for the welthe of the king*es* leige people as for the honestye of the sayde Crafte yt is now condescended and agreed that from hensforthe no parsons of ffelayship after he or they be made ffree of the sayd Companye shall presume to sett open any Shoppe of Barborye unto suche tyme as he or they be abled by the sayd Maister or gouernors *with*out he be of the clere value of his owne proper goods to the value of Ten*ne* m*arkes* sterlinge upon payne of fforfayture of xl[s] the one half to the Chamber of London and the other half to the Almesse of the sayd Crafte In wytnes whereof to this present Booke we have sett our Signes manuells the xiiijth Daye of Maye in the xxijnd yere of the reyne of our Sou*ereign* Lorde his reigne King Henrye the Eighte.[1]

Exd the xxvith Daye of Apryll Anno Dm: xv^{o} lvi^{m} [2] being the M^{r} of the Companye and Agreeing with the Original *per* me THOMAS KNOT.

[1] A.D. 1530.

[2] 2 Philip and 3 Mary.

APPENDIX J.

HARVEY'S GRANTS OF ARMS TO THE UNITED COMPANY OF BARBERS AND SURGEONS.[1]

To all and singular as well Kings heralds and Offycers of Armes as nobles gentilmen and others which these presents shall see or here[2] Willm Hervy Esquyer otherwyse called Clarencieulx principall herulde and Kinge of armes of the southe East and west parties of this Realme of Englande Sendethe due comendac*i*ons * * * And whereas in this citty of London thexperience and practise of the science & faculty of surgery is most requiset and dayly exersysed and experimented * * * And foreasmoche as within this said Cyttie of London there were two severall companies the one by the name of Barbors and the other by the name of Surgeons The Barbores beinge Incorporated and thother not and both occupied the arte of surgerye whervpon greate contention Dyd arise & for that yt [was] thought moste mete and necessarye that the said two Companyes shulde be vnited and made one bodye corporate to thentente that by ther vnyon and often assemblie togither the exercyse & knowlege of ther science myght appeare as well yn practice as yn speculacion not onelye to themselves but to others vnder them so that yt was thought most mete & convenient uppon graue & great consideracion To vnite and Jouyne the said two companyes into one as maye appeayre by acte of p*ar*l*i*ament in a° xxxij of henry theight with these wordes ffolowinge Be it enacted by the King * * * & for that it pleased the said Kinge henry theight not onely to vnytie and incorp[or]ate the two Companyies togyther by acte of Parlement but also hath Ratyfied & confermed the same by his Letters patentes under the great seale of Englond & wheras Thomas Gaele at this present beinge

[1] I. C. B., No. 101, 20, in the Heralds' College, London. [2] hear.

Mr Alexaunder Mason John Stondon and Robert Modesley governors of the said corporacion mystery and comunalty of Barbors & surgeones beinge Desyrous to have some Signes & tokines of honour added & augmented to their olde auncient armes for a perpetual memory as well of that famous prince kinge henry theight their founder & patronne as also a declaracion of the vnytinge of those t[w]o compaynes together hath instantly Required me the said Clarenciulx To consyder the premyses and to shew my indevour herein in consideracion wherof and findinge theyr request to be just & lawfull I the said Clarenciulx kinge of armes by power & authoritie to myn offyce annexed & graunted by Letters patents vnder the great Sealle of Englande for a testimony & further encrease of ther worshepes haue graunted & assigned vnto them for an augmentacion to ther old and aunscint Armes (which is sables a chevrone between thre flumes[1] argent) a chef paly argent & vert on a pale goles a lyon passant regardant gold betwene two spatters argent a roze gewles crowned golde & to ther creast vpon the healme an opinacus[2] golde standinge vpon a wreath argent and sables manteled gewles dubled argent as more playnely appereth depicted in this margent which armes with the said augmentacion helme & creaste I the said Clarenciulx kinge of armes haue Ratyfied confirmed gyven & graunted & by thes presents do ratyfie and conferme gyve & graunt unto the said Thomas ʒale at this present Master Alexander Mason John Stondon Robert Modesley governers of the corporacion mystery & cominallty of Barbors and Sourgeons & to ther successores & to the hole assistance company & felowshipe of the said corporacon mystery and comynalty of Barbors & surgeons within this cyttie of London and to ther successores for euer more and the same to vse beare

[1] fleams or lancets.

[2] An "Opinicus" is an heraldic beast having the body and legs of a lion, the neck and head of an eagle, the tail of a camel, and the wings of a griffin. My friend, Mr. Horace Noble, suggests that this animal was probably selected in allusion to the qualities required by a good surgeon, viz. the boldness of a lion, the keen vision of an eagle, the swiftness of a griffin, the patience of a camel. [Ed. note.]

and shew in shelde seall and banner banners or banerroles or standards penon or penons pensill or pensilles to ther honor & worshipes at their Lybertis & pleazure without Impediment let or intervpcion of any person or persones In wittines wherof I the said Clarenceulx kinge of armes have subscribed my name & set hervnto the seall of myn office & the seale of myn armes the xth day of July in the yere of our Lorde god 1561 and in the third yeare of the Reinge of our dread souvrayne Lady Elizabeth by the grace of god quene of England fraunce & Irland Defender of the ffeath etc.

The amended grant of arms to the united Company appears in the following form:

In consyderacion Whereof and for that it dothe appeyre a thyng most Requyset for the vnityng of theys two Companyes to gether and ffor that the Occupacion of barbors surgeons byeng Incorporatyd hath synce the xxx yere of kyng henry the Eygt vsed and borne armes that is to say sables a cheveron by twyne iij flewmes argent which armes were vnto them assinyd onely by the gyfte and assignement of clarenceoulx kyng of armes as by the patent thereof Dothe and maye more pleanly appere and since the vnityng of the sayd two companyes theys armes of the sayd Corporacion of barbores Surgeons hath byne vsed and none other yet notwythstandyng the late Kyng henry the eyght of ffamous memory assined and gave unto the Companye of the Surgeons onely a cougnisance w*hi*ch is a Spatter theron a Rose gules crouned gold ffor ther warrant in ffylde butt no Auctoritye by warrant ffor the bayryng of the same in Shylde as Armes and ffor that it pleased the same King henry the eyght not onely to vnyte and Incorporate theys two companyes together by acte of parliament but also hath Ratifyed and Confyrmed the same by Letters patents vnder the great Seale of England and so lately confyrmed by the quenes maigesty that now is.

And whereas thomas Galle in the thyrde yere of the quenes maigestyes Reigne that now is, byeng master, Allexaunder Mason, John Stondon and Robert Mudesley governores of the same Corporacion mysterye Comunaltye of barbores and Surgeons byeng Desyrus to haue some scynes and tokenes of honor Addyd and augmentyd to the olde and auncient armes of the barbour Surgeons not onely ffor a perpetuall memory as wele of that ffamous prynce Kyng henry the Eyght there ffounder and patron but also ffor a ffurther Declaracion of the unyting of theys two companyes together Dyd Instantly require the late Clarencieulx harvy to consider the primases[1] and to shew his Indevor then in Consyderacion whereof he ffyndyng there Request Just and lawfull Dyd graunt and gyue vnto them by the name of master and gouernors of the mysterye and Communaltye of barbores and Surgeons and to there Successours byeng mester and governours of the same Corporacion of barbores and Surgeons by his letters patents vnder his hande and scale bayryng date the xth of July in the yer of ouer lord God 1561 and in the thyrd yere of the Raynge of the quenes maigesty that now is Augmentacion in chyffe to ther olde and auncient armes wyth helme and creast to the same which chyffe was paly argent and vert on a pale gules a lyon passant regardant golde bytwyne two Spatters argent and on each a Dobell Rose gules and argent crowned gold and to theyr creast on a torsse sylver and sable a Opyinacus gold manteled gules Dobeled argent and further in the tyme that Robart balthrope Esquyer and Sargent of the Surgeons to the quenes maigestye then byeng Mester of the sayd mysterye and Communaltye of the barbores and Surgeons and George Vaughan Rychard hughes and George Corron governors of the same Corporacion the sayd Clarencieux harvy Dyd graunt vnto the sayd corporacion of barbors and Surgeons two Supporters to theyr armes which by ffore he had gyven vnto them that wais two lynxe in their proper Collor about there necks a croune with a cheyne argent as now pleanly apperyth by the sayd letters patents of armes.

Yet notwythstandyng vpon Dyvers and Soundrye Consyderacions shewin vnto vs the sayd Kinge of armes and at the Instante request of Richard Holmewood mester[2] of the sayd mystery and Cominalty, Nicolas Archenbolde Thomas Burston

[1] premises. [2] A.D. 1568.

and John ffelde governoures of the sayd corporacion wyth the sayd garter Clarencioulx and Norrey Kyngs of armes by power and actority to vs commytted vnder the greate Seale of England haue not onely at there Just and resonable request after consydered the primases but ffor our further testimonys and Increase of their worshype haue Ratifyed and Confyrmed theyr armes creastes and supporteres Biffore mencionyd in suche maner and fforme as here after folowys that is to saye quarterly the ffyrst sables a cheveron bytwyne iij flewmes argent the Second quarter perpale argent and vert on a Spatter of the fyrst a Dobele rose gules and argent crowned or over all on a crosse gules a lyon passant Regardant or, and to their creaste vpon the helme on a torsse argent and sable a Opinacus or, in his mouthe a . . . . [1] vert the flower sable manteled gules Dobeled argent suppurted *with* two lynxe in there proper coller[2] aboue there neckes a crowne *with* a chene argent as more clearly apperyth Depictyd in the margent which armes creaste and supporters we the sayd kyngs of armes haue confyrmed Ratyfyed gyven and grauntyd and by theys presents Do ratifye confyrme gyue and graunte vnto the sayd Rechard Holmewood master of the sayd mysterye and Comunalty nicolas Archenbold Thomas burston and John ffelde governours of the sayd Corporacion mystery and comunalty of barbores and Surgeons and to their Successors by the name of mester and governours and to the holl assistaunce company and ffeloshype of the sayd corporacion mystery and Comunaltye of barbors and Surgeons *with*in the Cyty of London and to their successors ffor ever more and the same to vse bayer and shew in shyld sealle banner or bannerrolls Standard or Standards penon or penons.

[1] I have been unable to decipher the name of the flower, but it is unimportant, as it is not mentioned by Dethick in the final and existing grant given in full in Appendix K.

[2] colour.

APPENDIX K.

GRANT OF ARMS BY SIR GILBERT DETHICK TO THE UNITED COMPANY OF BARBERS AND SURGEONS.

To all and singuler as well kinges heraudes and officers of Armes as nobles gentlemen and others to whome these presentes shall come be seene, heard read or vnderstand We *Sir* Gilbert Dethicke knighte also Garter principall kinge of Armes Robert Cooke Esquyr also Clarencieulx kinge of Armes of the Southe partes of England and Willm flower Esquyre also Norroy kinge of Armes of the northe partes of England sendithe gretinge in our Lord god euerlastinge forasmuche as aunciently from the beginninge the valiaunt and vertuose actes of excellent persones haue been commendid to the world and posteritie with sondry monumentes and remembraunce of their good Descartes Amongst the whiche the cheifest and most vsuall haue been the bearinge of shyldes called Armes which are none other thinges then demonstracions and tokens of prowisse and valoris diuersly distributid accordinge to the quallitie of the personnes meritinge the same To thentent that such as haue donn commendable seruice to their Prince or contry either in warre or peace or otherwyse by laudable and couragiouse Enterpryses or proceedinge of eny personne or personnes in the augmentacion of their estate or common weale of their Realme and Contrey might therby receiue due honor in their lyves and also deryve and continue the same successivly in their posterity and successores for euer And whereas in this Citye of London thexperience and practyse of the faculty and Science Chyrurgery is most requisit and dayly to be exersysed and experimented for the preservacion of many and by occasion of the practyse many expert psonnes be brought vpp and experimented to the reliefe succour & health of an infinyte number of personnes And forasmuche as within this Citye of London there were two seuerall Companyes one by the name of *Barbours Chirurgeones* and the other by the name of *Chyrurgiones* onely [the one] beinge incorporated and thother not. And both occupyeinge the arte of Chirurgery whervppon great controuersy did aryse and for that it was most meet and necessary that the sayd two companyes should be vnyted and made one sole body and so incorporatid to thentent that by their vnyon and often assembly together the exercyse and knowledge of their scyence and mystery might appere as well in practyse as in speculacon not onely to them selues but also to others vnder them so that it was thought most meet and conuenient vppon graue and great consideracion to vnyte and ioyne the said ij companyes in one which was donne as may appeer by an act of Parliament in Anno xxxij of henry theight with these wordes Be it enacted by the Kynge our Souuerayne Lord and the Lordes spirituall and temporall and the commones of the same that the sayd seuerall and distinct two companyes that is to say bothe the Barboures Chyrurgiones and the surgiones and euery personе of theim beinge a freeman of either of the said Companyes after the accustome of the sayd Cyty of London and their successores from hensforthe Immediatly be vnyted and made one entiere and whole body corperate and one ssociety perpetuall, which at all tymes hereafter shalbe called by the name of Masters and gouernours of the mistery and communalty of Barboures and Surgiones of London for euermore and by none other name. In consyderacon wherof and for that it dothe appeere A thynge most requysyt for the vnitinge of these two companyes togither: and for that the occupacion of the *Barboures Surgiones* beinge incorporate hathe since the tyme of Kynge henry the sixt vsid and borne Armes viz sable, a cheueron betweene three flewmes argent which Armes were vnto them assigned onely by the gifte and assignement of Clarencieulx kinge of Armes as by the patent thereof dothe and may most playnely appeere. And since the vnitinge of the sayde two companyes these Armes of the corporacion of Barboures Surgiones haue been vsed and none other yeat notwithstandinge of late kinge henry the eight of famous memory assigned and gaue vnto the company of Surgiones onely a recognisaunce which is A spatter thereon a Rose gules crowned gould with their warrant in field but no authority by warrant for bearynge the same in shyld as Armes, and for *that* it pleased the same Kynge henry the eight not onely to vnyte incorporate these two companyes togither by Acte of Parliament but also hath ratefyed and confirmed the same by his letters patentes vnder the great seall of England and also lately confirmed by the Quenes ma*jes*tie that nowe is And whereas Thomas Galle in the third yeare of the Quenes ma*jes*ties Reigne that now is beinge Mr Alexander Mason John Stondon and Rob*ert* Modesley gouernoures of the same Corporac*i*on, mistery and communalty of Barboures and surgiones beinge desyrouse to haue some signes and tokenes of honor added and augmented to the ould and auncient Armes of Barboures sourgiones not onely for a perpetuall memory as well of the famous prince Kinge henry the eight their founder and patron but also for a further declaracon of the vnytinge of these two companyes togither did instantly requyre the late Clarencieulx harvy to consyder the premisses and to shew his indeuore therin who fyndynge the request iust and lawefull did graunt and gyue vnto them by letters patentes vnder his hand and seale bearinge date the xth of July in the third yeare of the reigne of the Quenes ma*jes*tie that nowe is an augmentac*i*on in cheife to their ould and auncient Armes with healme and creaste to the same which cheife was paly Argent and vert one a pale geules a lyon passant gardant gold betwene two spatteres argent one eche a double rose geules and argent crowned gold And to their creaste one A torce silver and sable an Opinacus gold mantled geules doubled argent. And further in the tyme of Thomas Balthroppe Esquyre Sargent of the Quenes ma*jes*ties Surgions then beinge maister of the sayd mistery and communalty of Barboures and Surgiones and George Vaughan Richard hughes and George Correy gouernours of the same corporac*i*on the same Clarentieulx harvy did graunt vnto the sayd corporac*i*on two supporters to these Armes before gyven them which were two linx in ther propper colour about ther neckes a crowne with a cheyne argent pendent therat as now playnely doth appeere by the sayd L*ett*res patentes Yeat notwithstandinge forasmuche as it dothe playnely appeere vnto vs the sayd Garter Clarentieulx and Norroy kinges of Armes that the aforesayd Armes in some respectes were not onely contrary to the wordes of the corporacion of the sayd Barboures and Chirurgiones but that also in the same Patent of Armes ther are sondry other thinges contrary and not agreinge with the auncient Lawes & rules of Armes we the sayd kinges of Armes by power and aucthority to vs committed by l*ett*res patentes vnder the great seale of England haue confirmed geuen and graunted the aforesayde Armes creast and supporters heertofore mentioned to be borne in manner and forme heerafter specified viz quarterly the first Sables a chevron between three fleumes argent, the second quarter par pale argent and vert one a spatter of the first a double rose geules and argent crowned gold the third as the second the fourth as the first ouer all one a cross geules a Lyon passant gardant gold and to their creast vppon the heaulme one A torce argent and sable an Opinacus gold mantled geules Doubled argent supported with two linx in their propper colour about ther neckes a crowne with a cheyne argent pendent thereat as more playnely appeereth depicted in this margent.[1] which Armes creast and supporters and euery part and parcell therof we the sayd kinges of Armes haue ratefyed confirmed geuen and graunted vnto Richard Holmwoed maister of the sayd mistery and communalty nicholas Archenbold Thomas Burston and John feild gouernours of the sayd mistery corporation and communalty of Barbo*ur*s and surgiones and to their successores by the name of M*ast*ers and gouernours and the hole assistauntes company and felowshippe of the sayd corporac*i*on mistery and communalty of Barbours and surgiones within this city of London and to their successores for euermore And they the same to haue hold vse beare enioy and shewe forth in shyld seales banner or bannerroles standert or standertes penon or penones pencil or pencilles

[1] See the Frontispiece.

or otherwyse to ther honors and worshippes at all tymes and for euer hereafter at their lyberty and pleasures without the impediment lett molestacon or interruption of any other personne or personnes In Witnesse wherof wee the sayd Garter Clarentieulx and Norroy kinges of Armes haue signed *these* presentes with *our* handes and sett thereunto *our* seuerall seales of Armes the second daye of June in Anno 1569 and the eleuenth yere of the raigne of *our* souueragne Lady Elizabeth by the grace of god quene of England fraunce and Ireland Defender of the fayth etc.[1]

[1] Sir Gilbert Dethick's Guifts, 162, p. 99; in the College of Heralds, London.

APPENDIX L.

A SUMMARY OF SUCH THINGS AS THIS COMPANIE DOE BESECH TO HAUE CONTAINED IN *LETTERS* PATENTS FROM HIS *MAJESTIE* [JAMES 1ST, OCT. 5TH, 1604].

1. It recyted the acte of 32° Henry the Eighth, w*hi*ch incorporateth and uniteth the Companie of Barbors of London and the Companie of Chirurgeons of London into one body Corporat and comminalty, by the name of M*aste*rs, or Governors, and Co*mmin*alty of Barbers and Surgeons of London, And giveth power to Chose yearly fower M*aste*rs or Governors.

2nd. It ratifieth vnto them by the Kinges Majestie all grants, priviledges, powers, authorities, benefits, advantages, and other things whatsoever by forme*r* statutes or l*ette*rs patents graunted vnto them, by anie name or title whatsoe*v*er.

3rd. It giveth explanacion of form*er* graunts and farther supplies [1] in the *par*ts here ensuinge, viz.:

4th. That in respect of the greate enormytyes by vnskillful practizers of Surgerie, And for the suppressinge tryall and discovery of such, and well orderinge of the corporac*i*on,

5th. There shal be yearlie chosen xij of the Liu*er*y of that Companie as theie haue accustomed, w*hi*ch twelve shall Chose 4 of the Livery to be the M*aste*rs or Governors for one yeare, of w*hi*ch fower there shall twoo be professors of Surgery.

6th. That the fower Governors shall and maye take & vse, new search, examinac*i*on, imprest, correcc*i*on, and gou*ern*me*n*t of all freemen and others useinge the Misterie of Barbers or Surgeons w*i*thin London, the lib*er*ties, suburbes, or seaven myles compasse any waye.

7th. That the Gou*er*nors, or anie of [them] w*i*th one or twoe of thassistants, shall haue poure to enter into shoppes, howses,

[1] prays.

etc., of those useinge Barbery or Surgery within those lymmitts, as well to trye theire skill and experience as their instruments, vnguents, implaisters, medicynes, and recepts, and how theie do vse or misvsen, or not rightly vse or applye, the same.

8th. And to give tolleracion or allowance to the skillfull in such *par*ts wherein he shall so be found skillfull, so as theie submit themselves to the governm*en*t of the Companie, accordinge to his Ma*jesty*'s pleasure signified, And shall paie quarteridge as the freemen shall doe.

9th. And to suppresse the insufficient and vnskillfull, and to burne or destroy the vnguents and playsters, instruments, etc., w*hi*ch theie shall find corrupt, vnwholsom, or not fytteinge to be vsed.

10th. The M*aste*rs or Governors, by themselues or oth*e*rs, by warrants under three, or twoe of theire hand*es* and scales, to attach and imprison the contemptious infringer of theire ordinanc*es*, and the resister, refuser, or denier of search, etc.

11th. That such *per*son so to contynue imprisoned, and be received and deteyned by the g*a*olers, till hee shall haue submytted himself and become bounden to the M*aste*rs or Gouernors in [1] pound*es*, neu*er* to vse anie such vnwholsom or vnfytteinge vnguents, playsters, etc.

12th. It ordayneth and establisheth an assistance of xxvi of the said Company from tyme to tyme, to contynue for theire seu*er*al lives, except removed uppon iust occasion by the M*aste*rs and assistance, or the greater *par*t of them.

13th. The fower M*aste*rs or Governo*rs* to be yearely chosen oute of the xxvi.

14th. The now p*rese*nt fower M*aste*rs or Gou*er*nors, after theire yeare expired, to be and contynue of thassistance, to the ende that there shalbe alwayes xxvi besides the fower M*aste*rs or Governo*rs*, and no more, except for necessitie some of the M*aste*rs or Gou*er*nors shalbe chosen oute of the Liv*er*y, not beinge of thassistance, and then such, after his yeare expired, to contynue, notw*ith*standinge, of thassistance. Uppon death or remove of

[1] Left blank in the original.

anie of thassistance, another to be chosen by the M*aste*rs, or twoe of them, and the greater *par*t of Assistance.

15th. Item it giveth power of assemblyes of the M*aste*rs, Gou*er*nors, and Assistance, when theye shall appoynt; and the same to be called a Courte of Assistance.

16th. It sheweth the openinge, searinge, and embalmeinge of the dead corpes to be *proper*ly belongeinge to the science of Barbery and Surgery, and the same intruded into by Butchers, Taylors, Smythes, Chaundlers, and others of divers trades, unskillfull in Barbery or Surgery, and vnseemly and vnchristianlyke defaceing, disfiguringe, and dismemberinge the dead corpes, And so that by theire unskillfull searinge and embalmeinge the corpes corrupteth, and groweth *pres*entlie contagious, and ofensive to the place and *per*sons approachinge.

17th. Wherefore it *pro*hibiteth all *per*sons from so medling therein, Except such only of the said incorporacion as doe actually and vsually exercize the facultie of Barbor and Surgeon, and are allowed and approued accordeinge to the Statute.

18th. And giveth power to the M*aste*rs and Go*uer*nors to arrest and imprison such as doe to the contrary, till theire submission unto such fyne as shalbe payable by the said orders of the said Company confirmed, or to be confirmed accordinge to the statute of the 19th H. vii.

19th. For the better go*uer*nm*en*t of the said Companie, by reason of the increase and multitude of them, It giveth power to the M*aste*rs or Go*uer*nor*s* and Corte of assistance to constitute a yeomanry and rules for the better go*uer*nm*en*t of them and the comminaultye of the said Companie.

20th. For the better enforceinge of obedience to the rules and ordinance*s* of the said Companie, The said Charter giveth furth*er* power to the M*aste*rs or Go*uer*nor*s* to make ordinance*s* for the good of the Companye in and by theire Cortes of Assistance.

21st. And the Go*uer*nor*s* to ponish freemen foraynes and strangers *with*in the Lymmytts for offences contrary to theire ordinances made or to be made.

22nd. And to leveye fynes and amercemen*ts* by distresse, and Commytt offenders to p*ri*son till submission.

23rd. And in defalt of paim*ent* w*ith*in xx dayes, then to sell the distresse, and retayne to the value of the fyne and Chardges, and deli*uer* the rest to the owner.

24th. It freeth everye *per*son of this Society or incorporacion from constableship, watch, warde, all office or duety of bearinge arm*s*, inquests, Juryes, appearinge at musters, fyndeinge of men by sea or land, and taxe or assessm*ent* toucheinge such levies.

25th. Where a Charter of E. iv°[1] gave them power to purchase lands to the value of five marks *per* ann*um;* when theie weare onelye the Companie of Barbers; when by reason of the vniteinge of them and the Surgeons into one Company, and theire mayntenance and relief of the poore, and other occasion of Chardge and expence. This Charter confirmeth vnto them all lands graunted vnto them by anie name or tytle, and to receive oth*er* lands to the value of [2] pound*es* p*er* an*num*, not holden of his M*aje*s*tie in capite* or knight's service.

26th. In respect of the lykely occasion of the Kinges Majesty and his successors for his service in the warres of skillfull Surgeons whose skill is best knowen to the Surgeons of London, this Charter giveth poure to the M*aste*rs or Gou*ernors*, uppon notice given vnto them, to Chardge and empress w*i*thin the aforesaid Lymmytts hable men for such service, and to take from the vnhable to serve such instrum*ents* and stuffe of Surgery, as theie shall think fytt and necessary for the furnisheinge of them so impresst to service; and to such purpose to allowe or disallowe the *pro*vision of Surgery of the *per*sons imprest.

27th. And to such purpose, the M*aste*rs or Gouerno*rs*, by themselves or oth*er* hable *per*sons authorized to take view of the Chests, Caskets, and *pro*visions, from London to Lee, of anie *per*son so imprest, or to be ymployed, and to take away all Defective stuffe, and the same to destroy, that noe abuse be therby in the sayd seruice.

[1] King Edward the Fourth.

[2] Left blank in the original.

APPENDIX M.

VARIOUS REGULATIONS CONCERNING THE ANATOMY AND SURGERY LECTURES IN THE UNITED COMPANY OF BARBERS AND SURGEONS.

Jan. xiiijth 1566 [1]—Here was Mr doctor Julys & he made requeste y^{t} he myghte have the work of the anathomy these iiij or five years so y^{t} the coledge of the phicysions sholde not put hym from us.

Jan 17th 1574—It was agreed by the whole house that Mr Docto^{r} Smythe sholde worke upon Thanatomye ffor the space of thies iiij yeres next coming and yf he be sick or oute of Towne to take there choyse where they will.

11 July 1596 [2]—Dr Paddy appointed reader of the Anathomy Lectures.

23 Nov: 1609—Sr William Paddy kn*t* surrendered his place of reader of the Anathomyes Lecture and Mr Doctor Gwynn appointed in his place.

28 Mar: 1609—Dr Gwynn [M. Gwinne] read a Lecture upon a Body dissected.

17 Sept: 1612—This day it is ordered upon a motion by the Master propounded touching that one of the Colledge should read in this House the Weekly Lectures of Surgery on Tuesdays That the Master shall confer with Mr President of the Physicians College to see whether they will give consent that Doctor Davi[c]s or some other sufficient physician whom the Comp^{y} shall please, shall read the weekly lectures in our house. And if the president of the College shall not consent hereto then this house is to deal or to compound with some other of our own Company to read their Lecture in this House whereof the M*aste*rs are to make

[1] 1563, Dr. Will. Cunningham.

[2] Dec. 20th, 1577, "Mr. Thomas Hall to desect the Anatomies."

certificate unto the said Mr President or to take such order that the house may not anywise be charged towards the same Lecture.

6th Oct: 1612—Doctor Gwyn appointed reader of the Weekly Lectures with a stipend of £10 pr Annm.·

13 Decr 1627—Death of Doctor Gwyn reported to the company, weekly lectures to be read according to ancient custom by the surgeons of the Co approved according to law to begin with the antientest Mastr Mr Richd Mapes & soe after every Surgeon in his antiquity and degree in the Company.

9th April 1632—This Court taking into [cons*iderati*on the great care and pains of Mr Doctor [R] Andrews his agitac*i*ons & yearly reading of our lectures in time of the dissections of the public Anathomys for this four years past do now order that there shalbe given him £13 · 6 · 8 as of the free gift of this House for his pains therefore.

7 June 1632—Also as concerning reading Lectures in surgery this Court upon reading the order made for that purpose doth order Mr Mapes shall begin and read his lecture in surgery, & so every approved surgeon to read in his turn & ancientrye in the Companie.

6th Janry 1632—And so concerning the order for reading of lectures in Surgery by an approved surgeon of this Companie this court did againe deliberate upon the same & every one of the Assistants declared his opinion thereupon, & the plurality of voices was to have the lectures read by the approved surgeons of the house according to our ordinances & by a Doctor of Physick.

20th Novber 1632—Doctor Andrews appointed Weekly Lecturer in Surgery in consequence of a Letter from King Cha*rle*s.

28 Dec: 1632—Doctor Andrews excused Lecturr and Dr Read appointed.

28 Dec: 1637—Mem*oran*dum Upon the rising of the Cot of Assists it was concluded & agreed by the Examiners & Assists Surgeons that Mr Doctor [O.] Meverell an Ancient Physician of

the College shall be reader of an Anatomicall Lecture at the next public dissection to be held in the new erected Theatre.

8 Nov: 1638—D[r] [O.] Meverell excused Lect[r] & D[r] Pruieon appointed.

19 Aug[t] 1641—Also it is ordered that henceforward the Tuesdays Lectures shall be deliv[d] by the surgeons of this C[o] themselves and not by a Doctor & that the examiners shall meet & consider of the names.

23 Sept: 1641—Also that the Examiners are desired to meet & consider concerning Lectures on Tuesday next & that in regard of the present sickness this Court doth order that no Tuesday courts or Lectures be held till after the fortnight within the next Term.

30 Sept: 1641—This day the examiners taking into their consideration the manner of the reading of Lectures in surgery have thought fit & ordered that the surgery lecture should be read by approved surgeons only & the Lecture to begin by the first surgeon that is approved next to the Examiners & so every one by his turne to read the Tuesdays Lectures.

D[r] Pruieon to read the Six Public Anathomy Lectures this year.

27 Oct: 1645—This day M[r] Edward Arris acquainting this Court that a person a friend of his who desired his Name to be as yet concealed through his great desire of his increase of the knowledge of Chirurgery did by him freely offer to give unto this Corp[n] for ever the sum of £250 to the end & upon condition that a human body be once in every year hereafter publicly dissected and six Lectures thereupon read in this Hall if it may be had with convenience and the charges to be borne by this Company. And if no human body may be had nor conveniently dissected in one year, then the C[o] to distribute one half of the sum of the usual charges of a public Anatomy to our own poor & the other half to the poor of S[t] Sepulchre's. The said worthy overture is thankfully accepted by this court & it is ordered a Draught be drawn by our Clerk against the next Court of

Assist[s] for the performance thereof. And to that purpose a Rent charge of £20 per annum be granted out of our Lands at Holborn Bridge.

24 Nov: 1645—This Court taking into consideration in what manner the public Bodies hereafter shall be dissected and by whom that Anathomy which is now newly about to be established shall be performed doth think fit & so order that the present Mas[rs] of Anathomy or such others as shall be appointed by the two Mas[r] Surgeons for the time being and the more part of the Examiners shall performe the same and that the manner of dissections of every public Anathomy shall be such as they the s[d] two M[rs] or Governors for the time being & the Examiners & the more part of them shall direct.

13 Jan[ry] 1645—The Draught of the Instruments for establishing an Anathomy to be yearly hereafter betwixt the feasts of S[t] Michael the Archangel & of the Birth of our Lord God having been perused by the Comm[rs] on the part & behalf of the Company and of the Donor is read & allowed of. And do order that the present exam[rs] shall be the feoffees to be joined with the donor in the deed. And that the same be engrossed and sealed with the Common seal of an ordinary Court upon payment of the £240. And the Donors and Feoffees sealing the Counterp[t] th*ere* of.

6 Febr[y] 1645—This Court being informed from the Counsele that forasmuch as the Deed for the setting of the new Anatomy cannot be granted to all the Examiners as Feoffees Two of them being two of the present Governors unless the Deed be first made to a particular person & afterwards conveyed to them. This Court doth refer the manner of the Conveyance to the Donor his Councell whether it shall be granted to all the Examiners or to those Eight that are not now governors or to any other Brethren of this Company.

20th Feb: 1645—This Court doth agree that the Deed of an ann*uity* formerly granted to the use of the new public Anathomy be made for £24 upon the cons*iderati*on of £300. And it is promised by M[r] Arris on the behalf of the Donor that if the £300 shall be restored within 12 *years* he or his heirs shall grant unto this Comp*any* for the same use the like sum of £24 per annum out of some of his Lands or Teneme*nt*s. And do nominate & appoint for feoffees Mr Dunn Mr Collings Mr Kings Mr Pinder Mr Fleete Mr Arris Mr Boone & Mr Bennett.

24 Mar: 1645—This day Mr Edward Arris paid the sum of £300 to the use of this Compy and is the purchase money for the annuity of £24 per Annum for the use of the new Public Anatomy. Whereupon the Deed of Grant of the Sd Anny & for establishing the sd new Anatomy was sealed with the Common Seale & delivered to the Donor and the several Feoffees intrusted in the matter.

17 July 1646—Whereas £300 hath been worthily given to this House for the discharge of all expences to be laid out in & about a public Anathomy to be henceforth had yearly for ever between the Feasts of Michs & Xmas in every year. And for that Dr Prudion [Prujean] who formerly read the Anatomical Lectures hath desired to be excused from reading the lectures on the next Anathomyes to be dessected between Michs and Xmas next, this Court doth think fit that Dr [L. ?] Wright be desired to performe the same. And the Mr of the Anathomy for the time being when the said Anatomy shall be dissected do always in their several & respective times of Masrs of Anatomy dissect the sd Anatomy. And this Court doth think fit that the dissection of the sd Anatomy be of the Muscles of the Body. But that the manner thereof be left to the Judgmt of the readers and the Dissectors.

21 Sept 1646—Our Master acquainting the Court that Docr Prudion [Prujean] & divers other learned Physicians have recommended Doctor [Jno] Goddard as a man well qualified & very able to read the Anatomical Lectures, this Court doth order that Dr Prudion be requested to perform the Lectures on the next public Anatomy himself: but if he shall deny it, that then Dr Goddard or such other as Dr Prudion shall think more fitt be desired to read the Lectures.

7 Jan: 1646—Whereas this Court is well satisfied that

Dr Prudion is desirous to be excused from reading the next anatomical Lectures, this Court doth order that Doctor Goddard be desired to performe the same.

15 Ffeb 1646—This Court doth think fit & so order that the Tuesday lectures be again revived & read by surgeons freemen of this Compy in their turnes according to their Authority in the Livery. The Eldest Asst Surgeon to read the first lecture, & that to be on the first Tuesday in May next, and the other to be from thence monthly & no oftener viz The first Tuesday in every month. Provided nevertheless, that when as any such Tuesday shall not be within the time limited in & by an order of a Court of Assists of the 11 Aug 1563 in that behalfe or shall happen to be on any the days thereby excepted that then every such Tuesday be no Lecture day.

29 March: 1647—This Court doth explain the order of the last Court of Assists concerning the Tuesdays Lectures That it is the meaning of this Court and the Cort doth accordingly order That the sd Lectures be read as well by the ancient Masters Surgeons & Examrs in their course as by any others.

7 July 1647—This day Mr Coppinger moved this Court That inasmuch as two of our ancient Masters having been appointed to read the Tuesdays Lectures in their turns & had not read those lectures accordingly That this Court would be pleased to honor him so much as to permit him to read the next Lecture. This Court answered him that the next Lecture will not be till after Michs next & in the meantime it should be considered of.

23 Sept: 1647—Doctor Prudion excused reading the next Anatomicall Lectures & Dr Nurse appointed.

14 Jan: 1647—Dr Nurse appointed constant Anatomical reader.

8 Oct: 1649—This Court taking into con*siderati*on several werthy Physicians of whom one might be elected reader of the Anatomicall Lectures at the public dissections of this Co: do think fit that Dr Scarborough be elected thereunto, who being desired to come to this Court appeared during the sitting thereof & declared himself very willing to perform the same & rendered

thanks to this C° for their good opinion of him. Tuesdays lectures again revived.

12 Oct: 1649—Dr Scarborough elected Anatomical reader.

23 Oct: 1649—Surgery lectures revived & to be performed according to ancient practice the most ancient Masters reading the first lecture. *

30 June 1698—Ordered that there be an Anatomy Lecture called Gale's Anatomy. Dr [Clopton] Havers & Dr Hand being put in nomination for reading of the same Dr Havers was chosen for 3 years & to read on the 2d Tuesday, Wednesday & Thursday in July next by 3 of the Clock in the Aftn & to have 30s for his pains & the remr to be dysposed of by the Committee.

Oct 19th 1699—Ordered that two Doctors readers to this Society for the future shall be elected for no longer terme than 4 years only at one time. Dr Hand appointed reader in the room of Dr [E.] Tyson.

15 Apr 1708—Dr [T.] Wadsworth chosen reader of the Osteology Lecture.

16 Aug: 1711—The Court proceeded to the election of Readers for the Muscular Venter & Osteology Lectures & Dr Mead, Dr Freind & Wadsworth the present readers were unanimously chosen readers of the sd severall Lectures for the four ensuing years.

On the 24, 26 & 27 days of Dec 1711 was held at the Hall a public Lecture upon the Muscles performed by Dr R. Mead being Alderman Arris's gift.

7 July 1712—Dr Freind chosen reader of the Muscular Lecture.

Dr Comer [H. Colmer?] reader of the Venter Lecture.

18 Sept 1712—Dr Henry Plumtree chosen reader of the Muscular lecture in the room of Dr Mead.

Dr [J.] Douglas reader of the Osteology lecture.

15 Dec: 1715—Dr [J.] Douglas elected reader of the Muscular lecture.

Dr [W.] Wagstaffe upon the Viscera.

* 1662. Dr. Tearne. See "Pepys' Diary," Feb. 27, 1662-3.

13 Mar: 1716—D^r [W.] Barrowby elected reader of the Osteology Lecture.

6 Nov: 1717—Dr Plumtree chosen reader upon the muscles in the room of Dr Douglas resigned.

20 Aug: 1720—Dr Wagstaff chosen reader of the Muscular Lectures.

Dr Barrowby upon the Viscera.

Dr Jewin [James Jurin ?] upon the Bones.

30 Oct: 1721—Dr Jewin chosen reader upon the Viscera in the room of Dr Barrowby resigned.

Dr Chas Bale Osteology.

29 Mar: 1722—Dr Bale being in France, D^r [W.] Rutty elected reader of the Osteology lecture.

20 Aug: 1724—Dr Jewin elected reader upon the Muscles.

Dr Rutty upon the Viscera.

Dr Deodate [John Diodati] upon the Bones.

6 June 1727—D^r [E.] Wilmot elected Osteology Lect^r in room of Dr Deodate dece*ase*d.

15 Aug: 1728—Dr Rutty elected Muscular lecturer.

Dr Wilmot Viscera.

Dr Martlett [Lawrence Martel ?] Osteology.

7 April 1730—Dr Goldsmith [J^no Gouldsmith ?] chosen reader of the Osteology Lecture in the room of Dr Martlett resigned.

13 Aug: 1730—Dr Wilmot resigned & D^r [F.] Nicholls elected visceral Lecturer.

Dr Goldsmith elected Muscular Lecturer.

D^r [R.] Nesbett Osteology.

13 Aug: 1730—Mr Joshua Symmonds chosen Demonstrator or Teacher of Anatomy for 3 yeers.

5 Mar: 1730—Mr Symonds resigned & Dr Nourse elected in his room.

17 Aug: 1732—Dr Nesbitt chosen reader on the Muscles in the room of Dr Goldsmith dece*ase*d.

Dr Ruffiniar Osteology Lecturer.

5 Mar: 1733—Dr Nourse resigned the Demonstratorship.

17 July 1735—Dr Nicholls appointed Osteology Lect^{r} in the room of Dr Ruffiniar resigned.

21 Aug: 1735—Mr Ab^{m} Chovatt & Mr Peter Maccullock appointed Demonstrators of Anatomy.

19 Aug: 1736—Dr Nicholls elected Muscular Lecturer (Arris' Lecture).

Dr Nicholls elected Osteology Lecturer (Gale's Lecture).

Mr Chovett resigned the office of Demonstrator & Mr Macculloch & Mr Hawkins elected.

7 Dec: 1738—D^{r} [R.] Banks elected Viscera Lecturer.

16 Aug: 1739—Mr Macculloch elected Demonstrator.

10 July 1744—Mr W. Bromfield elected demonstrator in the room of Mr Macculloch dec*ease*d.

APPENDIX N.

A COPY OF THE BISHOP'S LICENCE TO A SURGEON.

A licence granted to Mr. Samuel Holditch, Chirurgeon.

GILBERT by the Providence of God Bishopp of London to all Xtian people to whom these *pre*sents shall come sendeth Greeting Whereas heretofore for the avoideing of many grievous accidents daylie appearing to many of his Maties louving subjects by the unskilful practicers of the Arte and Science of Chirurgery It was carefully provided by an especiall Act of Parliament made for the refotmacon thereof in *the* third yeare of the Reigne of our late Soveraigne Lord King Henry the Eight of famous memory That itt should not be lawfull for any person within this Realme of England to use or exercise the arte or science of Chirurgery except he were first examined approued & admitted according to the tenor of the said Statute Know ye therefore that wee the Bishopp aforesaid having received sufficient testimony (from John Frederick Esqr. Thomas Allen Abraham Clerke & Thomas Bowden M^rs.^ in Chirurgery heretofore approued & admitted according to Lawe to use and exercise the said Arte & Gouno^rs^ formerly of the mistery and cominaltie of Barber Chirurgeons of the Cittie of London incorporated) of the due examination and Tryall of Samuel Holditch a Freeman of the said mistery and cominalty & one of the cloathing of the said Corporation and findeing by the opinion of the said John Frederick Esqr. and Thomas Allen Abraham Clerke and Thomas Bowden that *the* said Samuel Holditch is a skilfull sufficient and able Chirurgeon and a very fitt man to use and exercise the said Arte and Science He being first examined by the examiners appointed and authorized according to Lawe for Examinacion and approbac*i*on of Chirurgeons (as by a testimoniall under the common scale of the said Corporacion a true coppie whereof remaineth in our principall regis*t*ry more att large may appeare) Doe now by these presents approue of the said Samuell Holditch to be an able & sufficient Chirurgeon (He being first solemnly sworn before John Exton Doctor of Lawes Surrogate of the right Wor^ll.^ Mr. Richard Chaworth Doctor of the Laws and our Chancellor to the Supremacy of the King's most excellent Matie And by these presents wee doe admitt him the said Samuel Holditch to use & exercise the said Arte or science of Chirurgery soe farre forth as by the Lawes and Statutes of this realme of England wee may lawfully admitt him In witnesse whereof wee have caused the seale of our said chancellor (w*hi*ch we use in this behalf) to be hereunto affixed Dated the fifth day of July in the yeare of our Lord One thousand six hundred sixtie one And in the first yeare of our consecracion

Ri Butler Regi*str*ar*us.

Jo Exton Ex^c.^

APPENDIX O.

TABLE OF FEES REQUIRED FOR THE VARIOUS GRADES IN THE UNITED COMPANY OF BARBERS AND SURGEONS.

Payments.	Surgeon's pay. £	s.	d.	Barber's pay. £	s.	d.
For the freedom by purchase . . .	10	10	0	6	6	0
For the Livery's fine	10	0	0			
For the fines for all offices to the Parlour door, the fines on the other side included, except the fine for the Lady's feast	25	0	0	25	0	0
For every examination for the great diploma	6	6	0			
For the fine for the four several offices of Master & the 3 Wardens of the Company which the Surgeons often pay, but the barbers never do, sometimes 30 guineas: but oftener . . .	40	0	0	40	0	0
For the fine for the Master and Stewards of Anatomy when called upon in turn & if they serve the expence is rather more	40	0	0			
	£131	16	0	£71	6	0

N.B. The Clerk & Beadle are not included.

Payments to the Poor's Box.	£	s.	d.	£	s.	d.
Paid by Everyone taking the freedom .	0	5	0	0	2	6
For the admission of a foreign brother surgeon	1	1	0			
For the Livery	0	5	0	0	5	0
For the diploma of Surgeon	0	10	6			
For the fine to the Parlour door . .	1	1	0	1	1	0
For Master & Stewards of Anatomy .	1	1	0			
	£4	3	6	£1	8	6

For expences of admitting foreign brothers.

	£	s.	d.
For the examination fee	7	7	0
The Poor's box	1	1	0
The Clerk's fee for diploma & bond	1	9	0
The Beadle's fee	0	5	0
	£10	2	0

APPENDIX P.

PROPOSALS FOR ESTABLISHING A SCHOOL OF ANATOMY AT SURGEON'S HALL, IN ORDER TO RAISE BOTH PROFIT & REPUTATION TO THE SURGEONS' COMPANY.

FIRST that a Professor be chosen out of the Members of the Company by the Master, Wardens & Court of Assistants. His Office shall be to teach Anatomy three days in the Week, throughout the whole Year to all such of any Profession, who shall enter with the Consent of the Master & Wardens. For which his Salary shall be £120 *per annum.*

The Price to each Pupil shall be Five Guineas for a Year: five Shillings of which shall be paid to the Beadle for attending all Lectures, & giving such other necessary Assistance as the Professor shall require, & the other Five Pounds shall be equally divided with the Company & the Professor.

That the two Masters, two Wardens & two Stewards of Anatomy that are annually appointed, each of them do their respective Duties by dissecting, demonstrating & reading twelve Lectures in the Year, on the four public Bodies allowed by Act of Parliament, otherwise on refusal to fine as by the By-law of the Company, on the said refusal the Professor to give the said public lecture. The Professor to make such preparations as shall be necessary for carrying on the Lectures, which shall be deposited in the Library from time to time.

The Professor may be at liberty to make what Advantage he can by any preparations over and above what shall be necessary for the Lectures, & all dissecting pupils to be for the Professor's sole advantage.

The Professor to be at equal Expence with the Company in

the purchase of any Bodies that may be necessary to be had over & above those allowed to the Company. That all Members of the Company may be present at every private or public lecture, paying ten shillings *per annum* each: the Masters, Wardens & Stewards of Anatomy excepted.

That each Person obtaining the Grand Diploma pay towards Anatomy two guineas, for which he shall have the Priviledge of attending all the Courses of Anatomy in that year. That each apprentice bound at the Hall pay one guinea towards Anatomy.

That each Person examined in order to qualify himself for the Navy, Army or *East India* company pay towards Anatomy five shillings.

That the Wardens & Stewards of Anatomy attend at the appointed Hour of giving every Lecture to demonstrate the Parts read upon, & to keep all Things in order during that Time.

At all Dissections etc: none to be present but the Members of the Court of Assistants and those employed in preparing the Parts for lecture.

The Number of Compleat courses in the Year will be three to be given by the Professor, besides twelve public Lectures by the Masters of Anatomy.

The Expences attending the foregoing plan:

	£	s.	d.
Professor's salary	120	0	0
Six adult subjects for the 3 courses at £2 2ˢ for muscles	12	12	0
Three small ditto at £1 1ˢ for the Viscera . .	3	3	0
" " " " Bloodvessels .	3	3	0
" " 10ˢ 6ᵈ " Nerves . .	1	11	6
Injections, glasses for preparations, subjects, etc. .	40	0	0
Allowance to the Beadle out of [say] 50 pupils at 5ˢ each.	12	10	0
Porter to clean & bury the flesh	5	5	0
Total of expences .	£198	4	6

	£	s.	d
By the foregoing plan it is supposed that Fifty pupils enter the first year at £5 5s each . .	262	10	0
Suppose out of the appointed Masters, Wardens & Stewards of Anatomy, three should refuse .	63	0	0
Suppose twelve Grand diplomas in the year at £2 2s each	25	4	0
Suppose twelve Apprentices in the year at £1 1s each	12	12	0
Suppose Fifty from the Navy, Army & East India Company at 5s each	12	10	0
Suppose 200 Members of the Company to pay 10s. each	100	0	0
	£475	16	0
Expences			
Total of Expences .	£198	4	6

	£	s.	d.
Clear profit to be divided between the Company & the Professor	277	11	6
One half of which is	138	15	9

APPENDIX Q.

REGULATIONS AS TO THE ANATOMY LECTURES AT SURGEONS' HALL.

At a Committee appointed to prepare conveniences for the Lectures the 12th July 1753.

Present—Mr Singleton
Mr M. Hawkins
Mr Nourse
Mr Pott Mr Crane Mr Hewitt
Mr Minors Mr Hunter.

To procure a proper table for the body in the dead room.

To make a door opposite the passage to the Theatre for the mob, & part off the passage to prevent them coming to the steps.

To have proper hatches or bars to separate ye members of ye Co: in ye Hall from the Court of Assistants & prevent them getting into ye Theatre till ye Court are seated.

To fix iron spikes in the Theatre to prevent the Mob getting over the outer rail.

To fix an iron chain from the top of the Theatre.

To take away the present Table in the Theatre & gett a new one after the model of Mr Hunter's or Mr Minors.'

To alter the reader's seat in the Theatre & make room to it.

To gett board to bring the Body in.

To line & make shelves to the closet in ye dissecting room.

To have new gowns for the Court of Assis^{ts}.

APPENDIX R.

SPEECH DELIVERED BY MR. GUNNING AT THE END OF HIS YEAR OF OFFICE AS MASTER, AT A QUARTERLY COURT OF ASSISTANTS HOLDEN AT THE THEATRE ON THE 1ST DAY OF JULY, 1790:—

Present: Messieurs Gunning, Lucas, Hawkins, Pyle, Warner, Watson, Minors, Harris, Pitts, Graves, Patch, Walker, Cooper, Wyatt, Hunter, Earle, Grindall, Long, Wathen.

The Minutes of the last Court of Assistants holden on the 1st day of April last were read and confirmed.

The Master reported that since the last meeting of the Court of Assistants he had with the concurrence of the Wardens, purchased the sum of £500 3 ℘ Ct. Consolidated Bk Annys on the Company's Account which at 73 and 7/8 ℘ Ct. and Commissn had cost the Sum of £370, and which purchase makes the whole Sum of Bank Annuities belonging to the Company to be £6300.

The Master from the Committee of Accounts reported that since the last report made on the 1st day of April last the Company's receipts including the Balance of £489 : 5 : 8, then in hand had amounted to £1084 : 6 : 2 & the expenditure including the money paid for the said £500. Bank Annuities to the sum of £774 : 18 : 7 and that the balance now in hand exclusive of £6300 three ℘ Cent Bank annuities & the Intt due thereon is £309 : 7 : 7 and that every Bill Tax & Salary is paid up to this time.

The Master read to the Court several observations relating to the affairs of the Company which being read it was moved and seconded that the same be entered upon the Minutes of this

Court and the question being put it was resolved in the affirmative and such observations are as follows.

Upon the point of taking leave of you Gentlemen on the expiration of my Office, it is necessary for me I think to tell you, that if I have been in any respect deficient in my duty, it has been owing rather to an Error in my Judgment than to any want of Industry, or intention to serve you.

By means of some regulations which were proposed by me and which you were pleased to approve of, I have the pleasure of informing you, that next Year the Sum saved in the Article of Dinners only will be found to be £170 & upwards when compared with the charge on that Account in the years 1788 and 1789, & if to this is added the sum of £50 or £60 being heretofore expended in a matter, which can never happen again, the whole will then amount to £230 independently of what may arise from retrenching all other unnecessary Expences, and this Sum will be nearly sufficient for your extraordinary expences should you come to any resolution of providing for them.

With respect to the Cash I have left very little in Bank, having funded all that I could; but what is of much greater moment, I have ordered all Bills Salaries & Taxes to be paid up to this day, that the Company may for once know what their necessary expences in one year really are.

Before I retire from my Office I will beg your indulgence for half an hour whilst I lay before you some Observations relating to the present state of the Company, & I am induced to do this from a hope that it may tend to bring forward a more particular enquiry into your affairs.

When I entered on my Office, I determined to make myself as well acquainted with the affairs of the Company as I could, & finding it difficult to obtain sufficient Information from those who had gone before me in this place, I was obliged to have recourse to your Books & papers & having spent as much time in reading & examining them as I could spare from my other avocations I shall think myself well rewarded if anything I shall suggest, may conduce in any degree to improve the present

System & enable us the better to answer the end of our Institution.

The remarks I intend to make shall be principally confined to the five following heads.

1st Committee — 2d the Hall — 3d Servants
4th Anatomy — 5th Charities.

Committee.

Our Books have been in general kept in a very irregular manner, tho' I am ready to acknowledge that our present Clerk has paid more attention to this point than any of his Predecessors.

No Entries have been ever made of what passed in the Court of Examiners & as to the Minutes of the Court of Assistants they have never been signed by the Master, and the drawing & posting them have been left almost entirely to the Clerk. It has not been usual to enter at large any Motion or proposal till it has been confirmed by a subsequent General Court; so that when any motion happens to be rejected nothing is to be found in your Books concerning it; & as a Motion tho' rejected at one time, may, in the whole, or in part, be of use at another, this practice may prove very prejudicial to us, & ought to be altered.

No punctuality has been observed in the discharge of your Bills & at present they are not so regularly paid as they ought to be. This made it very difficult to ascertain the precise expenditure of each year, as it often happens that a Bill for an Expence incurred in one year, has not been paid till a year or two after.

The nature of your Constitution Gentlemen, respecting your Accounts seems to be this Vizt. That the Clerk is responsible to the Governors of each year, the Governors to the Court of Assistants, and the Court of Assistants to the Company at large, and it appears from the Books of Accounts from the year 1745 to the year 1778, that the Master & Wardens, or Governors as they are sometimes called, did on quitting their Office crave an allowance for Sums expended by them, for that is the form of their prayer & those are their Words. Auditors from 3 to 8 were appointed on the annual change of Officers for the examination of Accounts & giving discharges.

They met in July to examine the Accounts of their Officers of the preceeding year, whose Duties terminated in the beginning of the same month; sometimes earlier in the month, sometimes later. Within a short space of time they deferred this Report till August & soon afterwards till September; at length it was deferred till the February & March of the succeeding year & sometimes later, & this Order, if it can be so called was observed till the Resolutions of 1778.

There were amongst others, two considerable defects in this plan. Your Officers were dismissed before their Accounts were passed & when properly speaking they were not responsible, you subjected them to an Enquiry. In fact these Officers or Governors or Servants, which ever you please to call them, ought not to have been dismissed till they had made up their Accounts; & there was no regularity in the Audit, late as it took place, whether on the 1st month or on ye 9th. To the above respecting your Accounts or Bills, I must add, that there is scarce any instance on your Books of Check, Complaint, or Defalcation, or prompt payment. The Bills were brought in late, paid late, & audited late; at last they scarcely made their appearance at all.

However and at length, whether there was a Suspicion or not that things were going on wrong with respect to their late Clerk to whom the conduct of almost every thing had been trusted I cannot tell, but the Court of Assistants came to certain resolutions on the 2d July 1778 which had they been duly observed would have saved the Company from the loss which soon followed. The resolutions were these; 1st That the Court of Examiners should be a standing Committee of Accounts with an Auditor General, the Master or Auditor with any two to be a Quorum.

2ndly That they should meet every Month at 1/2 an hour after 2 o'Clock that their business might be done before Dinner, so as not to interfere with the common business of the Day.

3rdly That the Clerk should lay before the Committee, all Bills whatever to be examined before they were paid, and an Order signed, if approved of, for their payment.

4thly That no repairs of the Hall should be directed without the consent of the Committee.

5thly That the Beadle should give an Account of what he had received for the Quarterage.

6thly That the Clerk should report at every Meeting what Sum of Money remained in his hands.

7thly That this Committee should be empowered to order Bonds to be purchased or any other public Securities they shall think proper with the Money over and above what may be thought necessary for the current Expences.

These Resolutions were in most instances very good, but they were deficient in two capital points: they did not direct an immediate payment of the Bills when they became due, or a comparing of the present Bills with those of the preceeding year that they might have check'd the excess had it become necessary. Your resolutions however had this further effect, that they lessened the responsibility of your Governors, and threw it upon the Examiners and their Committee at large.

Would you suppose, Gentlemen, however with the above regulations that within a space shorter than two years, the Company should have suffered their Clerk such was their blind confidence, to abscond with 8 or £900 of their Money the very Security for his Fidelity in his own possession.

To be short, Monies were received & suffered to remain in his hands; few or no Bills were brought in, & the Sum of £300 was even advanced by the Company on his own Bond. At length within the course of a few Months some reflection took place, the Monies were called for, and the confidential Servant was called for also; but he was not to be found.

Perhaps, Gentlemen, you will think that all this has nothing to do with the present Accounts, and that it is a painful and ungracious retrospect at the best.

I reply; that it is a transaction which should be always before your Eyes as no one till this deficiency happened was thought to stand on higher ground than the Servant now alluded to.

I assure you Gentlemen, I am astonished that this supineness in the responsible part of our Company, was not made the subject of an Enquiry.

It is easy enough to point out what ought to be done for y^e^ œconomical conduct of your affairs, & for your Security. The Bills should be sent in as soon as due. They should be examined compared & chequed and ordered for payment at every monthly Court; they should on that be immediately discharged, & at every quarterly Court an Account should be produced, & a balance struck. Any sum beyond what is necessary for the expence of the ensuing month should be funded, and no Monies should be left useless in the hands of a Banker. And these Directions differ very little from what you yourselves have enjoined.

I cannot however, dismiss this Article without proposing to you something further, you have done a great deal in regulating the number & expense of your Dinners but you may do still more, & I am confident the alteration will be extremely beneficial to you, it is, that the Court of Examiners, & your Committee of Accounts instead of meeting at 1/4 before 4 at the Tavern & dining almost immed^y^ afterwards should meet at their own Hall at 1/2 an hour after 5 precisely, an hour before the Examination begins, a period of time which will give them sufficient leisure for the inspection of your Accounts, that the 8 Dinners which you have allowed in the year for the Court of Examiners separately, sho^d^ be abolished, & that in lieu of them each of the Gentlemen who attend should receive the usual ffee of half a Guinea, which is paid to the Court of Assistants at large.

The utility of this is obvious. To many the half Guinea is

preferable to the Dinner. It would be rather a saving to the Company, as the charge for each Dinner would seldom be less than £6 and this could never exceed 5 Guineas & be sometimes less. Your Accounts would be well inspected, & you wo^{d} be less heated & sooner ready for your Examinations. Every thing that is really for the good of the Company you have the power to do, tho' nothing to its prejudice.

To the alteration of a Custom so much to your advantage, no solid objection can be made. You have encreased your own Fees, you have added others & you have raised the Emoluments & Gratifications of your Servants in points not quite clear & rather problematical.

The Meeting 4 times a year of the Court of Assistants, the governing part of our Society, seems to be quite sufficient to answer any solid purposes; and the proposed alteration, which is certainly for the good of the Company, being a matter which principally relates to the Court of Examiners, did they but agree to it, I should hope that the Court of Assistants at large could have no objection.

But if the Gentlemen should be of a contrary opinion & prefer the usual custom to the method now proposed, I should then think that a Committee of 3 should be appointed to meet quarterly on the Thursday preceeding the Meeting of the General Court, to inspect, regulate and check the Bills, & that their Members should be paid the usual ffee for such attendance.

I apprehend that such an appointment, even at this expence, would prove good management upon the whole, & that it would fully answer the purposes of its Institution.

If Gentlemen will not be prevailed on to do their duty without being paid for it (a custom very frequent now in much higher concerns than ours) let any reasonable allowance be made to them. I am confident it will be much to your Interest to invite them in this manner to do their duty. After having resolved in the year 1778 that the Court of Examiners which met every Month sho^{d} be a Committee of Accounts, why did they not pay some attention to this important matter.

How many Hundred Pounds (I am inclined to think 1500 or 2000) would have been saved by this means? But nobody as an Individual being interested in it & the attendance being inconvenient no regard was paid to this salutary measure. If the present ffee for such an attendance be tho^{t} insufficient let it be doubled, as I am convinced that the resolution is a wise one, & if properly pursued will produce the happiest consequences.

And now taking leave of the Accounts as far as they relate to ye Committee I propose that the Bills, if they are not paid monthly, should be paid quarterly at the furthest; that no money should be left with any Banker, nor more in the hands of your Clerk, or in your Chest, but what is necessary for one Months Current Expence.

You will by these means know the Expences incidental to each year, you will discharge them all within the year; and you will have your articles cheaper for prompt payment.

To this care of your Account succeeds the consideration of retrenching many unnecessary Expences, or at least the not suffering them to be continued.

The Expences incidental to your Committee have on an Average for the last 4 years amounted to about £420 or £430 ℘ Ann. For the succeeding four I hope and believe they will be within £250 each Year if we proceed with due care and caution.

I have further to recommend to you Gent. to insert the Summons in the Minutes of each Court of Assistants, & to suffer no Business to be transacted but what is expressly mentioned in the Summons; to enter at large whatever comes before you by way of Motion, & to keep Minutes of the transactions of the Court of Examiners. That all Minutes be signed by the Master in the course of the ensuing Week if not immediately. I recommend it to you likewise that a Book be made out containing all the rules & customs which are not printed but which ought to be known, that we may not be obliged to depend on any one for the information of the moment.

I will close this Article but with one observation more. I am confident Gents that you are insensibly led into many Expences from not having had time given you to consider them I therefore really think that no Donation, or Gratuity or repair, or the purchasing of ffurniture or any Expence of what kind soever sho[d] be incurred not only without a previous Notice being given in the Summons but that such measure should not operate till the determination of one General Court has been confirmed by that of the succeeding one; such a proceeding as this would sufficiently guard us against the ill effects arising from surprize.

Hall.

Respecting the Hall, Gentlemen, If there be no substantial repairs wanting (which a Surveyor should be appointed to examine & to make his report to you thereon) your annual expence on this head may amount to £250 including Taxes & what may be necessary after having put it into complete repair to preserve it so.

But this is but a part, & the smallest part of the charge for including the original Expence of 4 or £5,000 in the Building, this House may be estimated as costing you £500 ℘ Ann. & upwards.

You have in it a Theatre for your Lectures, a Room for a Library, a Committee room for your Court, a large room for the reception of your Communitys together with the necessary accommodations for your Clerk.

But how great soever your intentions were I am sorry to observe they have been but very ill executed.

Your Theatre is without Lectures, your Library Room without Books is converted into an Office for your Clerk, and your Committee Room is become his Eating parlour; and is not always used even in your Common Business & when it is thus made use of, it is seldom in a fit & proper state.

Every body on the least reflection must see & feel a great indecency in all this. The Court want these Rooms for themselves, I do not mean to have them shut up, but I wish to have the proper use of them, and that they should make the proper appearance. To continue to use them for other purposes than those for which they were originally designed will cause a great increase of expence in your furniture, your Coals, your Candles; but what is still more detrimental, the lower part of your House is by this means uninhabited, & I need not observe what damage the whole Building may receive from hence in a short space of time.

If your Committee Room is to be converted into an Eating Parlour, why should we not eat in it ourselves?—Your Dinners at the Tavern are exceedingly inconvenient & expensive & attended with a great loss of time.

You meet at the Hall, You adjourn to the Tavern, you return to the Hall again, & all this, when you have a House of your own inhabited at a great expence to you & where if you have not already all the conveniences you want you may at any time be furnished with them.

If Gentlemen, you make no better use of the Hall than what you have already done, you had better sell it, and apply the Money for the good of the Company in some other way. Your Hall has cost you a Sum during the last 4 years very little less than £1,200 & the Bills for repairs, independently of the Taxes and Ground Rent have amounted to upwards of £250 within the last two.

I could wish Gentlemen, on this head, that the laying out any Sum of Money beyond 5 or £10 may not be left to a Committee of 2 or 3 as has been usual I believe to the Master and Wardens chiefly—But that a Surveyor should be appointed, who in the Month of March every Year should inspect your Building and report its condition to the Master by whom it should be brought to the Court of Assistants, on the first Thursday in April every Year so that the subsequent General Court held in the beginning of July may be in good time to confirm & direct the immediate execution of such repairs as may be wanting The great business in matters of this kind is not to be led into unnecessary expences from precipitation.

Servants.

Respecting your Servants Gentlemen, The Clerk and the Beadle are objects of very great importance. I see with much surprize, the expence of this establishment increasing upon you very fast indeed. The original Salaries remain the same, and so do the Gratuities as they were settled formerly but of late you have appointed additional Servants, and allowed extra Gratuities. You have submitted to many new Articles of Expense, & as far as I can learn no perquisites have been asked, which have not been granted, without your shewing the least inclination to resist, and without taking any time to examine or compare. These Excesses appeared first in the year 1786 and have increased so rapidly since that time, that the perquisites now are in general one-third greater than they used to be in some cases twice as much; and these irregularities, enormous as they are, have been suffered to continue, without the least endeavour on your part to reform them.

Two Years before this period, the Establishment of the Clerk, including the Salary, Gratuity & perquisites of Coal, Candle, Wood, & Turnery amounted to £160 or £170 ℔ Ann. and even then the allowance for some of the Articles was too great. In the year 1786 an additional Sum of £25 was allowed to the Clerk for a Servant, so that his Establishment amounted then to £180 ℔ Ann. and upwards. In 1787 & 88 it amounted to £195 in 1788 and 89 to £225; in 89 and 90 to £230 nearly.

The original Establishment of your Clerk is as follows (viz[t.]) He accepts his office to hold it, and does hold it, during your pleasure only. He was at first allowed £60 ℔ Ann. and it was expressly stipulated that he should have no perquisite of any kind whatever. To this Stipend of £60 ℔ Ann. the Yearly Sum of £40 was added by way of Gratuity, then an Apartm[t] within the House was granted him, and an allowance of the two articles of Coal and Candle.

Besides the Apartm[t] which was at first granted to the Clerk he has now the use of the Library & Committee Room. To his perquisites of Coal & Candle that of Turnery is added & he has also an ample allow[ce] for a Serv[t.] & many other advantages.

The Coal and Candle are extended (deducting what is necessary for your Committee Rooms) greatly beyond the Expence of most private Families. What think you of a charge of £84 for Coals Candles Wood & Turnery during this last year on the Clerk's own account only? I am ready to acknowledge that his place ought to be a good one, sufficient to retain a Man of Credit and Ability in your service, but at the same time it ought, I think, to bear some proportion to your Income and his trouble. Perquisites are in all cases improper, and therefore it would be adviseable for you to give your Clerk a competent Salary, and allow him no more perquisites.

The Apartments in the Hall as they were originally granted must be equal to £70 ℘ Annum as Taxes and repairs of every kind are charged to your account.

Now what ought our Company whose whole expenditure sho[d] not exceed £900 ℘ Ann., and if the proposed alterations take place, can never greatly exceed that Sum; what, I say, ought such a Company to allow their Clerk for assisting them at their Meetings, paying their Bills & collecting their Revenue.

Is not a Stipend of 25 ℘ C[t.] with the advantages of such apartments rather too much? I really think it is.

Your Beadle also has been indulged in the same manner as your Clerk since to £33 : 5 ℘ Ann. being the Sum which was paid four years ago for the Beadle's Salary Gratuity and Dinners you have added a second and third Gratuity of £5 : 5 : 0 each; and besides these he is permitted to have many other advantages, of the propriety of which, little or no enquiry was ever made. I am of opinion that your Beadle ought to be put on a good footing, as good an one as your Clerk, allowance being made for the difference of their situations; and as you should pay your Beadle properly, so have you a right to his services. I could wish that the distinct duty of your Servants was a little better explained, as well as many other points relative to our Customs and Examinations, which I have mentioned before and which do not make an Article either in your Books or your printed Laws.

If the Beadle is paid at the rate of £35 ℘ Ann. which is rather at an higher rate than he stood 4 years ago, it is beyond a doubt that you pay him very liberally for his real Services & there is not the shadow of a pretence for such an extravagant augmentation in so short a space of time as from £33 to £44. He has perquisites of a doubtful nature, & his time is not wholly taken up in our Business.

Gentlemen, I cannot blame your Servants for asking, when they know you are always ready to give, the Evil originates with yourselves. If in giving a detail of facts, I appear to accuse, it is an accusation of our own remiss conduct, rather than theirs. However I hope that the Gentlemen who are to succeed me in office will finish these Enquiries, that they will sift everything to the bottom, that they will do their best towards cutting off all unnecessary Expences, and lastly that they ascertain what is the duty of our Servants, and proportion the reward to their real services only.

Anatomy.

The first thing that I shall remind you of on this head is, that you have a Fund for the support of Anatomical Lectures of £16 ℘ Ann., besides the Interest of £500, being what was appropriated to us, when our Company was separated from the Barbers, and this Fund is constantly increasing by means of the penalties which all our Members pay, who do not in their turns serve the Anatomical offices; but notwithstanding this encouragement I am sorry to observe that you have instituted Lecteres neither in Surgery, nor indeed in anatomy of any degree of importance; nor have you held out any Gratification or reward for rising Merit. Even the feeble attempts which you have already made, tho' found in the end to be totally inadequate and ineffectual, yet have been shamefully expensive.

Three Lectures in 1786 and 87 a period of one Year only, and given, I think upon the Bones, cost you upwards of £30 without the expence of subjects or dissection—That of the succeeding year in 1787 & 88 cost you upwards of £40. There were indeed 6 Lectures given in that year as there were in the succeeding one of 88 & 89, and their expences amounted to £46 and upwards. It would be scarcely credible if it did not appear upon your Books, with what needless Expences, & with what exactions it has been usual for every thing of this kind, to be deducted. In the present year a charge is made of £3 to Men who attend Executions, a mere Imposition; and what is more extraordinary, this allowance is made, at a time, when no Lectures at all have been given.

In 1782 you voted a Crown ⅌ head to all those Members of the Court of Assistants who should be present at these Lectures by way of encouraging the attendance of others by their own appearance; so that every Lecture, putting their right of doing this out of the question, would cost the company, if the whole number attended, an additional expence of 5 Guineas to what they had originally done, & what Man of sense would advise the company to be at the expence of 7 or 8 Guineas for each Lecture unless the Lectures were given on the most useful subjects in Surgery, and delivered by Gentlemen of the first experience in the profession.

I know it will be said that our Hall can never be a School of Anatomy the situation of the different Hospitals, the extent of the Town, together with its interference with private courses prohibit it.

I admit this mode of reasoning to be in most respects true, but there is a time of the year, in May & June, when the Anatomical Lectures should be over, & before the Students have left the Town, when a number not exceeding 15 Lectures in Surgery might be given gratis by Men of Experience in the profession, on those points which tend only to real practice and the establishment of good principles. Lectures so conducted could not fail of being highly useful to the Students as well as being honourable to ourselves.

Were the Gentlemen to give up their insignificant Fees on this occasion or to confine them altogether to the usual one of half a Guinea to be given to the Master and Wardens only, who should be under the necessity of being present at such Lectures, it would I think sufficiently answer the intended purposes.

And as an incitement to emulation it would be advisable for the Comp[y] to give every year a Gold Medal of the same value as that which is presented to your professor, to any Student who should produce the best Treatise in the Judgment of the majority of the Court of Examiners on any given subject.

The whole expence of this Establishment (all needless expences and exactions being removed) would not exceed the Sum of £140 ℘ Ann. The Lectures should be read by such Members from your Court of Assistants as are in practice, and are Surgeons to Hospitals, and for which they should be liberally paid.

I cannot dismiss this subject without making one more remark. From the construction of your Hall as well as from the Names and nature of the offices founded in your Body it plainly appears that one great object of your Institution was to raise a School for Anatomy. Why it failed in the begin[g] I cannot readily conceive, why it fails now, I can readily explain.

I find from your Books that a proposal was made by a very respectable Member of this Court for establishing a School of Anatomy here in 1775, about two years after your attempt began. This proposal was ordered to be printed, & then it was referred to the consideration of the next General Court who resolved that the then mode of conducting the Anatomical Lectures had not yet been allowed a proper length of time to prove its insufficiency.

I cannot approve of this resolution of the General Court, as there was certainly room to suspect that the mode was exceptionable in which the Anatomical Lectures were then conducted;

for if that had not been the case, why should any proposal of this kind have been made. Indeed whoever considers but the bare outline of your plan must be satisfied, that it could not be attended with success. For this business was put under the management of no less than 6 persons, 2 Masters, 2 Wardens & 2 Professors, & how could an Establishment of so mixed and complex a nature be properly carried on? and accordingly your Lectures were bad or your demonstrations were bad, or your parts were ill prepared, or ignorant persons were appointed who disgraced you by acting, or thought themselves hardly treated by being obliged to fine.

But tho' you rejected the proposal yet you took the blame and suspecting that your plan was capable of further improvement; you directed that your Lectures should be read at fixed times, between Michaelmas & Christmas and between Christmas and Lady day. The first period assigned was for those to be read on the Muscles, and the last for those on the Viscera. The Osteological Lectures were to be read in June. But still the mode of giving them remained unaltered & their number was undetermined.

The Committee of Anatomy which had been appointed to meet on this Business continued to proceed, and they appear no where to have been dismissed.

But your plan, with these improvements, did not answer, and after giving it the long Tryal of 13 years, in the year 1766 I think, you came to the following resolutions, viz$^{t.}$: That the present mode of giving the Lectures not having ans$^{d.}$ the intended purposes, a Professor should be appointed annually, to be rewarded with a Gold Medal, who should read his Lectures at fixed times, as specified in your resolution of 1775 and that the whole number to be given should be 15.

This was an improvement on the 2^{d} plan inasmuch as an Annual Professor was to be appointed of known ability (which took place) and a certain number of Lectures was stipulated to be given, which I believe did not take place; and this plan well executed would have been of real use to the Public.

But this plan was never likely to be well executed: 1st because no adequate reward was provided for the Professor.—2dly Because a Condition was annexed which was of itself sufficient to destroy the whole. The Condition was as follows: Vizt. That these Lectures should be so read, at such times, within the above mentioned period, as the Master & Wardens for the time being shod appoint.

This alteration, ostensibly I believe (for it does not appear that the proposed Lectures were read) took place for the present moment; but in the succeeding year of 1767, you determined, (really I believe without taking the trouble of informing yourselves of what had passed the year before) that the 12 Lectures as above mentioned (for you forgot the 3 Osteological) should be now varied, for that is the expression, and that such Professor for the time being should be obliged to give (an implication that they had not yet been given) such a number of 6 Lectures on the Muscles, and 6 on the Viscera, as the Master and Wardens for the time being shall from time to time appoint, and so this Business, for all the purposes of efficacy, has been given up, & in that whimsical state at the expiration of another 13 years it now continues; nor can I find that either the 12 or 15 Lectures have ever in any one year been given. In 1779 indeed 6 Lectures in Surgery were added to the usual anatomical ones, but whether they were given in that or any other year I cannot tell.

Should you revise these plans, Gentlemen, as you can hardly avoid to do, to preserve the Spirit, or rather to rouse the Spirit of your Institution, & indeed to apply properly the Income of Funds, created as it were for these specific purposes, should you direct 15 Lectures to be given only as above stipulated, on a solid & permanent foundation, after having determined on their nature; you will take care I hope never to lose sight of two Objects—the One that your Professors are properly qualified for the intended purpose; the Other that you provide for them an adequate reward.

I am apprehensive, Gentlemen, that you will think me rather

tedious on this subject, but I thought it necessary to give even an imperfect account, than no account at all.

Charity.

Respecting your Charities, I am sorry to observe these have decreased. They were 3 years ago at the rate of £93 & upwards, for the next £34, the succeeding one £38, and the present £67.

This Expence seems to have decreased in proportion as others have increased. I hope if Gentlemen are of opinion that the Company's affairs are in so flourishing a state as to make a further accumulation unnecessary, which some I know have thought; that they will be more liberal in this article.

The Widows and Children of Surgeons who may be left in a distress'd condition are much more proper objects of your attention than others which have been so much insisted upon, and which surely are objects of a very inferior consideration. These have the true claim to be benefited from the overflowing of a Fund, created in some respects by the Industry of their Fathers.

I could wish therefore that the sum of £80 ℘ Ann. at the least, might be put aside, to estsblish a Fund for this purpose, to be disposed of wholly by the majority of this Court. I think, however, that the persons in question, to whom this Charity is given, should attend once a year, with what Certificates may be proper, that we may be subject to no imposition, which some times are practised on such occasions.

Library.

Having finished my observations, Gentlemen, on five of the points before you, I must now mention what I early thought right and what many Gentlemen here agree with me in, which is, that this Society ought to be possessed of a Library, which when completed, should under certain Restrictions, be open to the Students in Surgery, for their Information.

I could wish that a Sum not exceeding £80 ℘ Annum might

be appropriated for the completion of this purpose, & I hope the Gentlemen will think this a proper way of disposing of their Money likewise.

Conclusion.

Gentlemen,

The savings on the whole in your Dinners to the amount of £150 ℔ Annum, & in other Articles, I trust to the amount of £50 more, by the reduction & limitation of unnecessary Charges, will go further than to enable you to appropriate the Annual Sum of £300 for the purposes of your Lectures your Charities & a Library, together with such a suitable reward as you shall think fit to bestow on merit. And if I am right in my estimate the annual expenditure of 1000 Guineas (your Hall once repaired) will allow even for contingent expences and will be found to be an establishment fully sufficient to answer all your intended purposes.

When Moneys are expended judiciously, & liberally for the public good, as well as for the private Emolument of the governing part of this Society, no Enquiry will be made even if we should exceed the bounds of what strictly speaking, we have a right to do.

But at present I am afraid, we are rather differently circumstanced, I recommend it therefore to the consideration of this Court, either at this, or at a short period of time from the present to appoint a Committee consisting of the Master and Wardens of the ensuing year, with any 3 others of the Court of Examiners, that the Gentlemen shall name, for the purpose of examining and retrenching the present expences where they have been found to exceed what is just and necessary, and of adopting and offering to your consideration either the above, or any other plans that may appear to them more eligible, for forwarding & effecting the above mentioned purposes, and that they will make a report of the same to the next General Court.

Believing further that when these plans are put into Execution, and when unaided by Royal Donations or the public, we have put ourselves on the footing of a liberal, of an humane, & learned Society and so beneficial a one to the public, from the proper Exertions of our own private Fund; we shall have a better pretence to ask for an extension of those powers from Parliament, which we have discovered a disposition to execute so much for the advantage of the Community.

Resolved that a Committee of Seven Members of this Court be appointed to consider of the above Observations and make their Report thereon to the next Court.

Resolved that the Master and Wardens Mr Warner Mr Grindall Mr Minors, & Mr Gunning be of that Committee and that any three of them do make a Quorum, And that such Committee do meet for the first time on Tuesday next the 6th instant at 6 o'Clock in the Evening, and that such Committee be paid the usual ffee of half a Guinea for each attendance.

This being the Day appointed by the Act of Parliament for the choice of a Master and Wardens for the ensuing year the Court proceeded to elect a Master; And Chas Hawkins Esqr one of His Majesty's Principal Serj^t^ Surgeons was in conformity to the Bye Laws [declared] to be unanimously chosen Master for the year ensuing.

The Court then proceeded to the election of Wardens for the year ensuing which Elections being by Ballot, upon the examination of the Votes by the Master Mr Wm Lucas appeared to be and was declared duly elected first or Upper Warden & Mr Edmund Pitts the second or Junior Warden To have hold & enjoy the said several & respective Offices of Master & Wardens for one whole year according to the said Act and the Bye Laws of this Corporation & the said Master & Wardens took the Oaths appointed by the Bye Laws to be taken by the Master & Wardens & their respective Seats in the Court accordingly.

Resolved that the thanks of this Court be given to the late Master & Wardens for their very diligent & punctual attendance during the past year and their strict & impartial discharge of their duty.

The Court then proceeded to elect a professor of Anatomy for the ensuing year when Mr John Abernethy was unanimously elected to that office.

The Court then proceeded to elect two Wardens & two Stewards of Anatomy for the ensuing year when Mr Wetherall & Mr Eden were unanimously elected Wardens & Mr Thos White & Mr Rice Benyon Stewards of Anatomy for the ensuing year.

Ordered that the Court do acquaint those Gentlemen of such their Elections to the aforesaid several Offices.

The Court then elected Wm Norris Henry Fearon Wm Breach, James Ward Robt Porter & George Davidson to be successively Stewards of Anatomy for the ensuing year in case any of the above Gentlemen shall die or pay their ffines.

Resolved that the disposal of the Bodies of all Murderers be left to the Master and Wardens for the ensuing year.

Ordered that a Gratuity of £40 to the Clerk and £10 to the Beadle be given to them for the past year.

Ordered that all Drafts on the Company's Bankers be signed by the Master and Wardens this day elected or any two of them.

Ordered that the Quarterage Book be made out and signed by the Master and Wardens.

The Clerk reported, that in pursuance of the directions of the last Court he had caused a Copy to be made of the Minute Book of the Court of Assistants and also Copies of all the accounts of all the receipts and payments on account of the Company from their separation from the Barbers up to this time which Books he produced; but that he had not been able to make out the other Book then ordered being an account of the receipts and paymts from the 1st day of July 1779 up to this time classed under the different heads as mentioned in the Resolution of the last Court because the Book from whence it must be taken had been some time in the possession of the Master for his perusal and since that had been copying as Ordered by the last Court but that he would make out the same before the next Court of Assistants.

Ordered that 20 Guineas be paid to the Clerk for the Expences which he has been at in preparing the Books agreeable to the directions of the Court of Assistants.

C. HAWKINS.

INDEX.

A A 3

Printed by Cassell & Company, Limited, La Belle Sauvage, London, E.C.

Illustrated, Fine Art, and other Volumes.

Abbeys and Churches of England and Wales, The: Descriptive, Historical, Pictorial. 21s.

After London; or, Wild England. By the late RICHARD JEFFERIES. *Cheap Edition*, 3s. 6d.

Along Alaska's Great River. By Lieut. SCHWATKA. Illustrated. 12s. 6d.

American Penman, An. By JULIAN HAWTHORNE. Boards, 2s.; cloth, 3s. 6d.

American Yachts and Yachting. Illustrated. 6s.

Animal Painting in Water Colours. With Eighteen Coloured Plates by FREDERICK TAYLER. 5s.

Arabian Nights Entertainments (Cassell's). With about 400 Illustrations. 10s. 6d.

Architectural Drawing. By PHENÉ SPIERS. Illustrated. 10s. 6d.

Art, The Magazine of. Yearly Volume. With several hundred Engravings, and Twelve Etchings, Photogravures, &c. 16s.

Behind Time. By G. P. LATHROP. Illustrated. 2s. 6d.

Bimetallism, The Theory of. By D. BARBOUR. 6s.

Bismarck, Prince. By C. LOWE, M.A. Two Vols. *Cheap Edition*. 10s. 6d.

Black Arrow, The. A Tale of the Two Roses. By R. L. STEVENSON. 5s.

British Ballads. 275 Original Illustrations. Two Vols. Cloth, 7s. 6d. each.

British Battles on Land and Sea. By the late JAMES GRANT. With about 600 Illustrations. Three Vols., 4to, £1 7s.; Library Edition, £1 10s.

British Battles, Recent. Illustrated. 4to, 9s. Library Edition, 10s.

British Empire, The. By SIR GEORGE CAMPBELL. 3s.

Browning, An Introduction to the Study of. By ARTHUR SYMONS. 2s. 6d.

Butterflies and Moths, European. By W. F. KIRBY. With 61 Coloured Plates. Demy 4to, 35s.

Canaries and Cage-Birds, The Illustrated Book of. By W. A. BLAKSTON, W. SWAYSLAND, and A. F. WIENER. With 56 Fac-simile Coloured Plates, 35s.

Cannibals and Convicts. By JULIAN THOMAS ("The Vagabond"). *Cheap Edition*, 5s.

Captain Trafalgar. By WESTALL and LAURIE. Illustrated. 5s.

Cassell's Family Magazine. Yearly Vol. Illustrated. 9s.

Celebrities of the Century: Being a Dictionary of Men and Women of the Nineteenth Century. 21s.; roxburgh, 25s.

Changing Year, The. With Illustrations. 7s. 6d.

Chess Problem, The. With Illustrations by C. PLANCK and others. 7s. 6d.

Children of the Cold, The. By Lieut. SCHWATKA. 2s. 6d.

China Painting. By FLORENCE LEWIS. With Sixteen Coloured Plates, and a selection of Wood Engravings. With full Instructions. 5s.

Choice Dishes at Small Cost. By A. G. PAYNE. *Cheap Edition*, 1s.

Christmas in the Olden Time. By Sir WALTER SCOTT. With charming Original Illustrations. 7s. 6d.

Cities of the World. Three Vols. Illustrated. 7s. 6d. each.

Civil Service, Guide to Employment in the. *New and Enlarged Edition*. 3s. 6d.

Civil Service.—Guide to Female Employment in Government Offices. Cloth, 1s.

Clinical Manuals for Practitioners and Students of Medicine. (*A List of Volumes forwarded post free on application to the Publishers.*)

Clothing, The Influence of, on Health. By FREDERICK TREVES, F.R.C.S. 2s.

Cobden Club, Some Works published for the:—

Writings of Richard Cobden. 6s.
Local Government and Taxation in the United Kingdom. 5s.
Displacement of Labour and Capital. 3d.
Free Trade versus Fair Trade. 5s.
Free Trade and English Commerce. By A. Mongredien. 6d.
Crown Colonies. 1s.
Popular Fallacies Regarding Trade. 6d.
Western Farmer of America. 3d.
Reform of the English Land System. 3d.
Fair Trade Unmasked. By G. W. Medley. 6d.
Technical Education. By F. C. Montague, M.A. 6d.
Our Land Laws of the Past. 3d.
The Caribbean Confederation. By C. S. Salmon. 1s. 6d.
Pleas for Protection Examined. By A. Mongredien. *New and Revised Edition*. 6d.
What Protection does for the Farmer. By J. S. Leadam, M.A. 6d.
The Old Poor Law and the New Socialism; or, Pauperism and Taxation. By F. C. Montague. 6d.
The Secretary of State for India in Council. 6d.
The National Income and Taxation. By Sir Louis Mallet. 6d.

Colonies and India, Our: How we Got Them, and Why we Keep Them. By Prof. C. RANSOME. 1s.
Colour. By Prof. A. H. CHURCH. *New and Enlarged Edition*, with Coloured Plates. 3s. 6d.
Columbus, Christopher, The Life and Voyages of. By WASHINGTON IRVING. Three Vols. 7s. 6d.
Commodore Junk. By G. MANVILLE FENN. 5s.
Cookery, Cassell's Shilling. The Largest and Best Work on the Subject ever produced. 1s.
Cookery, Cassell's Dictionary of. Containing about Nine Thousand Recipes. 7s. 6d.; roxburgh, 10s. 6d.
Cookery, A Year's. By PHYLLIS BROWNE. Cloth gilt or oiled cloth, 3s. 6d.
Cook Book, Catherine Owen's New. 4s.
Co-operators, Working Men: What They have Done, and What They are Doing. By A. H. DYKE-ACLAND, M.P., and B. JONES. 1s.
Countries of the World, The. By ROBERT BROWN, M.A., Ph.D., &c. Complete in Six Vols., with about 750 Illustrations. 4to, 7s. 6d. each.
Culmshire Folk. By the Author of "John Orlebar," &c. 3s. 6d.
Cyclopædia, Cassell's Concise. With 12,000 subjects, brought down to the latest date. With about 600 Illustrations, 15s.; roxburgh, 18s.
Cyclopædia, Cassell's Miniature. Containing 30,000 Subjects. Cloth, 3s. 6d.
Dairy Farming. By Prof. J. P. SHELDON. With 25 Fac-simile Coloured Plates, and numerous Wood Engravings. Demy 4to, 21s.
Dead Man's Rock. A Romance. By Q. 5s.
Decisive Events in History. By THOMAS ARCHER. With Sixteen Illustrations. Boards, 3s. 6d.; cloth, 5s.
Deserted Village Series, The. Consisting of *Éditions de luxe* of favourite poems by Standard Authors. Illustrated. Cloth gilt, 2s. 6d.

Goldsmith's Deserted Village. | **Wordsworth's Ode on Immortality, and Lines on Tintern Abbey.**
Milton's L'Allegro and Il Penseroso.
Songs from Shakespeare.

Dickens, Character Sketches from. FIRST, SECOND, and THIRD SERIES. With Six Original Drawings in each, by FREDERICK BARNARD. In Portfolio, 21s. each.
Diary of Two Parliaments. By H. W. LUCY. The Disraeli Parliament, 12s. The Gladstone Parliament, 12s.
Dog, The. By IDSTONE. Illustrated. 2s. 6d.
Dog, Illustrated Book of the. By VERO SHAW, B.A. With 28 Coloured Plates. Cloth bevelled, 35s.; half-morocco, 45s.
Dog Stories and Dog Lore. By Col. THOS. W. KNOX. 6s.
Domestic Dictionary, The. An Encyclopædia for the Household. Cloth, 7s. 6d.
Doré's Dante's Inferno. Illustrated by GUSTAVE DORÉ. *Popular Edition*, 21s.
Doré's Dante's Purgatorio and Paradiso. Illustrated by GUSTAVE DORÉ. *Popular Edition.* 21s.
Doré's Fairy Tales Told Again. With 24 Full-page Engravings by DORÉ. 5s.
Doré Gallery, The. With 250 Illustrations by GUSTAVE DORÉ. 4to, 42s.
Doré's Milton's Paradise Lost. With Full-page Drawings by GUSTAVE DORÉ. 4to, 21s.
Earth, Our, and its Story. By Dr. ROBERT BROWN, F.L.S. Vol. I., with Coloured Plates and numerous Wood Engravings. 9s.
Edinburgh, Old and New, Cassell's. With 600 Illustrations. Three Vols., 9s. each; library binding, £1 10s. the set.
Egypt: Descriptive, Historical, and Picturesque. By Prof. G. EBERS. Translated by CLARA BELL, with Notes by SAMUEL BIRCH, LL.D., &c. *Popular Edition*, in Two Vols., 42s.
"89." A Novel. By EDGAR HENRY. Cloth, 3s. 6d.
Electricity, Age of, from Amber Soul to Telephone. By PARK BENJAMIN, Ph.D. 7s. 6d.
Electricity, Practical. By Prof. W. E. AYRTON. Illustrated. 7s. 6d.
Electricity in the Service of Man. With nearly 850 Illustrations. 21s.
Encyclopædic Dictionary, The. A New and Original Work of Reference to all the Words in the English Language. Complete in Fourteen Divisional Vols., 10s. 6d. each; or Seven Vols., half-morocco, 21s each.
England, Cassell's Illustrated History of. With 2,000 Illustrations. Ten Vols., 4to, 9s. each. *New and Revised Edition.* Vols. I. and II., 9s. each.

English History, The Dictionary of. Cloth, 21s.; roxburgh, 25s.

English Literature, Library of. By Prof. HENRY MORLEY. Complete in 5 vols., 7s. 6d. each.

VOL. I.—SHORTER ENGLISH POEMS.
VOL. II.—ILLUSTRATIONS OF ENGLISH RELIGION.
VOL. III.—ENGLISH PLAYS.
VOL. IV.—SHORTER WORKS IN ENGLISH PROSE.
VOL. V.—SKETCHES OF LONGER WORKS IN ENGLISH VERSE AND PROSE.

English Literature, Morley's First Sketch of. *Revised Edition*, 7s. 6d.

English Literature, The Dictionary of. By W. DAVENPORT ADAMS. *Cheap Edition*, 7s. 6d.; roxburgh, 10s. 6d.

English Literature, The Story of. By ANNA BUCKLAND. *New and Cheap Edition.* 3s. 6d.

English Writers. An attempt towards a History of English Literature. By HENRY MORLEY, LL.D., Professor of English Literature, University College, London. Vols. I., II., III., and IV., 5s. each.

Æsop's Fables. With about 150 Illustrations by E. GRISET. *Cheap Edition.* cloth, 3s. 6d.; bevelled boards, gilt edges, 5s.

Etching: Its Technical Processes, with Remarks on Collections and Collecting. By S. K. KOEHLER. Illustrated with 30 Full-page Plates. Price £4 4s.

Etiquette of Good Society. 1s.; cloth, 1s. 6d.

Eye, Ear, and Throat, The Management of the. 3s. 6d.

Family Physician, The. By Eminent PHYSICIANS and SURGEONS. *New and Revised Edition.* Cloth, 21s.; roxburgh, 25s.

Fenn, G. Manville, Works by. *Popular Editions.* Boards, 2s. each; or cloth, 2s. 6d.

Dutch the Diver; or, a Man's Mistake.
My Patients.
The Parson o' Dumford.
Poverty Corner.
The Vicar's People. } In Cloth only.
Sweet Mace. }

Ferns, European. By JAMES BRITTEN, F.L.S. With 30 Fac-simile Coloured Plates by D. BLAIR, F.L.S. 21s.

Field Naturalist's Handbook, The. By Rev. J. G. WOOD & THEODORE WOOD. 5s.

Figuier's Popular Scientific Works. With Several Hundred Illustrations in each. 3s. 6d. each.

The Human Race.
World Before the Deluge.
Reptiles and Birds.
The Ocean World.
The Vegetable World.
The Insect World.
Mammalia.

Figure Painting in Water Colours. With 16 Coloured Plates by BLANCHE MACARTHUR and JENNIE MOORE. With full Instructions. 7s. 6d.

Fine-Art Library, The. Edited by JOHN SPARKES, Principal of the South Kensington Art Schools. Each Book contains about 100 Illustrations. 5s. each.

Tapestry. By Eugène Müntz. Translated by Miss L. J. Davis
Engraving. By Le Vicomte Henri Delaborde. Translated by R. A. M. Stevenson.
The English School of Painting. By E. Chesneau. Translated by L. N. Etherington. With an Introduction by Prof. Ruskin.
The Flemish School of Painting. By A. J. Wauters. Translated by Mrs. Henry Rossel.
The Education of the Artist. By Ernest Chesneau. Translated by Clara Bell. Non-illustrated.
Greek Archæology. By Maxime Collignon. Translated by Dr. J. H. Wright.
Artistic Anatomy. By Prof Duval. Translated by F. E. Fenton.
The Dutch School of Painting. By Henry Havard. Translated by G. Powell.

Five Pound Note, The, and other Stories. By G. S. JEALOUS. 1s.

Flower Painting in Water Colours. First and Second Series. With 20 Fac-simile Coloured Plates in each by F. E. HULME, F.L.S., F.S.A. With Instructions by the Artist. Interleaved. 5s. each.

Flower Painting, Elementary. With Eight Coloured Plates. 3s.

Flowers, and How to Paint Them. By MAUD NAFTEL. With Coloured Plates. 5s.

Forging of the Anchor, The. A Poem. By the late Sir SAMUEL FERGUSON, LL.D. With 20 Original Illustrations. Gilt edges, 5s.

Fossil Reptiles, A History of British. By Sir RICHARD OWEN, K.C.B., F.R.S., &c. With 268 Plates. In Four Vols., £12 12s.

France as It Is. By ANDRÉ LEBON and PAUL PELET. With Three Maps. Crown 8vo. cloth, 7s. 6d.

Franco-German War, Cassell's History of the. Two Vols. With 500 Illustrations. 9s. each.

Fresh-Water Fishes of Europe, The. By Prof. H. G. SEELEY, F.R.S. *Cheap Edition.* 7s. 6d.

Garden Flowers, Familiar. By SHIRLEY HIBBERD. With Coloured Plates by F. E. HULME, F.L.S. Complete in Five Series. Cloth gilt, 12s. 6d. each.

Gardening, Cassell's Popular. Illustrated. Complete in 4 Vols., 5s. each.

Geometrical Drawing for Army Candidates. By H. T. LILLEY, M.A. 2s.

Geometry, First Elements of Experimental. By PAUL BERT. 1s. 6d.
Geometry, Practical Solid. By Major ROSS. 2s.
Germany, William of. By ARCHIBALD FORBES. 3s. 6d.
Gladstone, Life of the Rt. Hon. W. E. By G. BARNETT SMITH. With Portrait. 3s. 6d.
Gleanings from Popular Authors. Two Vols. With Original Illustrations. 4to, 9s. each. Two Vols. in One, 15s.
Gold to Grey, From. Being Poems and Pictures of Life and Nature. By MARY D. BRINE. Illustrated. 7s. 6d.
Great Bank Robbery, The. A Novel. By JULIAN HAWTHORNE. Boards, 2s.
Great Industries of Great Britain. With 400 Illustrations. 3 Vols., 7s. 6d. each.
Great Northern Railway, The Official Illustrated Guide to the. 1s.; cloth, 2s.
Great Western Railway, The Official Illustrated Guide to the. 1s.; cloth, 2s.
Great Painters of Christendom, The, from Cimabue to Wilkie. By JOHN FORBES-ROBERTSON. Illustrated throughout. *Popular Edition*, cloth gilt, 12s. 6d.
Gulliver's Travels. With 88 Engravings by MORTEN. *Cheap Edition.* Cloth, 3s. 6d.; cloth gilt, 5s.
Gum Boughs and Wattle Bloom. By DONALD MACDONALD. 5s.
Gun and its Development, The. By W. W. GREENER. Illustrated. 10s. 6d.
Guns, Modern Shot. By W. W. GREENER. Illustrated. 5s.
Health at School. By CLEMENT DUKES, M.D., B.S. 7s. 6d.
Health, The Book of. By Eminent Physicians and Surgeons. Cloth, 21s.; roxburgh, 25s.
Health, The Influence of Clothing on. By F. TREVES, F.R.C.S. 2s.
Heavens, The Story of the. By Sir ROBERT STAWELL BALL, LL.D., F.R.S., Royal Astronomer of Ireland. Coloured Plates and Wood Engravings. 31s. 6d.
Heroes of Britain in Peace and War. In Two Vols., with 300 Original Illustrations. 5s. each; or One Vol., library binding, 10s. 6d.
Holy Land and the Bible, The. By the Rev. CUNNINGHAM GEIKIE, D.D. With Map. Two Vols. 24s.
Homes, Our, and How to Make them Healthy. By Eminent Authorities. Illustrated. 15s.; roxburgh, 18s.
Horse-Keeper, The Practical. By GEORGE FLEMING, LL.D., F.R.C.V.S. Illustrated. Crown 8vo, cloth, 7s. 6d.
Horse, The Book of the. By SAMUEL SIDNEY. With 28 *fac-simile* Coloured Plates. Demy 4to, 35s.; half-morocco, £2 5s.
Horses, The Simple Ailments of. By W. F. Illustrated. 5s.
Household Guide, Cassell's. With Illustrations and Coloured Plates. *New and Revised Edition*, complete in Four Vols., 20s.
How Dante Climbed the Mountain. By ROSE EMILY SELFE. With Eight Full-page Engravings by GUSTAVE DORÉ. 2s.
How Women may Earn a Living. By MERCY GROGAN. 1s.
Imperial White Books. In Quarterly Vols. 10s. 6d. per annum, post free; to subscribers separately, 3s. 6d. each.
India, Cassell's History of. By the late JAMES GRANT. With 400 Illustrations. 15s.
India: the Land and the People. By Sir JAMES CAIRD, K.C.B. 10s. 6d.
In-door Amusements, Card Games, and Fireside Fun, Cassell's. 3s. 6d.
Industrial Remuneration Conference. The Report of. 2s. 6d.
Insect Variety: its Propagation and Distribution. By A. H. SWINTON. 7s. 6d.
Irish Parliament, The, What it Was, and What it Did. By J. G. SWIFT MCNEILL, M.A., M.P. 1s.
Irish Parliament, A Miniature History of the. By J. C. HASLAM. 3d.
Irish Union; Before and After. By A. K. CONNELL, M.A. 2s. 6d.
John Parmelee's Curse. By JULIAN HAWTHORNE. 2s. 6d.
Kennel Guide, Practical. By Dr. GORDON STABLES. Illustrated. *Cheap Edition.* 1s.
Kidnapped. By R. L. STEVENSON. *Illustrated Edition.* 5s.
King Solomon's Mines. By H. RIDER HAGGARD. *Illustrated Edition.* 5s.
Khiva, A Ride to. By Col. FRED BURNABY. 1s. 6d.
Ladies' Physician, The. By a London Physician. 6s.
Lady Biddy Fane. By FRANK BARRETT. Three Vols. Cloth, 31s. 6d.
Lady's World, The. An Illustrated Magazine of Fashion and Society. Yrly. Vol. 18s.
Land Question, The. By Prof. J. ELLIOT, M.R.A.C. Including the Land Scare and Production of Cereals. 3s. 6d.
Landscape Painting in Oils, A Course of Lessons in. By A. F. GRACE. With Nine Reproductions in Colour. *Cheap Edition*, 25s.

Law, About Going to. By A. J. WILLIAMS, M.P. 2s. 6d.
Laws of Every Day Life, The. By H. O. ARNOLD-FORSTER. 1s. 6d.
Letts's Diaries and other Time-saving Publications are now published exclusively by CASSELL & COMPANY. (*A List sent post free on application.*)
Local Dual Standards. By JOHN HENRY NORMAN. Gold and Silver Standard Currencies. 1s.
Local Government in England and Germany. By the Rt. Hon. Sir ROBERT MORIER, G.C.B., &c. 1s.
London, Brighton, and South Coast Railway, The Official Illustrated Guide to the. 1s.; cloth, 2s.
London and North Western Railway, The Official Illustrated Guide to the. 1s.; cloth, 2s.
London and South Western Railway, The Official Illustrated Guide to the. 1s.; cloth, 2s.
London, Greater. By EDWARD WALFORD. Two Vols. With about 400 Illustrations. 9s. each. *Library Edition.* Two Vols. £1 the set.
London, Old and New. By WALTER THORNBURY and EDWARD WALFORD. Six Vols., each containing about 200 Illustrations and Maps. Cloth, 9s. each. *Library Edition.* Imitation roxburgh, £3.
Longfellow, H. W., Choice Poems by. Illustrated by his Son, ERNEST W LONGFELLOW. 6s.
Longfellow's Poetical Works. *Fine-Art Edition.* Illustrated throughout with Original Engravings. Royal 4to, cloth gilt, £3 3s. *Popular Edition.* 16s.
Luther, Martin: the Man and his Work. By PETER BAYNE, LL.D. Two Vols., 24s.
Marine Painting. By WALTER W. MAY, R.I. With 16 Coloured Plates. Cloth, 5s.
Mechanics, The Practical Dictionary of. Containing 15,000 Drawings. Four Vols. 21s. each.
Medicine, Manuals for Students of. (*A List forwarded post free on application.*)
Midland Railway, The Official Illustrated Guide to the. *New and Revised Edition.* 1s.; cloth, 2s.
Modern Europe, A History of. By C. A. FYFFE, M.A. Vol. I. From 1792 to 1814. 12s. Vol. II. From 1814 to 1848. 12s.
Music, Illustrated History of. By EMIL NAUMANN. Edited by the Rev. Sir F. A. GORE OUSELEY, Bart. Illustrated. Two Vols. 31s. 6d.
National Library, Cassell's. In Weekly Volumes, each containing about 192 pages. Paper covers, 3d.; cloth, 6d. (*A List of the Volumes already published sent post free on application.*)
Natural History, Cassell's Concise. By E. PERCEVAL WRIGHT, M.A., M.D., F.L.S. With several Hundred Illustrations. 7s. 6d.; roxburgh, 10s. 6d.
Natural History, Cassell's New. Edited by Prof. P. MARTIN DUNCAN, M.B., F.R.S., F.G.S. With Contributions by Eminent Scientific Writers. Complete in Six Vols. With about 2,000 high-class Illustrations. Extra crown 4to, cloth, 9s. each.
Nature, Short Studies from. Illustrated. *Cheap Edition.* 2s. 6d.
Neutral Tint, A Course of Painting in. With Twenty-four Plates by R. P. LEITCH. With full Instructions to the Pupil. 5s.
Nimrod in the North; or, Hunting and Fishing Adventures in the Arctic Regions. By Lieut. SCHWATKA. Illustrated. 7s. 6d.
Nursing for the Home and for the Hospital, A Handbook of. By CATHERINE J. WOOD. *Cheap Edition.* 1s. 6d.; cloth, 2s.
Oil Painting, A Manual of. By Hon. JOHN COLLIER. Cloth, 2s. 6d.
On the Equator. By H. DE W. Illustrated with Photos. 3s. 6d.
Orion the Gold Beater. A Novel. By SYLVANUS COBB, Junr. Cloth, 3s. 6d.
Our Own Country. Six Vols. With 1,200 Illustrations. Cloth, 7s. 6d. each.
Outdoor Sports and Indoor Amusements, Cassell's Book of. With about 900 Illustrations. *Cheap Edition.* 992 pages, medium 8vo, cloth, 3s. 6d.
Paris, Cassell's Illustrated Guide to. Cloth, 2s.
Parliaments, A Diary of Two. By H. W. LUCY. The Disraeli Parliament, 1874—1880. 12s. The Gladstone Parliament, 1881—1886. 12s.
Paxton's Flower Garden. By Sir JOSEPH PAXTON and Prof. LINDLEY. Revised by THOMAS BAINES, F.R.H.S. Three Vols. With 100 Coloured Plates. £1 1s. each.
Peoples of the World, The. By Dr. ROBERT BROWN. Complete in Six Volumes. With Illustrations. 7s. 6d. each.
Phantom City, The. By W. WESTALL. 5s.
Photography for Amateurs. By T. C. HEPWORTH. Illustrated, 1s.; or cloth, 1s. 6d.

Phrase and Fable, Dictionary of. By the Rev. Dr. BREWER. *Cheap Edition, Enlarged*, cloth, 3s. 6d.; or with leather back, 4s. 6d.

Picturesque America. Complete in Four Vols., with 48 Exquisite Steel Plates, and about 800 Original Wood Engravings. £2 2s. each.

Picturesque Canada. With about 600 Original Illustrations. Two Vols., £3 3s. each.

Picturesque Europe. Complete in Five Vols. Each containing 13 Exquisite Steel Plates, from Original Drawings, and nearly 200 Original Illustrations. £10 10s.; half-morocco, £15 15s.; morocco gilt, £26 5s. The POPULAR EDITION is now complete in Five Vols., 18s. each.

Pigeon Keeper, The Practical. By LEWIS WRIGHT. Illustrated. 3s. 6d.

Pigeons, The Book of. By ROBERT FULTON. Edited by LEWIS WRIGHT. With 50 Coloured Plates and numerous Wood Engravings. 31s. 6d.; half-morocco, £2 2s.

Pocket Guide to Europe (Cassell's). Size $5\frac{1}{8}$ in. × $3\frac{3}{4}$ in. Leather, 6s.

Poems, Representative of Living Poets, American and English. Selected by the Poets themselves. 15s.

Poets, Cassell's Miniature Library of the:—

Burns. Two Vols. Cloth, 1s. each; or cloth, gilt edges, 2s. 6d. the set.
Byron. Two Vols. Cloth, 1s. each; or cloth, gilt edges, 2s. 6d. the set.
Hood. Two Vols. Cloth, 1s. each; or cloth, gilt edges, 2s. 6d. the set.
Longfellow. Two Vols. Cloth, 1s. each; or cloth, gilt edges, 2s. 6d. the set.
Milton. Two Vols. Cloth, 1s. each; or cloth, gilt edges, 2s. 6d. the set.
Scott. Two Vols. Cloth, 1s. each; or cloth, gilt edges, 2s. 6d. the set.
Sheridan and Goldsmith. 2 Vols. Cloth, 1s. each; or cloth, gilt edges, 2s. 6d. the set.
Wordsworth. Two Vols. Cloth, 1s. each; or cloth, gilt edges, 2s. 6d. the set.
Shakespeare. Twelve Vols., half cloth, in box, 12s.

Popular Library, Cassell's. A Series of New and Original Works. Cloth, 1s. each.

The Russian Empire.
The Religious Revolution in the Sixteenth Century.
English Journalism.
Our Colonial Empire.
The Young Man in the Battle of Life.
John Wesley.
The Story of the English Jacobins.
Domestic Folk Lore.
The Rev. Rowland Hill.
Boswell and Johnson.
History of the Free-Trade Movement in England.

Post Office of Fifty Years Ago, The. 1s.

Poultry Keeper, The Practical. By LEWIS WRIGHT. With Coloured Plates and Illustrations. 3s. 6d.

Poultry, The Book of. By LEWIS WRIGHT. *Popular Edition.* With Illustrations on Wood, 10s. 6d.

Poultry, The Illustrated Book of. By LEWIS WRIGHT. With Fifty Exquisite Coloured Plates, and numerous Wood Engravings. Cloth, 31s. 6d.; half-morocco, £2 2s.

Pre-Raphaelites (The Italian) in the National Gallery. By COSMO MONKHOUSE. Illustrated. 1s.

Printing Machinery and Letterpress Printing, Modern. By FRED. J. F. WILSON and DOUGLAS GREY. Illustrated. 21s.

Queen Victoria, The Life and Times of. By ROBERT WILSON. Complete in 2 Vols. With numerous Illustrations, representing the Chief Events in the Life of the Queen, and Portraits of the Leading Celebrities of her Reign. Extra crown 4to, cloth gilt, 9s. each.

Queer Race, A. By W. WESTALL. 5s.

Rabbit-Keeper, The Practical. By CUNICULUS. Illustrated. 3s. 6d.

Red Library of English and American Classics, The. Stiff covers, 1s. each; cloth, 2s. each.

People I have Met.
The Pathfinder.
Evelina.
Scott's Poems.
Last of the Barons.
Adventures of Mr. Ledbury and his friend Jack Johnson.
Ivanhoe.
Oliver Twist.
Selections from Hood's Works.
Longfellow's Prose Works.
Sense and Sensibility.
Lytton's Plays.
Tales, Poems, and Sketches (Bret Harte).
Martin Chuzzlewit. Two Vols.
The Prince of the House of David.
Sheridan's Plays.
Uncle Tom's Cabin.
Deerslayer.
Eugene Aram.
Jack Hinton, the Guardsman.
Rome and the Early Christians.
The Trials of Margaret Lyndsay.
Edgar Allan Poe. Prose and Poetry, Selections from.
Old Mortality.
The Hour and the Man.
Washington Irving's Sketch-Book.
Last Days of Palmyra.
Tales of the Borders.
Pride and Prejudice.
Last of the Mohicans.
Heart of Midlothian.
Last Days of Pompeii.
Yellowplush Papers.
Handy Andy.
Selected Plays.
American Humour.
Sketches by Boz.
Macaulay's Lays and Selected Essays.
Harry Lorrequer.
Old Curiosity Shop.
Rienzi.
The Talisman.
Pickwick (Two Vols.).
Scarlet Letter.

Royal River, The: The Thames, from Source to Sea. With Descriptive Text and a Series of beautiful Engravings. £2 2s.

Russia. By Sir DONALD MACKENZIE WALLACE, M.A. 5s.

Russo-Turkish War, Cassell's History of. With about 500 Illustrations. Two Vols., 9s. each; library binding, One Vol., 15s.

Saturday Journal, Cassell's. Yearly Vols., 7s. 6d.

Science for All. Edited by Dr. ROBERT BROWN, M.A., F.L.S., &c. With 1,500 Illustrations. Five Vols., 9s. each.

Sea, The: Its Stirring Story of Adventure, Peril, and Heroism. By F. WHYMPER. With 400 Illustrations. Four Vols., 7s. 6d. each.

Section 558, or the Fatal Letter. A Novel. By JULIAN HAWTHORNE. Boards, 2s.; cloth, 2s. 6d.

Sent Back by the Angels. And other Ballads of Home and Homely Life. By FREDERICK LANGBRIDGE, M.A. 4s. 6d. *Popular Edition*, 1s.

Sepia Painting, A Course of. Two Vols., with Twelve Coloured Plates in each, and numerous Engravings. Each, 3s. Also in One Volume, 5s.

Shaftesbury, The Seventh Earl of, K.G., The Life and Work of. By EDWIN HODDER. With Portraits. Three Vols., 36s. *Popular Edition*, in One Vol., 7s. 6d.

Shakspere, The International. *Édition de luxe.*
"King Henry IV." Illustrated by Herr EDUARD GRÜTZNER. £3 10s.
"As You Like It." Illustrated by Mons. EMILE BAYARD. £3 10s.
"Romeo and Juliet." Illustrated by FRANK DICKSEE, A.R.A. £5 5s.

Shakspere, The Leopold. With 400 Illustrations, and an Introduction by F. J. FURNIVALL. Small 4to, cloth gilt, 7s. 6d.; half-morocco, 10s. 6d.; full morocco, £1 1s. *Cheap Edition* 3s. 6d.

Shakspere, The Royal. With Exquisite Steel Plates and Wood Engravings. Three Vols. 15s. each.

Shakespeare, Cassell's Quarto Edition. Edited by CHARLES and MARY COWDEN CLARKE, and containing about 600 Illustrations by H. C. SELOUS. Complete in Three Vols., cloth gilt, £3 3s.—Also published in Three separate Volumes, in cloth, viz.:—The COMEDIES, 21s.; The HISTORICAL PLAYS, 18s. 6d.; The TRAGEDIES, 25s.

Shakespeare, Miniature. Illustrated. In Twelve Vols., in box, 12s.; or in Red Paste Grain (box to match), with spring catch, lettered in gold, 21s.

Shakespearean Scenes and Characters. Illustrative of Thirty Plays of Shakespeare. With Thirty Steel Plates and Ten Wood Engravings. The Text written by AUSTIN BRERETON. Royal 4to, 21s.

Sketching from Nature in Water Colours. By AARON PENLEY. With Illustrations in Chromo-Lithography. 15s.

Skin and Hair, The Management of the. By MALCOLM MORRIS, F.R.C.S. 2s.

Sonnets and Quatorzains. By CHRYS, M.A. (Oxon). 5s.

Standards, Local-Dual. By JOHN HENRY NORMAN. 1s.

Steam Engine, The Theory and Action of the: for Practical Men. By W. H. NORTHCOTT, C.E. 3s. 6d.

Stock Exchange Year-Book, The. By THOMAS SKINNER. 12s. 6d.

Summer Tide, Little Folks Holiday Number. 1s.

Sunlight and Shade. With numerous Exquisite Engravings. 7s. 6d.

Surgery, Memorials of the Craft of, in England. With an Introduction by Sir JAMES PAGET. 21s.

Thackeray, Character Sketches from. Six New and Original Drawings by FREDERICK BARNARD, reproduced in Photogravure. 21s.

Thorah, The Yoke of the. A Novel. By SIDNEY LUSKA. Boards, 2s.; cloth, 3s. 6d.

Three and Sixpenny Library of Standard Tales, &c. All Illustrated and bound in cloth gilt. Crown 8vo. 3s. 6d. each.

- **Jane Austen and her Works.**
- **Mission Life in Greece and Palestine.**
- **The Romance of Trade.**
- **The Three Homes.**
- **Deepdale Vicarage.**
- **In Duty Bound.**
- **The Half Sisters.**
- **Peggy Oglivie's Inheritance.**
- **The Family Honour.**
- **Esther West.**
- **Working to Win.**
- **Krilof and his Fables.** By W. R. S. Ralston, M.A.
- **Fairy Tales.** By Prof. Morley.

Tot Book for all Public Examinations. By W. S. THOMSON, M.A. 1s.

Town Holdings. 1s.

Tragedy of Brinkwater, The. A Novel. By MARTHA L. MOODEY. Boards, 2s.; cloth, 3s. 6d.

Tragic Mystery, A. A Novel. By JULIAN HAWTHORNE. Boards, 2s.; cloth, 3s. 6d.

Treasure Island. By R. L. STEVENSON. Illustrated. 5s.

Tree Painting in Water Colours. By W. H. J. BOOT. With Eighteen Coloured Plates, and valuable instructions by the Artist. 5s.

Trees, Familiar. By G. S. BOULGER, F.L.S., F.G.S. Two Series. With Forty full-page Coloured Plates, from Original Paintings by W. H. J. BOOT. 12s. 6d. each.

Twenty Photogravures of Pictures in the Salon of 1885, by the leading French Artists. In Portfolio. Only a limited number of copies have been produced terms for which can be obtained of all Booksellers.

"Unicode": The Universal Telegraphic Phrase Book. Pocket and Desk Editions. 2s 6d. each.

United States, Cassell's History of the. By the late EDMUND OLLIER. With 600 Illustrations. Three Vols. 9s. each.

United States, The Youth's History of. By EDWARD S. ELLIS. Illustrated. Four Vols. 36s.

Universal History, Cassell's Illustrated. With nearly ONE THOUSAND ILLUSTRATIONS. Vol. I. Early and Greek History.—Vol. II. The Roman Period.—Vol. III. The Middle Ages.—Vol. IV. Modern History. 9s. each.

Vaccination Vindicated. An Answer to the leading Anti-Vaccinators. By JOHN C. MCVAIL, M.D., D.P.H. Camb. 5s.

Veiled Beyond, The. A Novel. By S. B. ALEXANDER. Cloth, 3s. 6d.

Vicar of Wakefield and other Works by OLIVER GOLDSMITH. Illustrated. 3s. 6d.; cloth, gilt edges, 5s.

Water-Colour Painting, A Course of. With Twenty-four Coloured Plates by R. P. LEITCH, and full Instructions to the Pupil. 5s.

What Girls Can Do. By PHYLLIS BROWNE. 2s. 6d.

Who is John Noman? A Novel. By CHARLES HENRY BECKETT. Boards, 2s.. Cloth, 3s. 6d.

Wild Birds, Familiar. By W. SWAYSLAND. Four Series. With 40 Coloured Plates in each. 12s. 6d. each.

Wild Flowers, Familiar. By F. E. HULME, F.L.S., F.S.A. Five Series. With 40 Coloured Plates in each. 12s. 6d. each.

Wise Woman, The. By GEORGE MACDONALD. 2s. 6d.

Woman's World, The. Yearly Volume. 18s.

World of Wit and Humour, The. With 400 Illustrations. Cloth, 7s. 6d.; cloth gilt, gilt edges, 10s. 6d.

World of Wonders, The. Two Vols. With 400 Illustrations. 7s. 6d. each.

World's Lumber Room, The. By SELINA GAYE. Illustrated. 2s. 6d.

Yule Tide. CASSELL'S CHRISTMAS ANNUAL. 1s.

ILLUSTRATED MAGAZINES.

The Quiver, for Sunday and General Reading. Monthly, 6d.
Cassell's Family Magazine. Monthly, 7d.
"Little Folks" Magazine. Monthly, 6d.
The Magazine of Art. Monthly, 1s.
The Woman's World. Monthly, 1s.
Cassell's Saturday Journal. Weekly, 1d.; Monthly, 6d.

*** *Full particulars of* CASSELL & COMPANY'S **Monthly Serial Publications** *will be found in* CASSELL & COMPANY'S COMPLETE CATALOGUE.

Catalogues of CASSELL & COMPANY'S PUBLICATIONS, which may be had at all Booksellers', or will be sent post free on application to the Publishers:—

CASSELL'S COMPLETE CATALOGUE, containing particulars of One Thousand Volumes.

CASSELL'S CLASSIFIED CATALOGUE, in which their Works are arranged according to price, from *Threepence to Twenty-five Guineas.*

CASSELL'S EDUCATIONAL CATALOGUE, containing particulars of CASSELL & COMPANY'S Educational Works and Students' Manuals.

CASSELL & COMPANY, LIMITED, *Ludgate Hill, London.*

Bibles and Religious Works.

Bible, The Crown Illustrated. With about 1,000 Original Illustrations. With References, &c. 1,248 pages, crown 4to, cloth, 7s. 6d.

Bible, Cassell's Illustrated Family. With 900 Illustrations. Leather, gilt edges, £2 10s.; full morocco, £3 10s.

Bible Dictionary, Cassell's. With nearly 600 Illustrations. 7s. 6d.; roxburgh, 10s. 6d.

Bible Educator, The. Edited by the Very Rev. Dean PLUMPTRE, D.D. With Illustrations, Maps, &c. Four Vols., cloth, 6s. each.

Bible Work at Home and Abroad. Yearly Volume, 3s.

Bible Talks about Bible Pictures. Illustrated by GUSTAVE DORÉ and others. Large 4to, 5s.

Bunyan's Pilgrim's Progress (Cassell's Illustrated). 4to. 7s. 6d.

Bunyan's Pilgrim's Progress. With Illustrations. *Popular Edition*, 3s. 6d.

Child's Life of Christ, The. Complete in One Handsome Volume, with about 200 Original Illustrations. Demy 4to, gilt edges, 21s.

Child's Bible, The. With 200 Illustrations. Demy 4to, 830 pp. *145th Thousand. Cheap Edition*, 7s. 6d.

Commentary, The New Testament, for English Readers. Edited by the Rt. Rev. C. J. ELLICOTT, D.D., Lord Bishop of Gloucester and Bristol. In Three Volumes, 21s. each.

Vol. I.—**The Four Gospels.**
Vol. II.—**The Acts, Romans, Corinthians, Galatians.**
Vol. III.—**The remaining Books of the New Testament.**

Commentary, The Old Testament, for English Readers. Edited by the Rt. Rev. C. J. ELLICOTT, D.D., Lord Bishop of Gloucester and Bristol. Complete in 5 Vols., 21s. each.

Vol. I.—**Genesis to Numbers.**
Vol. II.—**Deuteronomy to Samuel II.**
Vol. III.—**Kings I. to Esther.**
Vol. IV.—**Job to Isaiah.**
Vol. V.—**Jeremiah to Malachi.**

Dictionary of Religion, The. An Encyclopædia of Christian and other Religious Doctrines, Denominations, Sects, Heresies, Ecclesiastical Terms, History, Biography, &c. &c. By the Rev. WILLIAM BENHAM, B.D. Cloth, 21s.; roxburgh, 25s.

Doré Bible. With 230 Illustrations by GUSTAVE DORÉ. *Original Edition.* Two Vols., cloth, £8; best morocco, gilt edges, £15.

Early Days of Christianity, The. By the Ven. Archdeacon FARRAR, D.D., F.R.S.

LIBRARY EDITION. Two Vols., 24s.; morocco, £2 2s.
POPULAR EDITION. Complete in One Volume, cloth, 6s.; cloth, gilt edges, 7s. 6d.; Persian morocco, 10s. 6d.; tree-calf, 15s.

Family Prayer-Book, The. Edited by Rev. Canon GARBETT, M.A., and Rev. S. MARTIN. Extra crown 4to, cloth, 5s.; morocco, 18s.

Geikie, Cunningham, D.D., Works by:—

The Holy Land and the Bible. A Book of Scripture Illustrations gathered in Palestine. With Map. Two Vols. 24s.
Hours with the Bible. Six Vols. 6s. each
Entering on Life. 3s. 6d.
The Precious Promises. 2s. 6d.
The English Reformation. 5s.
Old Testament Characters. 6s.
The Life and Words of Christ. Illustrated. Two Vols., cloth, 30s. *Library Edition*, Two Vols., cloth, 30s. *Students' Edition*, Two Vols., 16s. *Cheap Edition*, in One Vol. 7s. 6d.

Glories of the Man of Sorrows, The. Sermons preached at St. James's, Piccadilly. By the Rev. H. G. BONAVIA HUNT, Mus.D., F.R.S.Edin. 2s. 6d.

Gospel of Grace, The. By a LINDESIE. Cloth, 2s. 6d.

Helps to Belief. A Series of Helpful Manuals on the Religious Difficulties of the Day. Edited by the Rev. TEIGNMOUTH SHORE, M.A., Chaplain in Ordinary to the Queen. Cloth. 1s. each.

CREATION. By the **Lord Bishop of Carlisle.**
MIRACLES. By the **Rev. Brownlow Maitland, M.A.**
PRAYER. By the **Rev. T. Teignmouth Shore, M.A.**
THE MORALITY OF THE OLD TESTAMENT. By the **Rev. Newman Smyth, D.D.**
THE DIVINITY OF OUR LORD. By the **Lord Bishop of Derry.**
THE ATONEMENT. By the **Lord Bishop of Peterborough.**

"Heart Chords." A Series of Works by Eminent Divines. Bound in cloth, red edges, 1s. each.

My Father. By the Right Rev. Ashton Oxenden, late Bishop of Montreal.
My Bible. By the Rt. Rev. W. Boyd Carpenter, Bishop of Ripon.
My Work for God. By the Right Rev. Bishop Cotterill.
My Object in Life. By the Ven. Archdeacon Farrar, D.D.
My Aspirations. By the Rev. G. Matheson, D.D.
My Emotional Life. By the Rev. Preb. Chadwick, D.D.
My Body. By the Rev. Prof. W. G. Blaikie, D.D.
My Soul. By the Rev. P. B. Power, M.A.
My Growth in Divine Life. By the Rev. Prebendary Reynolds, M.A.
My Hereafter. By the Very Rev. Dean Bickersteth.
My Walk with God. By the Very Rev. Dean Montgomery.
My Aids to the Divine Life. By the Very Rev. Dean Boyle.
My Sources of Strength. By the Rev. E. E. Jenkins, M.A., Secretary of the Wesleyan Missionary Society.

Holy Land and the Bible, The. A Book of Scripture Illustrations gathered in Palestine. By the Rev. CUNNINGHAM GEIKIE, D.D. Two Vols., demy 8vo, 1,120 pages, with Map. Price 24s.

"I Must." Short Missionary Bible Readings. By SOPHIA M. NUGENT. Enamelled cover, 6d.; cloth, gilt edges, 1s.

Life of Christ, The. By the Ven. Archdeacon FARRAR, D.D., F.R.S., Chaplain in Ordinary to the Queen.

ILLUSTRATED EDITION, with about 300 Original Illustrations. Extra crown 4to, cloth, gilt edges, 21s.; morocco antique, 42s.

LIBRARY EDITION. Two Vols. Cloth, 24s.; morocco, 42s.

POPULAR EDITION, in One Vol. 8vo, cloth, 6s.; cloth, gilt edges, 7s. 6d.; Persian morocco, gilt edges, 10s. 6d.; tree-calf, 15s.

Luther, Martin: his Life and Times. By PETER BAYNE, LL.D. Two Vols., demy 8vo, 1,040 pages, cloth, 24s.

Marriage Ring, The. By WILLIAM LANDELS, D.D. Bound in white leatherette, gilt edges, in box, 6s.; French morocco, 8s. 6d.

Moses and Geology; or, The Harmony of the Bible with Science. By the Rev. SAMUEL KINNS, Ph.D., F.R.A.S. Illustrated. *Cheap Edition.* 6s.

Protestantism, The History of. By the Rev. J. A. WYLIE, LL.D. Containing upwards of 600 Original Illustrations. Three Vols., 27s.; Library Edition, 30s.

Quiver Yearly Volume, The. With 250 high-class Illustrations. 7s. 6d. Also Monthly, 6d.

St. George for England; and other Sermons preached to Children. *Fifth Edition.* By the Rev. T. TEIGNMOUTH SHORE, M.A. 5s.

St. Paul, The Life and Work of. By the Ven. Archdeacon FARRAR, D.D., F.R.S., Chaplain in Ordinary to the Queen.

LIBRARY EDITION. Two Vols., cloth, 24s.; calf, 42s.

ILLUSTRATED EDITION, complete in One Volume, with about 300 Illustrations, £1 1s.; morocco, £2 2s.

POPULAR EDITION. One Volume, 8vo, cloth, 6s.; cloth, gilt edges, 7s. 6d.; Persian morocco, 10s. 6d.; tree-calf, 15s.

Secular Life, The Gospel of the. Sermons preached at Oxford. By the Hon. W. H. FREMANTLE, Canon of Canterbury. 5s.

Shall We Know One Another? By the Rt. Rev. J. C. RYLE, D.D., Bishop of Liverpool. *New and Enlarged Edition.* Cloth limp, 1s.

Twilight of Life, The. Words of Counsel and Comfort for the Aged. By JOHN ELLERTON, M.A. 1s. 6d.

Voice of Time, The By JOHN STROUD. Cloth gilt, 1s.

Educational Works and Students' Manuals.

Alphabet, Cassell's Pictorial. Size, 35 inches by 42½ inches. Mounted on Linen, with rollers. 3s. 6d.

Arithmetics, The Modern School. By GEORGE RICKS, B.Sc. Lond. With Test Cards. (*List on application.*)

Book-Keeping. By THEODORE JONES. FOR SCHOOLS, 2s.; or cloth, 3s. FOR THE MILLION, 2s.; or cloth, 3s. Books for Jones's System, Ruled Sets of, 2s.

Chemistry, The Public School. By J. H. ANDERSON, M.A. 2s. 6d.

Commentary, The New Testament. Edited by Bishop ELLICOTT. Handy Volume Edition. Suitable for School and general use.

St. Matthew. 3s. 6d.
St. Mark. 3s.
St. Luke. 3s. 6d.
St. John. 3s. 6d.
The Acts of the Apostles. 3s. 6d.
Romans. 2s. 6d.
Corinthians I. and II. 3s.
Galatians, Ephesians, and Philippians. 3s.
Colossians, Thessalonians, and Timothy. 3s.
Titus, Philemon, Hebrews, and James. 3s.
Peter, Jude, and John. 3s.
The Revelation. 3s.
An Introduction to the New Testament. 2s. 6d.

Commentary, Old Testament. Edited by Bishop ELLICOTT. Handy Volume Edition. Suitable for School and general use.

Genesis. 3s. 6d.
Exodus. 3s.
Leviticus. 3s.
Numbers. 2s. 6d.
Deuteronomy. 2s. 6d.

Copy-Books, Cassell's Graduated. Complete in 18 Books. 2d. each.

Copy-Books, The Modern School. Complete in 12 Books. 2d. each.

Drawing Copies, Cassell's "New Standard." Fourteen Books.

Books A to F, for Standards I. to IV. .. .. 2d. each.
" G, H, K, L, M, O, for Standards V. to VII. .. .. .. 3d. each.
" N, P, ... 4d. each.

Drawing Copies, Cassell's Modern School Freehand. First Grade, 1s.; Second Grade, 2s.

Electricity, Practical. By Prof. W. E AYRTON. 7s. 6d.

Energy and Motion: A Text-Book of Elementary Mechanics. By WILLIAM PAICE, M.A. Illustrated. 1s. 6d.

English Literature, A First Sketch of, from the Earliest Period to the Present Time. By Prof. HENRY MORLEY. 7s. 6d.

Euclid, Cassell's. Edited by Prof. WALLACE, M.A. 1s.

Euclid, The First Four Books of. In paper, 6d.; cloth, 9d.

French Reader, Cassell's Public School. By GUILLAUME S. CONRAD. 2s. 6d.

French, Cassell's Lessons in. *New and Revised Edition.* Parts I. and II., each 2s. 6d.; complete, 4s. 6d. Key, 1s. 6d.

French-English and English-French Dictionary. *Entirely New and Enlarged Edition.* 1,150 pages, 8vo, cloth, 3s. 6d.

Galbraith and Haughton's Scientific Manuals. By the Rev. Prof. GALBRAITH, M.A., and the Rev. Prof. HAUGHTON, M.D., D.C.L.

Arithmetic. 3s. 6d.
Plane Trigonometry. 2s. 6d.
Euclid. Books I., II., III. 2s. 6d. Books IV., V., VI. 2s. 6d.
Mathematical Tables. 3s. 6d.
Mechanics. 3s. 6d.
Natural Philosophy. 3s. 6d.
Optics. 2s. 6d.
Hydrostatics. 3s. 6d.
Astronomy. 5s.
Steam Engine. 3s. 6d.
Algebra. Part I., cloth, 2s. 6d. Complete, 7s. 6d.
Tides and Tidal Currents, with Tidal Cards, 3s.

Geometry, First Elements of Experimental. By PAUL BERT. Fully Illustrated. 1s. 6d.

Geometry, Practical Solid. By Major ROSS, R.E. 2s.

German of To-Day. By Dr. HEINEMANN. 1s. 6d.

German-English and English-German Dictionary. 3s. 6d.

German Reading, First Lessons in. By A. JAGST. Illustrated. 1s.

Handbook of New Code of Regulations. By JOHN F. MOSS. 1s.; cloth, 2s.

Historical Course for Schools, Cassell's. Illustrated throughout. I.—Stories from English History, 1s. II.—The Simple Outline of English History, 1s. 3d. III.—The Class History of England, 2s. 6d.

Historical Cartoons, Cassell's Coloured. Size 45 in. × 35 in. 2s. each. Mounted on canvas and varnished, with rollers, 5s. each.

Latin-English Dictionary, Cassell's. Thoroughly revised and corrected, and in part re-written by J. R. V. MARCHANT, M.A. 3s. 6d.

Latin-English and English-Latin Dictionary. By J. R. BEARD, D.D., and C. BEARD, B.A. Crown 8vo, 914 pp., 3s. 6d.
Latin Primer, The New. By Prof. J. P. POSTGATE. 2s. 6d.
Laws of Every-Day Life. For the Use of Schools. By H. O. ARNOLD-FORSTER. 1s. 6d.
Lay Texts for the Young, in English and French. By Mrs. RICHARD STRACHEY. 2s. 6d. [1s. 6d.
Little Folks' History of England. By ISA CRAIG-KNOX. With 30 Illustrations.
Making of the Home, The: A Book of Domestic Economy for School and Home Use. By Mrs. SAMUEL A. BARNETT. 1s. 6d.
Marlborough Books.

- **Arithmetic Examples.** 3s.
- **Arithmetic Rules.** 1s. 6d.
- **French Exercises.** 3s. 6d.
- **French Grammar.** 2s. 6d.
- **German Grammar.** 3s. 6d.

Mechanics and Machine Design, Numerical Examples in Practical. By R. G. BLAINE, M.E. With Diagrams. Cloth, 2s. 6d.
Music, An Elementary Manual of. By HENRY LESLIE. 1s.
Popular Educator, Cassell's. *New and Thoroughly Revised Edition.* Illustrated throughout. Complete in Six Vols., 5s. each; or in Three Vols., half calf, 42s. the set.
Readers, Cassell's "Higher Class":—"The World's Lumber Room," Illustrated, 2s. 6d.; "Short Studies from Nature," Illustrated, 2s. 6d.; "The World in Pictures." (Ten in Series.) Cloth, 2s. each.
Readers, Cassell's Readable. Carefully graduated, extremely interesting, and illustrated throughout. *(List on application.)*
Readers, Cassell's Historical. Illustrated throughout, printed on superior paper, and strongly bound in cloth. *(List on application.)*
Readers for Infant Schools, Coloured. Three Books. Each containing 40 pages, including 8 pages in colours. 4d. each.
Reader, The Citizen. By H. O. ARNOLD-FORSTER. With Preface by the late Rt. Hon. W. E. FORSTER, M.P. 1s. 6d.
Readers, The Modern Geographical. Illustrated throughout, and strongly bound in cloth. *(List on application.)*
Readers, The Modern School. Illustrated. *(List on application.)*
Reading and Spelling Book, Cassell's Illustrated. 1s.
School Bank Manual, A. By AGNES LAMBERT. 6d.
Shakspere Reading Book, The. By H. COURTHOPE BOWEN, M.A. Illustrated. 3s. 6d. Also issued in Three Books, 1s. each.
Shakspere's Plays for School Use. 5 Books. Illustrated. 6d. each.
"Slöjd," as a means of Teaching the Essential Elements of Education. By EMILY LORD. 6d.
Spelling, A Complete Manual of. By J. D. MORELL, LL.D. 1s.
Technical Manuals, Cassell's. Illustrated throughout:—

- **Handrailing and Staircasing.** 3s. 6d.
- **Bricklayers, Drawing for.** 3s.
- **Building Construction.** 2s.
- **Cabinet-Makers, Drawing for.** 3s.
- **Carpenters & Joiners, Drawing for.** 3s. 6d.
- **Gothic Stonework.** 3s.
- **Linear Drawing & Practical Geometry.** 2s.
- **Linear Drawing and Projection.** The Two Vols. in One, 3s. 6d.
- **Metal-Plate Workers, Drawing for.** 3s.
- **Machinists & Engineers, Drawing for.** 4s. 6d.
- **Model Drawing.** 3s.
- **Orthographical and Isometrical Projection.** 2s.
- **Practical Perspective.** 3s.
- **Stonemasons, Drawing for.** 3s.
- **Applied Mechanics.** By Sir R. S. Ball, LL.D. 2s.
- **Systematic Drawing and Shading.** By Charles Ryan. 2s.

Technical Educator, Cassell's. Illustrated throughout. Popular Edition. Four Vols., 5s. each.
Technology, Manuals of. Edited by Prof. AYRTON, F.R.S., and RICHARD WORMELL, D.Sc., M.A. Illustrated throughout.

- **The Dyeing of Textile Fabrics.** By Prof. Hummel. 5s.
- **Watch and Clock Making.** By D. Glasgow. 4s. 6d.
- **Steel and Iron.** By Prof. W. H. Greenwood, F.C.S., M.I.C.E., &c. 5s.
- **Spinning Woollen and Worsted.** By W. S. McLaren, M.P. 4s. 6d.
- **Design in Textile Fabrics.** By T. R. Ashenhurst. 4s. 6d.
- **Practical Mechanics.** By Prof. Perry, M.E. 3s. 6d.
- **Cutting Tools Worked by Hand and Machine.** By Prof. Smith. 3s. 6d.

A Prospectus on application.

Test Cards, Cassell's Combination. In sets, 1s. each.
Test Cards, Cassell's Modern School. In sets, 1s. each.

A Copy of **Cassell and Company's Complete Catalogue** *will be forwarded post free on application.*

Books for Young People.

"Little Folks" Half-Yearly Volume. With 200 Illustrations, with Pictures in Colour. Boards, 3s. 6d.; or cloth gilt, 5s.

Bo-Peep. A Book for the Little Ones. With Original Stories and Verses. Illustrated throughout. Yearly Volume. Boards, 2s. 6d.; cloth gilt, 3s. 6d.

Every-day Heroes. By LAURA LANE. Illustrated. Cloth, 2s. 6d.

Legends for Lionel. New Picture Book by WALTER CRANE. 5s.

Flora's Feast. A Masque of Flowers. Penned and Pictured by WALTER CRANE. With 40 pages in Colours. 5s.

The New Children's Album. Fcap. 4to, 320 pages. Illustrated throughout. 3s. 6d.

The Tales of the Sixty Mandarins. By P. V. RAMASWAMI RAJU. With an Introduction by Prof. HENRY MORLEY. Illustrated. 5s.

Sunday School Reward Books. By Popular Authors. With Four Original Illustrations in each. Cloth gilt, 1s. 6d. each.

- **Seeking a City.**
- **Rhoda's Reward; or, "If Wishes were Horses."**
- **Jack Marston's Anchor.**
- **Frank's Life-Battle; or, The Three Friends.**
- **Rags and Rainbows: a Story of Thanksgiving.**
- **Uncle William's Charge; or, The Broken Trust.**
- **Pretty Pink's Purpose; or, The Little Street Merchants.**

"Golden Mottoes" Series, The. Each Book containing 208 pages, with Four full-page Original Illustrations. Crown 8vo, cloth gilt, 2s. each.

- **"Nil Desperandum."** By the Rev. F. Langbridge, M.A.
- **"Bear and Forbear."** By Sarah Pitt.
- **"Foremost if I Can."** By Helen Atteridge.
- **"Honour is my Guide."** By Jeanie Hering (Mrs. Adams-Acton).
- **"Aim at a Sure End."** By Emily Searchfield.
- **"He Conquers who Endures."** By the Author of "May Cunningham's Trial," &c.

The "Proverbs" Series. Consisting of a New and Original Series of Stories by Popular Authors, founded on and illustrating well-known Proverbs. With Four Illustrations in each Book, printed on a tint. Crown 8vo, 160 pages, cloth, 1s. 6d. each.

- **Fritters; or, "It's a Long Lane that has no Turning."** By Sarah Pitt.
- **Trixy; or, "Those who Live in Glass Houses shouldn't throw Stones."** By Maggie Symington.
- **The Two Hardcastles; or, "A Friend in Need is a Friend Indeed."** By Madeline Bonavia Hunt.
- **Major Monk's Motto; or, "Look Before you Leap."** By the Rev. F. Langbridge.
- **Tim Thomson's Trial; or, "All is not Gold that Glitters."** By George Weatherly.
- **Ursula's Stumbling-Block; or, "Pride comes before a Fall."** By Julia Goddard.
- **Ruth's Life-Work; or, "No Pains, no Gains."** By the Rev. Joseph Johnson.

The "Cross and Crown" Series. Consisting of Stories founded on incidents which occurred during Religious Persecutions of Past Days. With Illustrations in each Book. 2s. 6d. each.

- **By Fire and Sword: a Story of the Huguenots.** By Thomas Archer.
- **Adam Hepburn's Vow: a Tale of Kirk and Covenant.** By Annie S. Swan.
- **No. XIII; or, The Story of the Lost Vestal.** A Tale of Early Christian Days. By Emma Marshall.
- **Strong to Suffer: A Story of the Jews.** By E. Wynne.
- **Heroes of the Indian Empire; or, Stories of Valour and Victory.** By Ernest Foster.
- **In Letters of Flame: A Story of the Waldenses.** By C. L. Matéaux.
- **Through Trial to Triumph.** By Madeline B. Hunt.

The World's Workers. A Series of New and Original Volumes by Popular Authors. With Portraits printed on a tint as Frontispiece. 1s. each.

- **The Earl of Shaftesbury.** By Henry Frith.
- **Sarah Robinson, Agnes Weston, and Mrs. Meredith.** By E. M. Tomkinson.
- **Thomas A. Edison and Samuel F. B. Morse.** By Dr. Denslow and J. Marsh Parker.
- **Mrs. Somerville and Mary Carpenter.** By Phyllis Browne.
- **General Gordon.** By the Rev S. A Swaine.
- **Charles Dickens.** By his Eldest Daughter.
- **Sir Titus Salt and George Moore.** By J. Burnley.
- **Florence Nightingale, Catherine Marsh, Frances Ridley Havergal, Mrs. Ranyard ("L. N. R.")** By Lizzie Alldridge.
- **Dr. Guthrie, Father Mathew, Elihu Burritt, Joseph Livesey.** By the Rev. J. W. Kirton.
- **Sir Henry Havelock and Colin Campbell, Lord Clyde.** By E. C. Phillips.
- **Abraham Lincoln.** By Ernest Foster.
- **David Livingstone.** By Robert Smiles.
- **George Muller and Andrew Reed.** By E. R. Pitman.
- **Richard Cobden.** By R. Gowing.
- **Benjamin Franklin.** By E. M. Tomkinson.
- **Handel.** By Eliza Clarke.
- **Turner the Artist.** By the Rev. S. A. Swaine.
- **George and Robert Stephenson.** By C. L. Matéaux.

Five Shilling Books for Young People. With Original Illustrations. Cloth gilt, 5s. each.

- **The Palace Beautiful.** By L. T. Meade.
- **"Follow my Leader;" or, the Boys of Templeton.** By Talbot Baines Reed.
- **For Fortune and Glory; a Story of the Soudan War.** By Lewis Hough.
- **Under Bayard's Banner.** By Henry Frith.
- **The Champion of Odin; or, Viking Life in the Days of Old.** By J. Fred. Hodgetts.
- **Bound by a Spell; or, the Hunted Witch of the Forest.** By the Hon. Mrs. Greene.
- **The King's Command. A Story for Girls.** By Maggie Symington.
- **The Romance of Invention.** By Jas. Burnley.

Three and Sixpenny Books for Young People. With Original Illustrations. Cloth gilt, 3s. 6d. each.

- **The Cost of a Mistake.** By Sarah Pitt.
- **A World of Girls: A Story of a School.** By L. T. Meade.
- **On Board the "Esmeralda;" or, Martin Leigh's Log.** By John C. Hutcheson.
- **Lost among White Africans: A Boy's Adventures on the Upper Congo.** By David Ker.
- **In Quest of Gold; or, Under the Whanga Falls.** By Alfred St. Johnston.
- **For Queen and King; or, the Loyal 'Prentice.** By Henry Frith.
- **Perils Afloat and Brigands Ashore.** By Alfred Elwes.
- **Freedom's Sword: A Story of the Days of Wallace and Bruce.** By Annie S. Swan.

The "Boy Pioneer" Series. By EDWARD S. ELLIS. With Four Full-page Illustrations in each Book. Crown 8vo, cloth, 2s. 6d. each.

- **Ned in the Woods.** A Tale of Early Days in the West.
- **Ned on the River.** A Tale of Indian River Warfare.
- **Ned in the Block House.** A Story of Pioneer Life in Kentucky.

The "Log Cabin" Series. By EDWARD S. ELLIS. With Four Full-page Illustrations in each. Crown 8vo, cloth, 2s. 6d. each.

- **The Lost Trail.**
- **Camp-Fire and Wigwam.**
- **Footprints in the Forest.**

The "Great River" Series. (Uniform with the "Log Cabin" Series.) By EDWARD S. ELLIS. Illustrated. Crown 8vo, cloth, bevelled boards, 2s. 6d. each.

- **Down the Mississippi.**
- **Lost in the Wilds.**
- **Up the Tapajos: or, Adventures in Brazil.**

The "Chimes" Series. Each containing 64 pages, with Illustrations on every page, and handsomely bound in cloth, 1s.

- **Bible Chimes.** Contains Bible Verses for Every Day in the Month.
- **Daily Chimes.** Verses from the Poets for Every Day in the Month.
- **Holy Chimes.** Verses for Every Sunday in the Year.
- **Old World Chimes.** Verses from old writers for Every Day in the Month.

Sixpenny Story Books. All Illustrated, and containing Interesting Stories by well-known Writers.

- **The Smuggler's Cave.**
- **Little Lizzie.**
- **The Boat Club.**
- **Luke Barnicott.**
- **Little Bird.**
- **Little Pickles.**
- **The Elchester College Boys.**
- **My First Cruise.**
- **The Little Peacemaker.**
- **The Delft Jug.**

Cassell's Picture Story Books. Each containing 60 pages of Pictures and Stories, &c. 6d. each.

- **Little Talks.**
- **Bright Stars.**
- **Nursery Toys.**
- **Pet's Posy.**
- **Tiny Tales.**
- **Daisy's Story Book.**
- **Dot's Story Book.**
- **A Nest of Stories.**
- **Good Night Stories.**
- **Chats for Small Chatterers.**
- **Auntie's Stories.**
- **Birdie's Story Book.**
- **Little Chimes.**
- **A Sheaf of Tales.**
- **Dewdrop Stories.**

Illustrated Books for the Little Ones. Containing interesting Stories. All Illustrated. 1s. each.

- **Indoors and Out.**
- **Some Farm Friends.**
- **Those Golden Sands.**
- **Little Mothers and their Children.**
- **Our Pretty Pets.**
- **Our Schoolday Hours.**
- **Creatures Tame.**
- **Creatures Wild.**
- **Up and Down the Garden.**
- **All Sorts of Adventures.**
- **Our Sunday Stories.**
- **Our Holiday Hours.**

Shilling Story Books. All Illustrated, and containing Interesting Stories.

- **Seventeen Cats.**
- **Bunty and the Boys.**
- **The Heir of Elmdale.**
- **The Mystery at Shoncliff School.**
- **Claimed at Last, and Roy's Reward.**
- **Thorns and Tangles.**
- **The Cuckoo in the Robin's Nest.**
- **John's Mistake.**
- **Diamonds in the Sand.**
- **Surly Bob.**
- **The History of Five Little Pitchers.**
- **The Giant's Cradle.**
- **Shag and Doll.**
- **Aunt Lucia's Locket.**
- **The Magic Mirror.**
- **The Cost of Revenge.**
- **Clever Frank.**
- **Among the Redskins.**
- **The Ferryman of Brill.**
- **Harry Maxwell.**
- **A Banished Monarch.**

Cassell's Children's Treasuries. Each Volume contains Stories or Poetry, and is profusely Illustrated. Cloth, 1s. each.

Cook Robin, and other Nursery Rhymes.
The Queen of Hearts.
Old Mother Hubbard.
Tuneful Lays for Merry Days.
Cheerful Songs for Young Folks.
Pretty Poems for Young People.
The Children's Joy.
Pretty Pictures and Pleasant Stories.
Our Picture Book.
Tales for the Little Ones.
My Sunday Book of Pictures.
Sunday Garland of Pictures and Stories.
Sunday Readings for Little Folks.

"Little Folks" Painting Books. With Text, and Outline Illustrations for Water-Colour Painting. 1s. each.

Fruits and Blossoms for "Little Folks" to Paint.
The "Little Folks" Illuminating Book.
Pictures to Paint.
The "Little Folks" Proverb Painting Book.

Eighteenpenny Story Books. All Illustrated throughout.

Wee Willie Winkie.
Ups and Downs of a Donkey's Life.
Three Wee Ulster Lassies.
Up the Ladder.
Dick's Hero; and other Stories.
The Chip Boy.
Raggles, Baggles, and the Emperor.
Roses from Thorns.
Faith's Father.
By Land and Sea.
The Young Berringtons.
Jeff and Leff.
Tom Morris's Error.
Worth more than Gold.
"Through Flood—Through Fire;" and other Stories.
The Girl with the Golden Locks.
Stories of the Olden Time.

The "World in Pictures" Series. Illustrated throughout. 2s. 6d. each.

A Ramble Round France.
All the Russias.
Chats about Germany.
The Land of the Pyramids (Egypt).
Peeps into China.
The Eastern Wonderland (Japan).
Glimpses of South America.
Round Africa.
The Land of Temples (India)
The Isles of the Pacific.

Two-Shilling Story Books. All Illustrated.

Stories of the Tower.
Mr. Burke's Nieces.
May Cunningham's Trial.
The Top of the Ladder: How to Reach it.
Little Flotsam.
Madge and her Friends.
The Children of the Court.
A Moonbeam Tangle.
Maid Marjory.
The Four Cats of the Tippertons.
Marion's Two Homes.
Little Folks' Sunday Book.
Two Fourpenny Bits.
Poor Nelly.
Tom Heriot.
Aunt Tabitha's Waifs.
In Mischief Again.
Through Peril to Fortune.
Peggy, and other Tales.
The Magic Flower Pot.
School Girls.

Half-Crown Books.

Little Hinges.
Margaret's Enemy.
Pen's Perplexities.
Notable Shipwrecks.
Golden Days.
Wonders of Common Things.
At the South Pole.
Truth will Out.
Pictures of School Life and Boyhood.
The Young Man in the Battle of Life. By the Rev. Dr. Landels.
The True Glory of Woman. By the Rev. Dr. Landels.
The Wise Woman. By George Macdonald.
Soldier and Patriot (George Washington).

Picture Teaching Series. Each book Illustrated throughout. Fcap. 4to, cloth gilt, coloured edges, 2s. 6d. each.

Through Picture-Land.
Picture Teaching for Young and Old.
Picture Natural History.
Scraps of Knowledge for the Little Ones.
Great Lessons from Little Things.
Woodland Romances.
Stories of Girlhood.
Frisk and his Flock.
Pussy Tip-Toes' Family.
The Boy Joiner and Model Maker.
The Children of Holy Scripture.

Library of Wonders. Illustrated Gift-books for Boys. Paper, 1s.; cloth, 1s. 6d.

Wonders of Acoustics.
Wonderful Adventures.
Wonders of Animal Instinct.
Wonders of Architecture.
Wonderful Balloon Ascents.
Wonders of Bodily Strength and Skill.
Wonderful Escapes.
Wonders of Water.

The "Home Chat" Series. All Illustrated throughout. Fcap. 4to. Boards, 3s. 6d. each; cloth, gilt edges, 5s. each.

Home Chat.
Peeps Abroad or Folks at Home.
Decisive Events in History.
Around and About Old England.
Half-Hours with Early Explorers.
Paws and Claws.

Books for the Little Ones. Fully Illustrated.

A Dozen and One; or, The Boys and Girls of Polly's Ring. By Mary D. Brine. Full of Illustrations. 5s.
The Merry-go-Round. Poems for Children. Illustrated throughout. 5s.
Rhymes for the Young Folk. By William Allingham. Beautifully Illustrated. 3s. 6d.
The Little Doings of some Little Folks. By Chatty Cheerful. Illustrated. 5s.
The Sunday Scrap Book. With One Thousand Scripture Pictures. Boards, 5s.; cloth, 7s. 6d.
Daisy Dimple's Scrap Book. Containing about 1,000 Pictures. Boards, 5s.; cloth gilt, 7s. 6d.
The History Scrap Book. With nearly 1,000 Engravings. 5s.; cloth, 7s. 6d.
The Little Folks' Out and About Book. By Chatty Cheerful. Illustrated. 5s.
Myself and my Friends. By Olive Patch. With numerous Illustrations. Crown 4to. 5s.
A Parcel of Children. By Olive Patch. With numerous Illustrations. Crown 4to. 5s.
Little Folks' Picture Album. With 168 Large Pictures. 5s.
Little Folks' Picture Gallery. With 150 Illustrations. 5s.
The Old Fairy Tales. With Original Illustrations. Boards, 1s.; cloth, 1s. 6d.
My Diary. With Twelve Coloured Plates and 366 Woodcuts. 1s.
Happy Little People. By Olive Patch. With Illustrations. 5s.
"Little Folks" Album of Music, The. Illustrated. 3s. 6d.
Cheerful Clatter. Nearly One Hundred Full-page Pictures. 3s. 6d.
Twilight Fancies. Full of charming Pictures. Boards, 2s. 6d.
Happy Go Lucky. 2s.
Daisy Blue Eyes. 2s.
Good Times. 1s. 6d.
Jolly Little Stories. 1s. 6d.
Our Little Friends. 1s. 6d.
Daisy Dell's Stories. 1s. 6d.
Little Toddlers. 1s. 6d.
Wee Little Rhymes. 1s. 6d.
Little One's Welcome. 1s. 6d.
Little Gossips. 1s. 6d.
Ding Dong Bell. 1s. 6d.
The Story of Robin Hood. With Coloured Illustrations. 2s. 6d.
The Pilgrim's Progress. With Coloured Illustrations. 2s. 6d.

Books for Boys.

Commodore Junk. By G. Manville Fenn. 5s.
The Black Arrow. A Tale of the Two Roses. By R. L. Stevenson. 5s.
Dead Man's Rock. A Romance. By Q. 5s.
A Queer Race. By W. Westall. 5s.
Captain Trafalgar. A Story of the Mexican Gulf. By W. Westall. Illustrated. 5s.
Kidnapped. By R. L. Stevenson. Illustrated. 5s.
King Solomon's Mines. By H. Rider Haggard. 5s.
Treasure Island. By R. L. Stevenson. With Full-page Illustrations. 5s.
Ships, Sailors, and the Sea. By R. J. Cornewall-Jones. Illustrated. 5s.
The Phantom City. By W. Westall. 5s.
Famous Sailors of Former Times, History of the Sea Fathers. By Clements Markham. Illustrated. 2s. 6d.
Modern Explorers. By Thomas Frost. Illustrated. 5s.
Wild Adventures in Wild Places. By Dr. Gordon Stables, M.D., R.N. Illustrated. 5s.
Jungle, Peak, and Plain. By Dr. Gordon Stables, R.N. Illustrated. 5s.
O'er Many Lands, on Many Seas. By Gordon Stables, R.N. Illustrated. 5s.
At the South Pole. By W. H. G. Kingston. *New Edition.* Illustrated. 2s. 6d.

Books for all Children.

Cassell's Robinson Crusoe. With 100 striking Illustrations. Cloth, 3s. 6d.; gilt edges, 5s.
Cassell's Swiss Family Robinson. Illustrated. Cloth, 3s. 6d.; gilt edges, 5s.
Sunny Spain: Its People and Places, with Glimpses of its History. By Olive Patch. Illustrated. 5s.
Rambles Round London Town. By C. L. Matéaux. Illustrated. 5s.
Favorite Album of Fun and Fancy, The. Illustrated. 3s. 6d.
Familiar Friends. By Olive Patch. Illustrated. Cloth gilt, 5s.
Odd Folks at Home. By C. L. Matéaux. With nearly 150 Illustrations. 5s.
Field Friends and Forest Foes. By Olive Patch. Profusely Illustrated. 5s.
Silver Wings and Golden Scales. Illustrated. 5s.
Little Folks' Holiday Album. Illustrated. 3s. 6d.
Tiny Houses and their Builders. Illustrated. 5s.
Children of all Nations. Their Homes, their Schools, their Playgrounds. Illustrated. 5s.
Tim Trumble's "Little Mother." By C. L. Matéaux. Illustrated. 5s.

CASSELL & COMPANY, Limited, Ludgate Hill, London, Paris, New York & Melbourne.

www.ingramcontent.com/pod-product-compliance
Lightning Source LLC
Chambersburg PA
CBHW031820270326
41932CB00008B/479

* 9 7 8 3 3 3 7 3 6 5 8 4 4 *